Advanced Structured Materials

Volume 27

Series Editors

Andreas Öchsner
Lucas F. M. da Silva
Holm Altenbach

For further volumes:
http://www.springer.com/series/8611

J. M. P. Q. Delgado
Antonio Gilson Barbosa de Lima
Marta Vázquez da Silva
Editors

Numerical Analysis of Heat and Mass Transfer in Porous Media

Springer

Editors
J. M. P. Q. Delgado
Faculdade de Engenharia
Laboratorio de Fisica das
Universidade do Porto
Rua Dr. Roberto Frias
4200-465 Porto
Portugal

Antonio Gilson Barbosa de Lima
Department of Mechanical Engineering
Federal University of Campina Grande
Campina Grande
Brazil

Marta Vázquez da Silva
Norte, CESPU, Department of Sciences
Instituto Superior de Ciências da Saúde
Rua Central de Gandra 1317
4585-116 Gandra
Portugal

ISSN 1869-8433 ISSN 1869-8441 (electronic)
ISBN 978-3-642-30531-3 ISBN 978-3-642-30532-0 (eBook)
DOI 10.1007/978-3-642-30532-0
Springer Heidelberg New York Dordrecht London

Library of Congress Control Number: 2012940958

Printed on acid-free paper

Springer is part of Springer Science+Business Media (www.springer.com)

Preface

In the last years, several heat and mass transfer textbooks have been published. The study of heat and mass transfer is a subject of continuous discussion, since the knowledge gained about the processes involving the heating/cooling of fluids or solids, the change in the concentration of a substance, the rate at which these phenomena occur and the determining factors in its development, will allow to predict the behaviour of many materials, the equipment design and to predict the performance of equipment in operation.

Thermal energy is the fraction of internal energy of a body that can be transferred due to a difference in temperatures itself and the process is called heat transfer. Mass transfer can be understood as the spatial motion of matter. However, "mass transfer" generally refers to the movement of a particular component in a system of several components, either by molecular diffusion or by convection. In most cases, it simultaneouly occurs in heat and mass transfer, so that these topics are often addressed simultaneously.

A porous medium refers to a solid having empty spaces that are filled with a fluid. Generally, many of these pores are interconnected so that transport of mass and heat is possible through the pore. This process is, in the vast majority of cases, a faster transport process than through the solid matrix. A wide variety of materials can be studied as porous media, such as rocks, soils, plant and animal tissues, paper and other packaging materials, etc.

The book *Numerical Analysis of Heat and Mass Transfer in Porous Media* was designed with the purpose of providing the scientific community with the latest developments in the field of heat and/or mass transfer in porous media, with special emphasis on a numerical/computational approach of the physical situations.

The main benefit of the book is that it discusses the majority of the topics related to numerical transport phenomenon in engineering (including state of the art and applications) and presents some of the most important theoretical and computational developments in porous media and transport phenomenon domain, providing a self-contained major reference that is appealing to scientists, researchers and engineers. At the same time, these topics will encounter a variety of scientific and engineering disciplines, such as chemical, civil, agricultural, mechanical engineering, etc. The book is divided into several chapters that intend to be a resume of the current state of knowledge for the benefit of professional colleagues.

J. M. P. Q. Delgado
A. G. Barbosa de Lima
M. Vázquez da Silva

Contents

Numerical Analysis of Mass Transfer Around a Sphere Buried in Porous Media: Concentration Contours and Boundary Layer Thickness

J. M. P. Q. Delgado and M. Vázquez da Silva

Abstract The mass transfer process between a moving fluid and a slightly soluble or fast reacting sphere buried in a packed bed, with "uniform velocity", was obtained numerically, for solute transport by both advection and diffusion. Fluid flow in the granular bed around the sphere was assumed to follow Darcy's law and the elliptic Partial Differential Equations (PDE), resulting from a differential material balance on the solute in an elementary control volume, was studied and solved numerically over the "whole range" of values of the relevant parameters (Peclet number and Schmidt number). The numerical solutions gave the concentration contour plots and concentration boundary layer thickness as a function of the relevant parameters. For each concentration level, the width and downstream length of the corresponding contour surface were determined. General expressions are presented to predict contaminant "plume" size downstream of the polluting source. An important feature of this work is the detailed discussion of the finite differences method adopted, with emphasis on the High-Resolution Schemes (HRS) used in the discretization of the convection term of the PDE.

Keywords Mass transfer · Active sphere · Porous media · Diffusion · Dispersion · High-resolution schemes

J. M. P. Q. Delgado (✉)
LFC–Building Physics Laboratory, Civil Engineering Department,
Faculty of Engineering, University of Porto, Porto, Portugal
e-mail: jdelgado@fe.up.pt

M. Vázquez da Silva
Centro de Estudos de Fenómenos de Transporte, Faculdade de Engenharia Da,
Universidade Do Porto, Rua Dr. Roberto Frias, s/n, 4200-465 Porto, Portugal
e-mail: mvazquez@fe.up.pt

M. Vázquez da Silva
Departamento de Ciências, Instituto Superior de Ciências da Saúde, Norte CESPU,
Rua Central de Gandra, n° 1317, 4585-116 Gandra, Portugal

J. M. P. Q. Delgado et al. (eds.), *Numerical Analysis of Heat and Mass Transfer in Porous Media*, Advanced Structured Materials 27,
DOI: 10.1007/978-3-642-30532-0_1, © Springer-Verlag Berlin Heidelberg 2012

1 Introduction

There are many processes, both in industry and in the environment, in which a lump of solid matter, buried in a granular bed of small inert particles, interacts with the fluid flowing around it, through the interstices of the bed. This work deals with mass transfer from a sphere buried in a packed bed of inert particles through which a uniform stream of gas flows, as sketched in Fig. 1.

Early attempts [5, 24] to predict the mass transfer coefficient were based on the analogy with fluid flow through an isolated sphere, for which the correlations of Ranz and Marshall [32] and Brian and Hales [4] can be applied. However, the two situations are very different from a hydrodynamic point of view and, as it happens, the flow field around the buried sphere is easier to treat, since there is no hydrodynamic boundary layer and no fluid separation (and therefore no wake) on the leeward side of the sphere.

The interest on this problem was reinforced by the need to predict the rate of combustion of char particles immersed in beds of sand (or ash), where combustion takes place mostly in the dense phase [3]. For a long time it has been known [38] that mass transfer in the dense phase of such beds occurs at the same rate as in a fixed bed of the same particles, but with gas flowing at the velocity of incipient fluidization. This may be seen as a corollary of the two-phase theory of fluidization [12], which is supported by experimental evidence [8, 31].

La Nauze et al. [21] attempted a solution of the mass transfer problem, proposing a mechanism that combines the processes of transient "dipping" and surface renewal, but their approach did not consider the detailed fluid mechanics of the gas flow. Prins et al. [31] tried to establish a correlation of dimensionless groups, based on a large amount of experimental data, but, since the number of independent variables involved is large, this result has limited applicability.

Coelho and Guedes de Carvalho [8] were the first to produce a full analytical treatment of the problem, although their analysis was restricted to a thin mass transfer boundary (i.e., to high Peclet numbers); their approach took into account the important role of transverse dispersion in mass transfer. The resulting asymptotic expression, valid for high Peclet numbers, was

$$\frac{Sh'}{\varepsilon} = \left[1.28Pe' + 0.141(d/d_1)Pe'^2\right]^{1/2} \tag{1}$$

where Sh' is the Sherwood number, ε is the bed voidage, d and d_1 are, respectively, the inert particle and the "active sphere" diameters and Pe' is the Peclet number. A primed is used in $D'_m = D_m/\tau$ and in all dimensionless groups containing D'_m, being D_m and D'_m, respectively, the molecular and the effective molecular diffusion coefficients and τ the tortuosity.

That result was then combined with the well known result $Sh'/\varepsilon = 2$ for molecular diffusion from a buried sphere into a stagnant medium (i.e., for $Pe' = 0$) and the general expression was proposed

$$\frac{Sh'}{\varepsilon} = \left[4 + 0.576Pe'^{0.78} + 1.28Pe' + 0.141(d/d_1)Pe'^2\right]^{1/2} \tag{2}$$

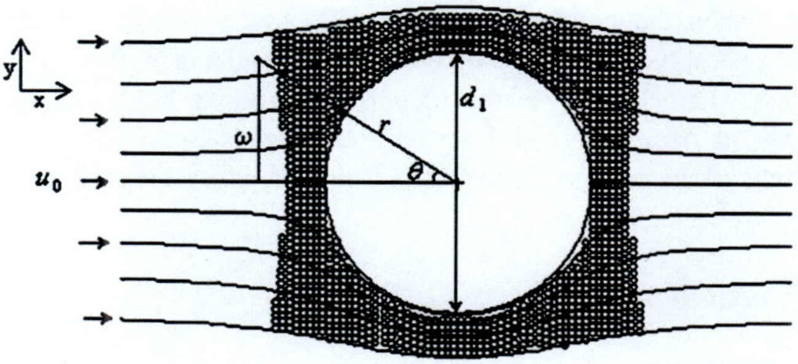

Fig. 1 Flow around an "active sphere" immersed in a porous media

Middleman [25] quotes this result as an empirical correlation, but it is worth emphasizing that the expressions obtained for the limiting conditions of both small and large values of Pe' were derived from first principles, without the introduction of any adjustable parameters. It is nevertheless true that for intermediate values of Pe', Eq. (2) represents an empirical correlation, which has been successfully tested against a huge number of experimental data [8, 19, 29].

The solution of this problem is of interest in the analysis of different physical situations. For example, in the combustion of char, in fluidised beds of sand, it may be useful to know how far does the cloud of CO stretch downstream of the burning particles, to see if it reaches the bubbles [3, 29]. Also in the understanding of certain mineralogical processes, such as the dissolution of plagioclase grains, during digenesis [28, 36], modelling the transport of aqueous Al by diffusion and convection around the grains may be required. Finally, in the contamination of groundwater [15] by pools or drops of NAPLs (Non-Aqueous Phase Liquids) or by compacted buried waste, for example, it may be necessary to estimate the distance from the "contaminant source" beyond which the levels of contaminant are expected to fall below some "safe" limit.

In all such processes, a region of higher solute concentration develops in the immediate vicinity of the "active mass", with consequent solute migration away from it, by diffusion and convection, in the moving fluid (such as fast reaction at the surface of the "active mass", there will be a local depletion of some of the reagents, resulting in diffusion/convection towards it). It is often important to estimate the distribution of solute concentration in the interstitial fluid, around and downstream of the "active mass". This region, especially rich in solute, is very often called "diffusion cloud" or "contaminant plume".

Prandtl [30] introduced in the early years of the last century the concept of a boundary layer for fluid flow past a solid. Over the years, for the case of a thin boundary layer (high Peclet numbers), several analytical steady and unsteady state mass transfer solutions for concentration boundary layer thickness have been presented in the literature [33]. However, there is a lack of correlations for concentration boundary layer thickness of spheres buried in a porous media, at low fluid velocities, which motivated this work.

The influence of the rheological properties on the boundary layer thickness was studied in detail by different scientists. For example, Nebbali and Bouhadef [26]

showed that, when the permeability and/or kinematic viscosity is increased, the velocity gradients at the duct walls decrease, which results in increase of the boundary layer thickness. Cheng [6] showed that increasing the power-law index, decreasing the Lewis number or decreasing the buoyancy ratio leads to an increase of the fluid velocity, which means that the concentration boundary layer thickness is also enlarged.

2 Mathematical Analysis

The situation considered in this work is sketched in Fig. 1, where the flow field is depicted around a sphere of diameter d_1 buried in a bed of inert particles of diameter d (with $d \ll d_1$); the granular bed is assumed to have uniform voidage, ε, and to be of "infinite" extent, with liquid flowing through it at a uniform interstitial velocity u_0, away from the sphere.

The equations describing the flow field are obtained by coupling Darcy's law ($\mathbf{u} = -K\, grad\, p$, being $\mathbf{u}$ the interstitial velocity vector, K the permeability and p the pressure) with the continuity equation ($div\mathbf{u} = 0$) to yield Laplace's equation ($\nabla^2\phi = 0$) for the flow potential, $\phi = Kp$.

The problem has axial symmetry (coordinates x and y, in Fig. 1), but it is convenient to express the potential and stream functions in spherical coordinates $(r,\ \theta)$, respectively by [10]:

$$\phi = -u_0\left[1 + \frac{1}{2}\left(\frac{a}{r}\right)^3\right] r\cos\theta \tag{3}$$

$$\psi = \frac{u_0}{2}\left[1 - \left(\frac{a}{r}\right)^3\right] r^2\sin^2\theta \tag{4}$$

where a is the "active sphere" radius and ψ is the stream function. The corresponding velocity components are

$$u_r = \frac{\partial\phi}{\partial r} = -u_0\cos\theta\left[1 - \left(\frac{a}{r}\right)^3\right] \tag{5}$$

$$u_\theta = \frac{1}{r}\frac{\partial\phi}{\partial\theta} = u_0\sin\theta\left[1 + \frac{1}{2}\left(\frac{a}{r}\right)^3\right] \tag{6}$$

Making use of the potential and stream lines it is possible to perform a steady state material balance on the solute in a differential element of a "stream tube" to obtain [20]

$$\frac{\partial c}{\partial\phi} = \frac{\partial}{\partial\phi}\left(D_L\frac{\partial c}{\partial\phi}\right) + \frac{\partial}{\partial\psi}\left(D_T\omega^2\frac{\partial c}{\partial\psi}\right) \tag{7}$$

where ω is distance to the flow axis, D_{L} and D_{T} are, respectively, the longitudinal and transverse dispersion coefficients and c is the solute concentration. The steady state may not be observed strictly, but it represents an important limiting condition.

The boundary conditions to be used in the integration of Eq. (7) must be such that: (*i*) far from the sphere, the solute concentration is equal to the background concentration, $c = c_0$; (*ii*) on the surface of the sphere, the solute concentration in the fluid equals the equilibrium value, $c = c^*$; (*iii*) the concentration field is axisymmetric.

Making use of the following dimensionless variables

$$C = \frac{c - c_0}{c^* - c_0} \tag{8}$$

$$U = \frac{u}{u_0} = \frac{\left(u_r^2 + u_\theta^2\right)^{1/2}}{u_0} \tag{9}$$

$$R = \frac{r}{a} \tag{10}$$

$$\Phi = \frac{4}{3}\frac{\phi}{u_0 d_1} \tag{11}$$

$$\Psi = \frac{\psi}{u_0 d_1^2} \tag{12}$$

$$Pe' = \frac{u_0 d_1}{D'_m} \tag{13}$$

Eq. (7) may be re-arranged to result in

$$\frac{\partial C}{\partial \Phi} = \frac{\partial}{\partial \Phi}\left(A\frac{\partial C}{\partial \Phi}\right) + \frac{\partial}{\partial \Psi}\left(B\frac{\partial C}{\partial \Psi}\right) \tag{14}$$

with

$$A = \frac{4}{3Pe'}\frac{D_L}{D'_m} \tag{15a}$$

$$B = \frac{3}{16}\frac{R^2 \sin^2\theta}{Pe'}\frac{D_T}{D'_m} \tag{15b}$$

The boundary conditions in dimensionless form are

$$\Phi \to -\infty, \Psi \geq 0 \quad C \to 0 \tag{16a}$$

$$\Phi \to +\infty, \Psi \geq 0 \quad C \to 0 \tag{16b}$$

$$|\Phi| \leq 1, \Psi = 0 \quad C = 1 \tag{16c}$$

$$|\Phi| > 1, \Psi = 0 \quad \frac{\partial C}{\partial \Psi} = 0 \tag{16d}$$

$$\forall_\Phi, \Psi \to +\infty \quad C \to 0 \tag{16e}$$

Equation (14) is to be solved numerically over the ranges of interest in terms of the Peclet and Schmidt numbers, and the solution must satisfy the boundary conditions expressed by Eqs. (16a–16e).

3 Numerical Method

Equation (14) was solved numerically using a finite-difference method similar to that described by Guedes de Carvalho and Alves [20]. A second-order Central Differencing Scheme (CDS) in a general non-uniform grid was adopted for the discretization of the diffusive terms [14] on the right hand side (r.h.s) of Eq. (14). The convection term on the left hand side (l.h.s) of Eq. (14) was also discretized with central differences but, for high values of Pe', the mass transfer process is dominated by advection and the CDS scheme is not bounded, leading to the appearance of unphysical oscillations in the solution near sharp gradients of the convected variable or even to divergence of the numerical method. This undesirable occurrence is typical of the CDS scheme for convection-dominated transport conditions and is an issue quite familiar to the Computational Fluid Dynamics (CFD) community [14]. To obviate this unbounded property of higher-order schemes in highly convective flows, several approaches have been proposed in the literature. The initial attempt was the first-order Upwind Differencing Scheme (UDS) proposed by Courant et al. [9]. This scheme ensures that a physically realistic solution is always attained, but it is only first-order accurate and, thus, it is not recommended nowadays [16]. The HYBRID [35] and the POWER-LAW schemes [27] were proposed to increase the accuracy of the UDS scheme, while preserving boundedness; in practice, they hardly ever offer better accuracy than the simple application of the first-order upwind scheme, especially for multidimensional cases. A better alternative is to achieve either second-order accuracy by extrapolating the advected variable from values at two upstream cells (the Linear Upwind Differencing Scheme, LUDS, of [34]), or by using two upstream and one downstream nodes (the third-order QUICK scheme proposed by [22]). The use of these two higher order schemes has been widely reported in the literature but, whenever there are strong gradients, the problem of unboundedness arises and a more stable alternative is needed. Other approaches have been proposed to enforce boundedness while preserving higher-order accuracy, such as flux limiters [37] or HRS [23]. This was the methodology adopted in the present work for the treatment of the convection term.

The discretized equation resulting from the finite-difference approximation of Eq. (14) leads to

$$\frac{C_{i+1/2,j} - C_{i-1/2,j}}{\Phi_{i+1/2} - \Phi_{i-1/2}} = \frac{A_{i+1/2}\frac{C_{i+1,j}-C_{i,j}}{\Phi_{i+1}-\Phi_{i}} - A_{i-1/2}\frac{C_{i,j}-C_{i-1,j}}{\Phi_{i}-\Phi_{i-1}}}{\Phi_{i+1/2} - \Phi_{i-1/2}} + \frac{B_{j+1/2}\frac{C_{i,j+1}-C_{i,j}}{\Psi_{j+1}-\Psi_{j}} - B_{j-1/2}\frac{C_{i,j}-C_{i,j-1}}{\Psi_{j}-\Psi_{j-1}}}{\Psi_{j+1/2} - \Psi_{j-1/2}} \tag{17}$$

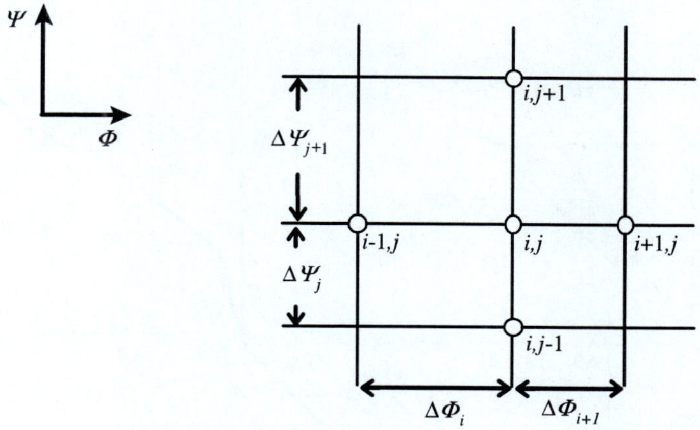

Fig. 2 Sketch of the computational grid near point (i, j) and identification of important variables

where the unknown $C_{i+1/2,j}$ and $C_{i-1/2,j}$ values have to be conveniently interpolated from the calculated node values (represented as circles in Fig. 2) using a HRS to ensure convective stability.

3.1 High-Resolution Schemes

A high-resolution scheme is a combination of different schemes in such a way that boundedness is preserved. In this work the Convergent and Universally Bounded Interpolation Scheme for the Treatment of Advection (CUBISTA) scheme of Alves et al. [1] was adopted to interpolate the $C_{i+1/2,j}$ and $C_{i-1/2,j}$ values needed in the discretized form of the convection term appearing on the l.h.s. of Eq. (17). The Normalized Variable Approach (NVA) of Leonard [23] was adopted, according to which a general differencing scheme of order 3 or less can be expressed as

$$C_{i+1/2,j} = f(C_{i-1,j}, C_{i,j}, C_{i+1,j}) \tag{18}$$

The NVA uses an appropriate upwind biased normalization, and Eq. (18) can be re-written in compact form as

$$\widehat{C}_{i+1/2,j} = g(\widehat{C}_{i,j}) \tag{19}$$

where

$$\widehat{C}_{m,j} = \frac{C_{m,j} - C_{i-1,j}}{C_{i+1,j} - C_{i-1,j}} \quad (m = i-1, i, i+1/2, i+1) \tag{20}$$

since, by definition of Eq. (20), $\widehat{C}_{i-1,j} = 0$ and $\widehat{C}_{i+1,j} = 1$. The unknown dimensionless variable $\widehat{C}_{i+1/2,j}$ is then a function of a single variable.

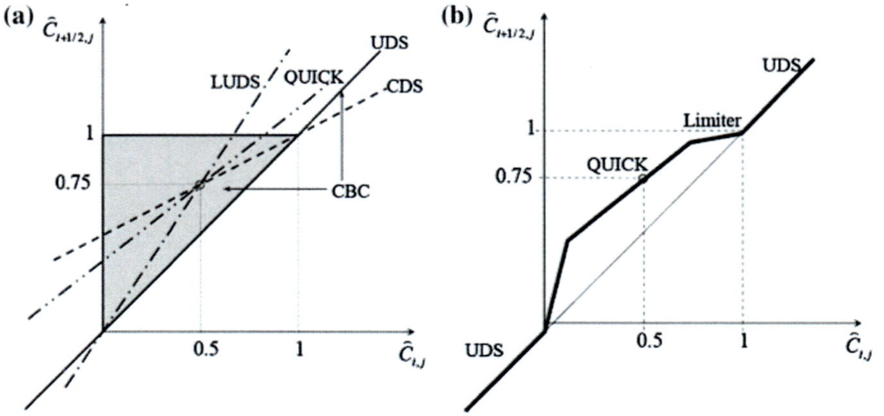

Fig. 3 **a** Normalized variable diagram (NVD) for different interpolating schemes showing the convective boundedness criterion (CBC). (Differencing schemes: UDS≡upwind differencing scheme, LUDS≡linear upwind differencing scheme, QUICK≡quadratic upwind, CDS≡central differencing scheme). **b** The CUBISTA high-resolution scheme on the NVD

The occurrence of unphysical oscillations in the solution can be avoided if the interpolated value of $C_{i+1/2,j}$ lies within bounds imposed by the neighbouring nodes $C_{i,j}$ and $C_{i+1,j}$. Gaskell and Lau [17] formulated the Convection Boundedness Criterion (CBC), which can be conveniently illustrated in the Normalized Variable Diagram (NVD) shown in Fig. 3(a).

According to the CBC, a scheme is bounded if it lies within the shadowed area shown in Fig. 3a, or alternatively along the line of unit slope which represents the upwind scheme. In Fig. 3a the lines corresponding to other commonly used higher-order differencing schemes are also shown for comparison. It may be seen that the only basic (linear) scheme that fully satisfies the CBC is the UDS, but, as already stated, it is only first-order accurate. It can be demonstrated [11] that a second-order scheme necessarily passes through the point $(1/2, 3/4)$ and a third-order scheme, at that point, must have a slope equal to the QUICK scheme $(= 3/4)$.

The creation of a HRS in the context of the NVA is straightforward. First, a generally unbounded, but accurate, basic differencing scheme is selected. Representing that scheme in the NVD, the unbounded regions are easily identified. Piecewise adjustments are performed so that those regions are avoided and the CBC restrictions are respected. In Fig. 3b is shown the CUBISTA scheme that is a bounded version of the QUICK scheme. The CUBISTA scheme may be represented in the context of the NVA by the following piecewise linear function [1]:

$$\widehat{C}_{i+1/2,j} = \begin{cases} \frac{7}{4}\widehat{C}_{i,j} & 0 \le \widehat{C}_{i,j} < \frac{3}{8} \\ \frac{3}{4}\widehat{C}_{i,j} + \frac{3}{8} & \frac{3}{8} \le \widehat{C}_{i,j} \le \frac{3}{4} \quad \text{QUICK} \\ \frac{1}{4}\widehat{C}_{i,j} + \frac{3}{4} & \frac{3}{4} < \widehat{C}_{i,j} \le 1 \\ \widehat{C}_{i,j} & \text{elsewhere} \quad \text{UDS} \end{cases} \tag{21}$$

It is computationally advantageous to express Eq. (21) as a combination of linear equations of the form

$$\widehat{C}_{i+1/2,j} = A_{i+1/2} + B_{i+1/2}\widehat{C}_{i,j} \tag{22}$$

which may be re-dimensioned to finally express $C_{i+1/2,j}$ and $C_{i-1/2,j}$ as a function of its neighbouring node values:

$$C_{i+1/2,j} = A_{i+1/2}C_{i+1,j} + B_{i+1/2}C_{i,j} + (1 - A_{i+1/2} - B_{i+1/2})C_{i-1,j} \tag{23}$$

where the $A_{i+1/2}$ and $B_{i+1/2}$ coefficients are calculated from Eq. (22), depending on the local value of $\widehat{C}_{i,j}$, calculated by Eq. (20), using the concentration field values obtained in the previous iteration.

Substitution of Eq. (23) in (17), with A and B given by Eqs. (15a, 15b), respectively, leads to the final form of the discretized equation, which can be expressed in compact form as

$$E\,C_{i,j} = F\,C_{i-2,j} + G\,C_{i-1,j} + H\,C_{i+1,j} + I\,C_{i,j-1} + J\,C_{i,j+1} \tag{24}$$

with

$$E = \frac{A_{i+1/2}}{\Delta\Phi_{i+1}\Delta\Phi_{i+1/2}} + \frac{A_{i-1/2}(1-\Delta\Phi_i)}{\Delta\Phi_i\Delta\Phi_{i+1/2}} + \frac{B_{i+1/2}}{\Delta\Phi_{i+1/2}} + \\ + \frac{B_{j+1/2}}{\Delta\Psi_{j+1}\Delta\Psi_{j+1/2}} + \frac{B_{j-1/2}}{\Delta\Psi_j\Delta\Psi_{j+1/2}} \tag{25a}$$

$$F = \frac{1 - A_{i-1/2} - B_{i-1/2}}{\Delta\Phi_{i+1/2}} \tag{25b}$$

$$G = \frac{A_{i-1/2}}{\Delta\Phi_i\Delta\Phi_{i+1/2}} - \frac{1 - A_{i+1/2} - B_{i-1/2} - B_{i+1/2}}{\Delta\Phi_{i+1/2}} \tag{25c}$$

$$H = \frac{A_{i+1/2}(1-\Delta\Phi_{i+1})}{\Delta\Phi_{i+1}\Delta\Phi_{i+1/2}} \tag{25d}$$

$$I = \frac{B_{j-1/2}}{\Delta\Psi_j\Delta\Psi_{j+1/2}} \tag{25e}$$

$$J = \frac{B_{j+1/2}}{\Delta\Psi_{j+1}\Delta\Psi_{j+1/2}} \tag{25f}$$

Equation (24) was solved iteratively using the Successive Over-Relaxation (SOR) method [14] as described in a previous work [20].

Although the convective term was discretized with the CÜBISTA high-resolution scheme (which has a theoretical third-order accuracy), it is expected that Eq. (24) is second-order accurate, due the CDS scheme used for the discretization of the diffusive terms.

3.2 Boundary Conditions

For a given value of Pe', the domain of integration was chosen so that $C < 10^{-4}$ along the boundary lines corresponding to $\Phi = \Phi_{min}$ and $\Psi = \Psi_{max}$, i.e., $C_{1,j} < 10^{-4}$ and $C_{i,N_\Psi} < 10^{-4}$. This is a satisfactory substitute for boundary conditions expressed by equations (16a, 16e).

The boundary condition corresponding to Eq. (16b) was not used as such, because that would have required the use of very large values of Φ_{max} (see Fig. 4). Instead, a moderate value of Φ_{max} was adopted (typically $\Phi_{max} = 5$), with the corresponding grid value of C being estimated from the three neighbouring concentrations on the left obtained in the previous iteration. A second order interpolating polynomial was used for that purpose. Finally, the specification of the condition of symmetry, given by Eq. (16d), was approximated by means of a second order interpolating polynomial [2], so the concentration on the nodes along the axis of symmetry (j index equal to 1) is

$$C_{i,1}^{(itern+1)} = \frac{C_{i,2}^{(itern)}(\Psi_3 - \Psi_1)^2 - C_{i,3}^{(itern)}(\Psi_2 - \Psi_1)^2}{(\Psi_3 - \Psi_1)^2 - (\Psi_2 - \Psi_1)^2} \tag{26}$$

where the index *itern* represents the iteration number.

3.3 Computational Mesh

For the situation under study, an orthogonal mesh, in the (Φ, Ψ) coordinate system, is adequate (see Fig. 4). Special care was taken to ensure proper refinement in the regions where the highest concentration gradients were expected. The computational domain used in the calculations ranged from $\Phi = -10^8$ to 10^9 and from $\Psi = 0$ to 10^{12} in all the tested cases (Pe'varied from 10^{-3} to 10^5), in order to adequately describe boundary conditions (16a–16e). It should be emphasized that the computational mesh needs to be very refined near the soluble sphere and coarse away from it (where the concentration gradients are smaller), given the great extent of the computational domain. This means that the length scale of the computational mesh varies over several orders of magnitude and this requires that the numerical method be both accurate and robust, especially for highly advective flow conditions. For all numerical simulations, mesh refinement studies were undertaken in order to obtain accurate results. Typically, three (or, in some cases, more) computational meshes were used, to be able to apply Richardson's extrapolation technique [14] in order to estimate the numerical accuracy. For all results reported in this work, the accuracy is better than 0.1 %. We note that Fig. 4 is simply an illustration of the transformation from the physical to the computational domain. The values of Φ_{min}, Φ_{max} and Ψ_{max} chosen depend on Pe' values

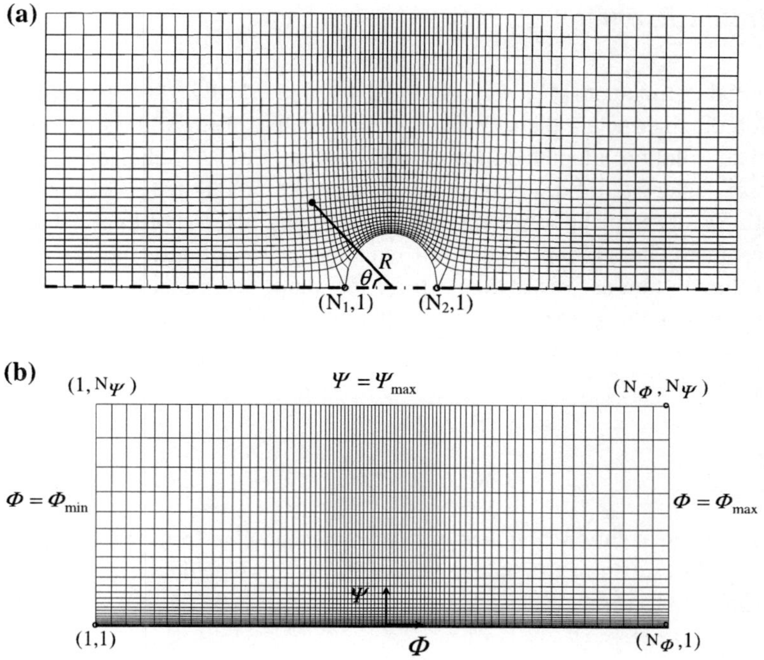

Fig. 4 Sketch of domain of computation. **a** Physical domain; **b** Computational mesh

and d/d_1. The mesh represented is rather coarse, and actually the numerical results were obtained in much more refined, and longer, meshes. Typically, the refined meshes used in the calculations had an order of 10^5 grid points.

It should be noted that we always started our calculations with a zero concentration field on a coarse grid. A converged solution could be obtained very quickly (O(10 s) in a desktop PC with a 1.4 GHz AMD® processor) and then we proceeded to a finer grid, by doubling the number of grid points in each direction. Instead of restarting the calculations with a zero concentration field in this new finer grid, we simply interpolated the solution obtained in the coarse-level grid, leading to a significant decrease in the time of CPU required to attain convergence. This fully automated procedure was repeated until the finest mesh calculations were performed. The use of Richardson's extrapolation to the limit allowed us to obtain very accurate solutions (with errors in the computed Sh' value below 0.1 %). A more elaborate multi-grid technique could have been implemented to further increase the convergence rate, but we found that this simple technique was adequate to obtain mesh-independent solutions in affordable CPU times.

3.4 Initial Estimate

The initial estimate of $C_{i,j}$ was obtained by numerically solving the parabolic equation obtained when the second term on the l.h.s. of Eq. (14) is neglected; the numerical solution adopted followed the implicit Crank–Nicholson method [2]. The term neglected in Eq. (14) represents transport by longitudinal dispersion and, for high Pe', its contribution is expected to be negligible. Accordingly, under those circumstances, integration of the full equation gave virtually the same concentration distribution as was obtained from the simplified equation. For lower Pe', the solution of the parabolic equation merely gives an initial estimate for the iterative process.

3.5 Results

The converged solution obtained yields values of $C_{i,j}$, from which the overall mass-transfer rate from the 'active sphere', $\dot{n}$, could be calculated and expressed by means of an average Sherwood number,

$$Sh' = \frac{kd_1}{D'_m} = \left[\frac{\dot{n}}{\pi\, d_1^2}(c^* - c_0)\right]\frac{d_1}{D'_m} \tag{27}$$

With reference to Fig. 4, the value of $\dot{n}$ will have to be equal to the net amount of solute leaving the domain of integration along the line joining the grid points $(N_\Phi, 1)$ and (N_Φ, N_Ψ), since $c \cong c_0$ all along the upper boundary of that domain. In the discretized form this may be expressed as

$$\dot{n} \equiv 2\,\pi\,\varepsilon\left\{\left(c - D_L\frac{\partial c}{\partial \varphi}\right)_{N_\Phi,1}\frac{\Delta\psi_2}{2} + \sum_{j=2}^{N_\Psi - 1}\left[\left(c - D_L\frac{\partial c}{\partial \varphi}\right)_{N_\Phi,j}\frac{\Delta\psi_{j+1}+\Delta\psi_j}{2}\right]\right\} \tag{28}$$

or, after some manipulation

$$\frac{Sh'}{\varepsilon} = Pe'\left\{\left(C - A\frac{\partial C}{\partial \Phi}\right)_{N_\Phi,1}\Delta\Psi_2 + \sum_{j=2}^{N_\Psi - 1}\left[\left(C - A\frac{\partial C}{\partial \Phi}\right)_{N_\Phi,j}(\Delta\Psi_{j+1}+\Delta\Psi_j)\right]\right\} \tag{29}$$

An alternative route to the evaluation of $\dot{n}$ is by integration of the diffusional flux all around the surface of the sphere

$$\dot{n} = \varepsilon\sum_{i=N_1+1}^{N_2-1} -D_T a\sin(\theta_i)\frac{3}{2}u_0\sin(\theta_i)\left(\frac{\partial c}{\partial \psi}\right)_{i,1} 2\pi a^2\left[\frac{\cos(\theta_{i-1}) - \cos(\theta_{i+1})}{2}\right] \tag{30}$$

which in the dimensionless discretized form results in

$$\frac{Sh'}{\varepsilon} = -\frac{3}{8}\sum_{i=N_1+1}^{N_2-1}\left[1+\frac{3}{2}\frac{Pe'(d/d_1)}{Pe_T(\infty)}\sin(\theta_i)\right]\sin^2(\theta_i)\left(\frac{\partial C}{\partial \Psi}\right)_{i,1}\left[\frac{\cos(\theta_{i-1})-\cos(\theta_{i+1})}{2}\right] \tag{31}$$

where $Pe_T(\infty)$ is the asymptotic value of Pe_T when Reynolds number tends to infinite.

4 Mass Transfer

In our calculations, for any chosen grid, the value of Sh'/ε was calculated numerically by both methods above and if the results differed by more than 1 %, the grid was refined until satisfactory agreement was reached. This gives additional confirmation that the numerical solutions had been refined until grid independence was achieved. Numerical solutions were worked out for many pairs of Pe' and d/d_1, in the ranges $0.1 \leq Pe' \leq 10^5$ and $0 \leq d/d_1 \leq 0.2$.

4.1 Case I: Molecular Diffusion (low $Pe'(d/d_1)$ values)

For $Pe'(d/d_1)$ lower than 0.1, dispersion is the direct result of molecular diffusion, with $D_T = D_L = D'_m$. The numerical solutions for this condition were worked out taking $d/d_1 \to 0$ in Eqs. (14, 29, 31) and the values obtained are shown as dots in the plot of Fig. 5.

As expected $Sh'/\varepsilon \to 2$ when $Pe' \to 0$, whereas for high Pe', the asymptote for convection with molecular diffusion across a thin boundary layer is observed, $Sh'/\varepsilon = \left[4Pe'/\pi\right]^{1/2}$, which is consistent with the result of La Nauze (1984), corrected by Guedes de Carvalho and Coelho [18].

The quadratic mean of the two asymptotes is

$$\left(\frac{Sh'}{\varepsilon}\right)_{quadmean} = \left[4+\frac{4}{\pi}Pe'\right]^{1/2} \tag{32}$$

and values of Sh'/ε calculated from this equation differ at most by 12 % from the corresponding numerical solution obtained in the present work. An improved approximation, inspired by the form of Eq. (32), is

$$\frac{Sh'_{md}}{\varepsilon} = \left[4+\frac{4}{5}(Pe')^{2/3}+\frac{4}{\pi}Pe'\right]^{1/2} \tag{33}$$

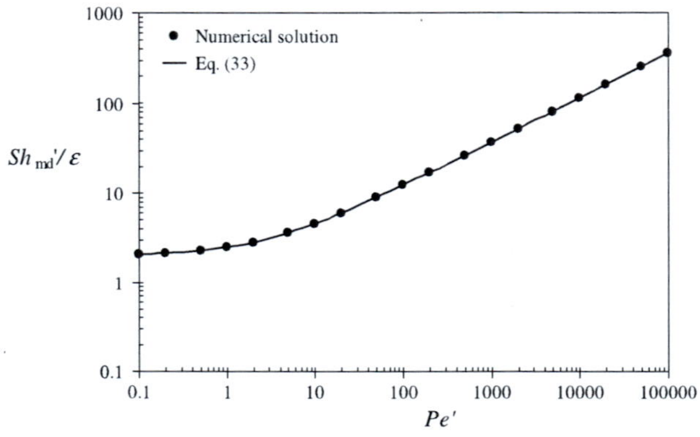

Fig. 5 Dependence of Sh'_{md}/ε on Pe' for soluble sphere

which is represented as a full line in Fig. 5, and gives values of Sh'_{md}/ε that differs less than 1 % from those obtained by the numerical solution described above. It should be stressed that Eq. (33) is a general result, valid for a gas or liquid flowing around a sphere buried in a packed bed of inert particles, under conditions for which molecular diffusion dominates over convective dispersion. For that reason, we shall use the symbol Sh'_{md} to represent the value of Sh' given by Eq. (33).

4.2 Case II: Intermediate to High $Pe'(d/d_1)$ Values

As $Pe'(d/d_1)$ is increased above 0.1, values of D_T and D_L start to deviate from D'_m and this situation has to be taken into account in the integration of Eq. (14). For gas flow, through beds of approximately isometric particles [7, 39], it is generally accepted, for the entire range of values of $Pe'(d/d_1)$, that good approximated values are given by

$$D_L = D'_m + \frac{ud}{Pe_L(\infty)} \quad \text{or} \quad \frac{D_L}{D'_m} = 1 + \frac{Pe'(d/d_1)}{Pe_L(\infty)}\frac{u}{u_0} \tag{34}$$

$$D_T = D'_m + \frac{ud}{Pe_T(\infty)} \quad \text{or} \quad \frac{D_T}{D'_m} = 1 + \frac{Pe'(d/d_1)}{Pe_T(\infty)}\frac{u}{u_0} \tag{35}$$

with $Pe_L(\infty) = 2$ and $Pe_T(\infty) = 12$, for flow.

The numerical solution of Eq. (14), with D_L and D_T given by Eqs. (34, 35), was worked out by Guedes de Carvalho and Alves [20], who observed that the values of Sh' obtained from the numerical solution are well represented (within 2 % deviation in Sh') by

$$\frac{Sh'}{\varepsilon} = \frac{Sh'_{md}}{\varepsilon}.\eta \tag{36}$$

with,

$$\eta = \left(1 + \frac{Pe'}{9}\frac{d}{d_1}\right)^{1/2} \tag{37}$$

Now, for liquids, since Eqs. (34) and (35) are still good approximations, at low to intermediate values of $Pe'(d/d_1)$ (up to 1, for D_L and up to 10, for D_T), Eq. (36) will be adequate, but only for this narrow range of $Pe'(d/d_1)$. Beyond that range, more accurate values of D_T and D_L are required, in the numerical solution of Eq. (14).

Values of D_T for liquid flow were reported by Delgado [13], in what seems to be the only available study on the influence of Schmidt number, *Sc*, over D_T. His data, for the range of values of $Pe'(d/d_1)$ of interest in the present study, shown an empirical correlation of D_T as a function of the Schmidt number. For $Sc \leq 550$,

$$\frac{D_T}{D'_m} = 1 + \frac{Pe'(d/d_1)}{12} - \left(\frac{Sc}{1500}\right)^{4.8} [Pe'(d/d_1)]^{4.83 - 1.3\log_{10}(Sc)} \tag{38}$$

For $Sc > 550$, the transverse dispersion coefficient is found to be independent of the Schmidt number, and the correlation reduces to:

$$\frac{D_T}{D'_m} = 1 + \frac{Pe'(d/d_1)}{12} - 8.1 \times 10^{-3}[Pe'(d/d_1)]^{1.268} \tag{39}$$

In the case of D_L, it is fortunate that its value is not needed with accuracy, since, for $Pe'(d/d_1) > 1$, the boundary layer for mass transfer around the sphere is thin, provided by the fact that the approximate condition $d_1/d > 10$ is observed. Indeed, for $Pe' > 10$, the boundary layer is thin and the term with D_L, in Eq. (14), may be neglected [8], numerical computations were undertaken in the present work that confirm the independence of Sh$'$ from D_L, for $Pe' > 10$.

Equation (14) was solved numerically for different values of Sc, being the values of D_T given by the corresponding fitted curve (Eqs. 38 or 39).

These numerical simulations results, let us to prepare plots of η *vs.* $Pe'(d/d_1)$, for given values of d/d_1, in a similar fashion to what was done by Guedes de Carvalho and Alves [20]. The results of the numerical computations are shown as points in Fig. 6 and an expression was sought to describe the functional dependence observed, with good accuracy. The following equation is proposed for $Sc \leq 550$:

$$\eta = \left[1 + \frac{Pe'(d/d_1)}{12} - \left(\frac{Sc}{1500}\right)^{4.8}\left(\frac{4}{3}Pe'\frac{d}{d_1}\right)^{4.83 - 1.3\log_{10}(Sc)}\right]^{1/2} \tag{40}$$

For $Sc > 550$, since D_T is independent from *Sc*, the value of Sh' becomes also independent from *Sc*, being possible to obtain a simplified version of the above correlation, by merely substituting the *Sc* value by 550

$$\eta = \left[1 + \frac{Pe'(d/d_1)}{9} - 1.16 \times 10^{-2}[Pe'(d/d_1)]^{1.268}\right]^{1/2} \tag{41}$$

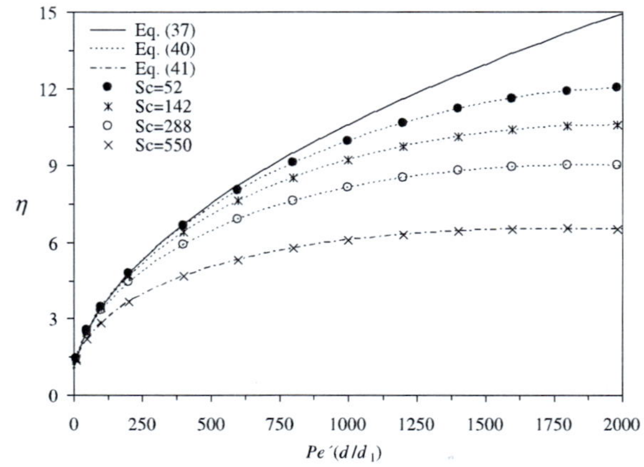

Fig. 6 Dependence of η on $Pe'(d/d_1)$ for different values of Sc. The points were obtained from the numerical solution of Eq. (14)

In the plots of Fig. 6, the solid lines represent Eqs. (40) or (41) and it may be seen that they describe the results of the numerical computations with very good accuracy.

It will be noticed that the enhancement factor is dependent on Sc only if $Sc \leq 550$, which is in accordance to the fact that mass transfer rates around the sphere depend strongly on D_T for small values of Sc, as shown recently by Delgado [13] in a detailed study on dispersion in liquids.

4.3 Concentration Contours

In the present section attention is restricted to those situations for which $Pe'(d/d_1)$ is low (i.e., $Pe'(d/d_1) < 0.1$), in which case $D_T \cong D_L \cong D'_m$; the concentration profiles, in dimensionless form, are then only a function of Pe', as may be seen from Eq. (17). This is a simplification which restricts the applicability of the results to low fluid velocities, but it is an important first step towards establishing the methodology for a more general solution.

The results of the numerical computation were used to draw concentration contour plots for a wide range of values of Pe', as illustrated in Figs. 7, 8, 9 (note that the radius of the 'active sphere' is $a = d_1/2$). It is worth recalling that for $Pe' = 0$ the situation is that of pure diffusion around the sphere and the well known theoretical result for the concentration contours, $C = 1/R$, is applied. As the value of Pe' is increased, the contour surfaces of constant concentration are gradually distorted, as if "blown" by the flowing fluid.

For low values of Pe' (up to $Pe' \cong 20$), the concentration contour surfaces have an ovoid shape, elongated towards the downstream side, for all values of C, as shown in Fig. 7. For $Pe' > 20$, and low enough values of C, the ovoid shape is still observed as it can be seen in Fig. 8. However, for higher values of C, the pronounced curvature of the flow lines, in the vicinity of the sphere, will influence

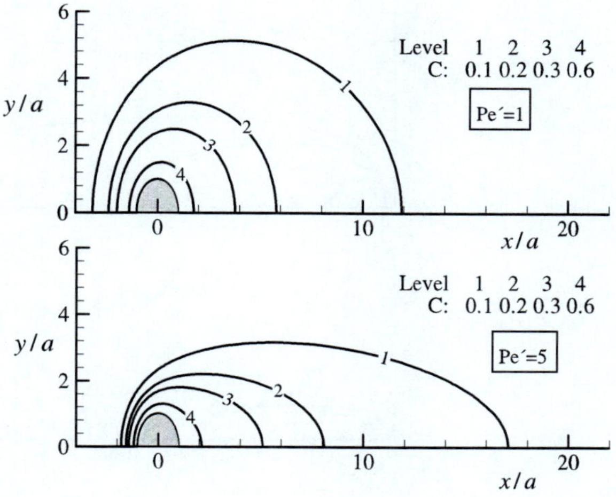

Fig. 7 Concentration contours for two low values of Pe'; the soluble sphere corresponds to the shaded area

the shape of the concentration contour surfaces; looking at Fig. 9, the plot for $Pe' = 200$ resembles the shape of a "comet".

The range $Pe' > 20$ is of great practical interest, for example in studies of groundwater contamination and it is interesting to look at the concentration contours, plotted as in Fig. 9, with different linear scales for y/a and x/a, to cover significant ranges of both variables. The figure shows that the variation of y/a with x/a, along the contour surfaces, goes through a first maximum, close to the sphere and then through a second maximum downstream of the sphere (obviously, for C close to 1, only one maximum is observed).

The parameters X_C (maximum value of x/a for concentration level C) and Y_C (maximum value of y/a for concentration level C, downstream of the sphere) are important in giving some crucial information on the size of the contaminant plume, for a specific value of C. X_C measures the downstream reach of contaminant and its definition is unequivocal. Y_C measures the lateral spread of contaminant, but its definition may lead to ambiguity due to the possible inflection of the contour line.

From the results of the many numerical simulations in the present work, values of X_C *vs.* Pe' and Y_C *vs.* Pe' were obtained for a wide range of values of C, and they are shown as dots in the plots of Figs. 10 and 11. A close examination of the trends followed by each set of points at constant C, in the plot of Fig. 10, reveals two asymptotes for x/a, namely

$$X_C = \frac{1}{C} \quad \text{for} \quad Pe' \to 0 \tag{42}$$

and

$$X_C = \frac{1}{C}\sqrt{\frac{Pe'}{\pi}} \quad \text{for} \quad Pe' \to \infty \text{ and low enough } C\ (C<0.1) \tag{43}$$

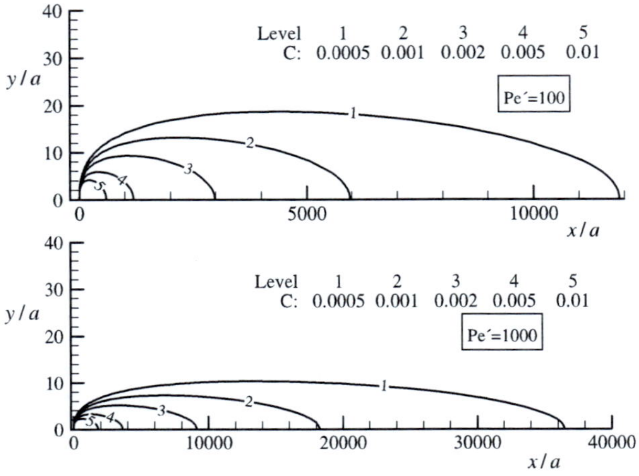

Fig. 8 Concentration contours far from the sphere for two high values of Pe'

Fig. 9 Concentration contours near the sphere for two high values of Pe'; the soluble sphere corresponds to the shaded area

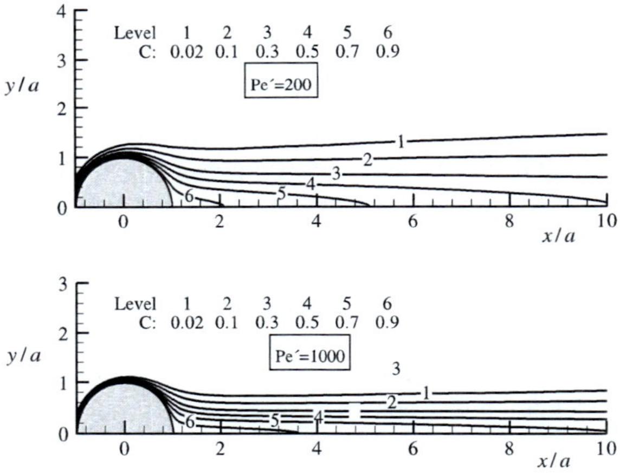

Taking the asymptotes as starting guide lines, an effort was then made to obtain a general approximate expression for the function $X_C(Pe', C)$, which would represent the "numerical points" with good accuracy over the whole range of Pe' and C. Led by intuition and after a few attempts, the function

$$X_C = \sqrt{\left(\frac{1}{C}\right)^2 + \frac{(1-C)^{5/3}}{C^2}\frac{Pe'}{\pi}} \tag{44}$$

was selected and it is represented by the solid lines in Fig. 10. It may be seen that the agreement with the numerical results is very good, throughout; the individual point values do not deviate from the lines by more than 10 %, over the entire range of Pe' and C.

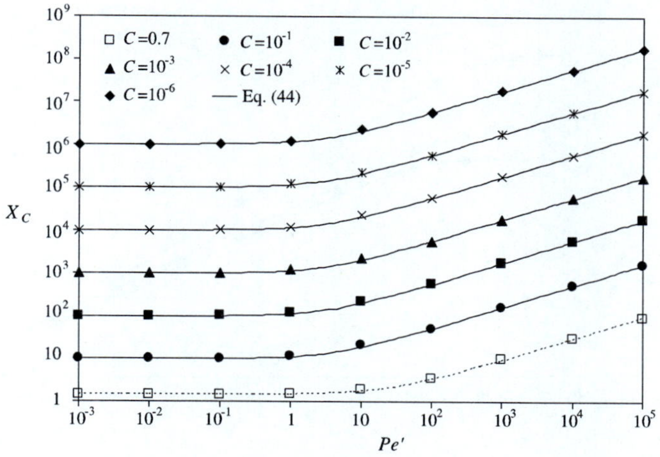

Fig. 10 Variation of X_C with Pe' for different values of C. The points represent the numerical solution and the solid lines correspond to Eq. (44)

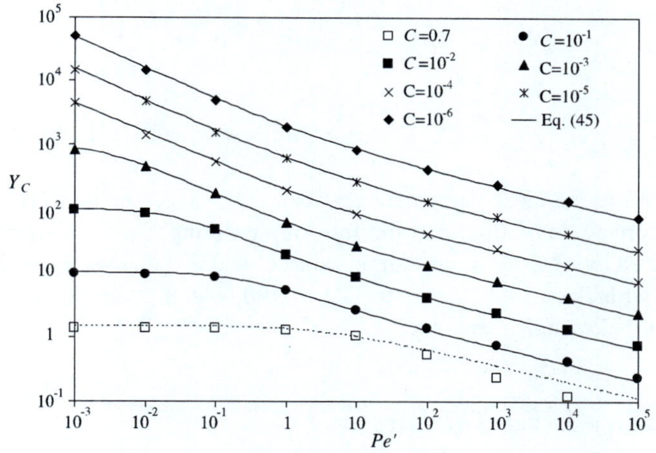

Fig. 11 Variation of Y_C with Pe' for different values of C. The points represent the numerical solution and the solid lines correspond to Eq. (45)

Adopting a similar approach to establish an approximate analytical expression for the functional relationship between Y_C, Pe' and C, led to

$$Y_C = \frac{1}{\sqrt{\frac{1}{\left(\frac{1}{C}\right)^2} + \frac{(1-C^2)Pe'}{\frac{12/5}{C} + \frac{7/5}{C}Pe'^{0.5}}}} \tag{45}$$

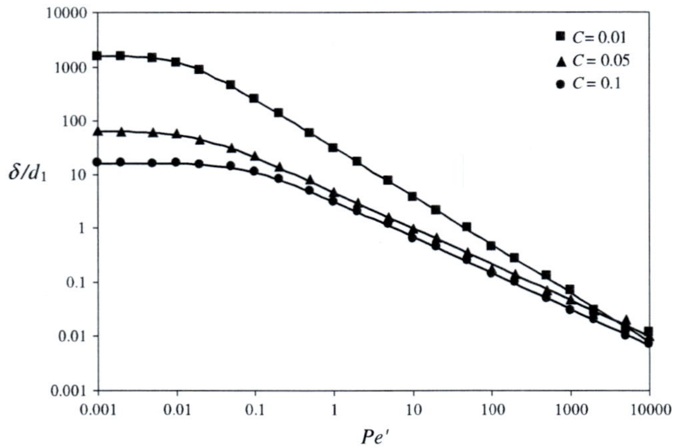

Fig. 12 Dependence of δ/d_1 on Pe' for different values of C. The points represent the numerical solution and the solid lines correspond to Eqs. (52) to (54)

which has the asymptotes

$$Y_C = \frac{1}{C} \quad \text{for}\, Pe' \to 0 \tag{46}$$

$$Y_C = \sqrt{\frac{3}{2}\frac{1}{C}}\frac{1}{Pe'^{0.25}} \quad \text{for} \quad Pe' \to \infty \,\text{and low enough}\, C\, (C < 0.1) \tag{47}$$

Figure 11 shows the close agreement between the "exact" point values, provided by the numerical simulation, and the lines representing Eq. (45), over the entire range of Pe'. Only for the rather large value $C = 0.7$ (taken as an example), the agreement with Eq. (45) is poor, for $Pe' > 100$, but the values of Y_C are then significantly below unity, anyway.

4.4 Boundary Layer Thicknesses

The numerical solution of Eq. (14) gives the concentration field and, from it, the dimensionless concentration boundary layer thickness, δ/d_1, for $C = 0.01$, $C = 0.05$ and $C = 0.1$.For very low $Pe'(d/d_1)$ values, dispersion is the direct result of molecular diffusion, with $D_T = D_L = D'_m$. The plot of Fig. 12 reveal three well-known asymptotes, for high values of Pe' :

$$\left(\frac{\delta}{d_1}\right)_{C=0.1} = \frac{\pi}{Pe'^{2/3}} \tag{48}$$

$$\left(\frac{\delta}{d_1}\right)_{C=0.05} = \frac{\pi^{4/3}}{Pe'^{2/3}} \tag{49}$$

$$\left(\frac{\delta}{d_1}\right)_{C=0.01} = \frac{\pi^3}{Pe'^{0.9}} \tag{50}$$

and for low values of Pe', $Pe' \to 0$, Fig. 12 reveals the following asymptote, for all values of C,

$$\left(\frac{\delta}{d_1}\right)_C = \frac{1}{2\pi}\frac{1}{C^2} \tag{51}$$

Taking the asymptotes as starting guide lines, an effort was then made to obtain a general approximate expression, which would represent the "numerical points" with good accuracy over the whole range of Pe'. The function obtained were

$$\left(\frac{\delta}{d_1}\right)_{C=0.1} = \frac{1}{\left[4\pi^2C^4 + \frac{Pe'^{4/3}}{\pi^2}\right]^{0.5}} \tag{52}$$

$$\left(\frac{\delta}{d_1}\right)_{C=0.05} = \frac{1}{\left[4\pi^2C^4 + \frac{Pe'^{4/3}}{\pi^{8/3}}\right]^{0.5}} \tag{53}$$

$$\left(\frac{\delta}{d_1}\right)_{C=0.01} = \frac{1}{\left[4\pi^2C^4 + \frac{Pe'^{9/5}}{\pi^6}\right]^{0.5}} \tag{54}$$

and the numerical values do not deviate by more than 7 % from the values given by Eqs. (52–54), over the entire range of values of Pe'.

5 Conclusions

The problem of solute migration from an active sphere buried in a packed bed of inert particles, through which fluid flows with uniform velocity (far from the sphere), was treated in detail. The PDE resulting from the differential mass balance was solved numerically over a wide range of values of the relevant parameters, and general expressions, given as Eqs. (33, 37, 40, 41), were obtained to relate the Sherwood number with the Peclet number, Schmidt number and d/d_1.

The PDE describing mass transfer in the interstitial fluid was solved numerically to obtain concentration contour plots of the solute, around the sphere, for a very wide range of values of Peclet number. The concentration contour surfaces were then inspected to determine the maximum downstream stretch and approximate maximum lateral spread of the contaminant, for each concentration level. Equations (44) and (45) were found to relate those two parameters accurately with Peclet number and dimensionless solute concentration and provide useful formulae for prediction of contaminant plume sizes.

Finally, the partial differential equation was solved numerically and general expressions, given as Eqs. (52–54), were obtained to relate the concentration boundary layer thickness with the Peclet number. The theory developed applies only in the range $Pe' < 0.2$, over which dispersion is the result of only advection and molecular diffusion.

Nomenclature

a	"Active sphere" radius
A	Dimensionless parameter, Eq. (15a)
B	Dimensionless parameter, Eq. (15b)
c	Solute concentration
c^*	Equilibrium solute concentration
c_0	Background concentration
C	Normalized concentration
d	Inert particles diameter
d_1	"Active sphere" diameter
D_L	Longitudinal dispersion coefficient
D_m	Molecular diffusion coefficient
D'_m	Effective molecular diffusion coefficient, $D'_m = D_m/\tau$
D_T	Transverse dispersion coefficient
itern	Iteration number
k	Mass transfer coefficient
K	Permeability in Darcy's law
n	Overall mass-transfer rate from the "active sphere"
N_Ψ	Number of elements on direction Ψ
N_Φ	Number of elements on direction Φ
p	Pressure
r	Spherical coordinate
u	Interstitial velocity value
$\boldsymbol{u}$	Interstitial velocity vector
u_0	Interstitial velocity of the fluid
u_r	Radial component of the interstitial velocity, Eq. (5)
u_θ	Radial component of the interstitial velocity, Eq. (6)
x	Axial coordinate
X_C	Maximum value of x/a for concentration level C
y	Axial coordinate
Y_C	Maximum value of y/a for concentration level C
E	Dimensionless parameter, Eq. (25a)
F	Dimensionless parameter, Eq. (25b)
G	Dimensionless parameter, Eq. (25c)
H	Dimensionless parameter, Eq. (25d)
I	Dimensionless parameter, Eq. (25e)
J	Dimensionless parameter, Eq. (25f)

Dimensionless groups

C	Dimensionless concentration, Eq. (8)
Pe	Peclet number, $Pe = u_0 d_1 / D_m$
Pe_L	Longitudinal Peclet number, $Pe_L = u_0 d_1 / D_L$
Pe_T	Transversal Peclet number, $Pe_T = u_0 d_1 / D_T$
Pe'	Peclet number based on effective molecular diffusion coefficient, $Pe' = u_0 d_1 / D'_m$
R	Dimensionless "active sphere" radius, Eq. (10)
S	Schmidt number, $S_c = \mu / (\rho D_m)$
Sh'	Sherwood number based on effective molecular diffusion coefficient, $Sh' = k d_1 / D'_m$
Sh'_{md}	Sherwood number based on effective molecular diffusion coefficient for a molecular diffusion regime
U	Dimensionless interstitial velocity, Eq. (9)

Greek letters

δ	Boundary layer thickness
ρ	Density
Φ	Dimensionless potential function, Eq. (11)
Ψ	Dimensionless stream function, Eq. (12)
ω	Distance to the axis, Fig. 1
μ	Dynamic viscosity
η	Enhancement factor, Eqs. (37, 40, 41)
ϕ	Flow potential
θ	Spherical coordinate
ψ	Stream function
ε	Bed voidage
τ	Tortuosity

Acronyms

CBC	Convection Boundedness Criterion
CDS	Central Differencing Scheme
CFD	Computational Fluid Dynamics
CUBISTA	Convergent and Universally Bounded Interpolation Scheme for the Treatment of Advection
HRS	High-Resolution Scheme
LUDS	Linear Upwind Differencing Scheme
NAPL	Non-Aqueous Phase Liquids
NVA	Normalized Variable Approach
NVD	Normalized Variable Diagram
PDE	Partial Differential Equations
QUICK	Quadratic Upwind

SOR Successive Over-Relaxation
UDS Upwind Differencing Scheme

Subscripts and superscripts

i, m Horizontal location in the computational mesh
j Vertical location in the computational mesh
max Maximum
min minimum
md Molecular diffusion
$'$ A primed is used in $D'_m = D_m/\tau$ and in all dimensionless groups containing, D'_m

References

1. Alves, M.A., Oliveira, P.J., Pinho, F.T.: A convergent and universally bounded interpolation scheme for the treatment of advection. Int. J. Numer. Meth. Fl. **41**, 47–75 (2003)
2. Anderson, J.D.: Computational Fluid Dynamics. McGraw-Hill, New York (1995)
3. Avedesian, M.M., Davidson, J.F.: Combustion of carbon particles in a fluidised bed. Trans. Inst. Chem. Eng. **51**, 121–131 (1973)
4. Brian, P.L.T., Hales, H.B.: Effects of transpiration and changing diameter on heat and mass transfer to spheres. AIChE J. **15**, 419–425 (1969)
5. Chakraborty, R.K., Howard, J.R.: Combustion of single carbon particles in fluidized beds of high-density alumina. J I Energy **54**, 48–54 (1981)
6. Cheng, C.Y.: Double diffusion from a vertical wavy surface in a porous medium saturated with a non-Newtonian fluid. Int. Commun. Heat Mass **34**, 285–294 (2007)
7. Coelho, M.N., Guedes de Carvalho, J.R.: Transverse dispersion in granular beds Part I- Mass transfer from a wall and the dispersion coefficient in packed beds. Chem. Eng. Res. Des. **66**, 165–177 (1988)
8. Coelho, M.N., Guedes de Carvalho, J.R.: Transverse dispersion in granular beds Part II- Mass transfer from large spheres immersed in fixed or fluidised beds of small inert particles. Chem. Eng. Res. Des. **66**, 178–189 (1988)
9. Courant, R., Isaacson, E., Rees, M.: The solution of nonlinear hyperbolic differential equations by finite differences. Commun. Pur. Appl. Math. **5**, 243–255 (1952)
10. Currie, I.G.: Fundamental Mechanics of Fluids. McGraw-Hill, New York (1993)
11. Darwish, M.S., Moukalled, F.: Normalized variable and space formulation methodology for high-resolution schemes. Numer. Heat Transfer, Part B **26**, 79–96 (1994)
12. Davidson, J.F., Harrison, D.: Fluidised Particles. Cambridge University Press, Cambridge (1963)
13. Delgado, J.M.P.Q.: Longitudinal and transverse dispersion in porous media. Chem. Eng. Res. Des. **85**, 386–394 (2007)
14. Ferziger, J.H., Peric, M.: Computational Methods for Fluid Dynamics. Springer-Verlag, Berlin (1996)
15. Fetter, C.W.: Contaminant Hydrogeology. Prentice-Hall, Upper Sadle River (1999)
16. Freitas, C.J.: Policy statement on the control of numerical accuracy. J. Fluid Eng-T. ASME **115**, 339–340 (1993)
17. Gaskell, P.H., Lau, A.K.C.: Curvature compensated convective transport: SMART a new boundedness preserving transport algorithm. Int. J. Numer. Meth. Fl. **8**, 617–641 (1988)
18. Guedes de Carvalho, J.R.F., Coelho, M.A.N.: Comments on mass transfer to large particles in fluidized beds of smaller particles. Chem. Eng. Sci. **41**, 209–210 (1986)

19. Guedes de Carvalho, J.R., Pinto, A.R., Pinho, C.T.: Mass transfer around carbon particles burning in fluidised beds. Trans. IChemE **69**, 63–70 (1991)
20. Guedes de Carvalho, J.R.F., Alves, M.A.: Mass transfer and dispersion around active sphere buried in a packed bed. AIChE J. **45**, 2495–2502 (1999)
21. La Nauze, R.D., Jung, K., Kastl, J.: Mass transfer to large particles in fluidized beds of smaller particles. Chem. Eng. Sci. **39**, 1623–1633 (1984)
22. Leonard, B.P.: A stable and accurate convective modelling procedure based on quadratic upstream interpolation. Comput. Method Appl. M. **19**, 59–98 (1979)
23. Leonard, B.P.: Simple high-accuracy resolution program for convective modelling of discontinuities. Int. J. Numer. Meth. Fl. **8**, 1291–1318 (1988)
24. Leung, L.A., Smith, I.W.: Role of fuel reactivity in fluidized-bed combustion. Fuel **58**, 354–360 (1979)
25. Middleman, S.: An Introduction to Mass and Heat Transfer. Wiley, New York (1998)
26. Nebbali, R., Bouhadef, K.: Numerical study of forced convection in a 3D flow of a non-Newtonian fluid through a porous duct. Int. J. Numer. Method H. **16**, 870 (2006)
27. Patankar, S.V.: Numerical Heat Transfer and Fluid Flow (Series in Computational Methods in Mechanics and Thermal Science). Hemisphere Pub Corp, Washington (1980)
28. Phillips, O.M.: Flow and Reactions in Permeable Rocks. Cambridge University Press, Cambridge (1991)
29. Pinto, A.R., Guedes de Carvalho, J.R.: Transverse dispersion in granular beds Part III- Mass transfer around particles dispersed in granular beds of inerts and the combustion of carbon particles in beds of sand. Trans. IChemE **68**, 503–509 (1990)
30. Prandtl, L.: Eine beziehung zwishen wärmeaustausch und strömungwiderstand der flüssigkeiten. Phys. Z. **11**, 1072 (1910)
31. Prins, W., Casteleijn, T.P., Draijer, W., Van Swaaij, W.P.M.: Mass transfer from a freely moving single sphere to the dense phase of a gas fluidized bed of inert particles. Chem. Eng. Sci. **40**, 481–497 (1985)
32. Ranz, WE., Marshall, WR. Jr: Evaporation from drops. Chem. Eng. Prog. **48**, 141–146 (Part I) and 173–180 (Part II) (1952)
33. Schlichting, H.: Boundary Layer Theory. McGraw-Hill, New York (1979)
34. Shyy, W.: A study of finite difference approximations to steady-state convection-dominated flow problems. J. Comput. Phys. **57**, 415–438 (1985)
35. Spalding, D.B.: A novel finite-difference formulation for differential expressions involving both first and second derivatives. Int. J. Numer. Meth. Eng. **4**, 551–559 (1972)
36. Stoessell, R.K.: Mass transport in sandstones around dissolving plagioclase grains. Geology **15**, 295–298 (1987)
37. Sweby, P.K.: High resolution schemes using flux limiters for hyperbolic conservation-laws. SIAM J. Numer. Anal. **21**, 995–1011 (1984)
38. Van Heerden, C., Nobel, A.P.P., Krevelen, D.W.: Mechanism of heat transfer in fluidised beds. Ind. Eng. Chem. **45**, 1237–1242 (1953)
39. Wilhelm, R.H.: Progress towards the a priori design of chemical reactors. Pure Appl. Chem. **5**, 403–421 (1962)

Mass Transport in Porous Media With Variable Mass

Alfio Grillo, Chiara Giverso, Marco Favino, Rolf Krause, Michael Lampe and Gabriel Wittum

Abstract We present a theoretical and numerical study of mass transport in a porous medium saturated with a fluid and characterised by an evolving internal structure. The dynamics of the porous medium and the fluid as well as their reciprocal interactions are described at a coarse scale, so that the fundamental tools of Mixture Theory and Continuum Mechanics can be used. The evolution of the internal structure of the porous medium, which is here primarily imputed either to growth or to mass exchange with the fluid, is investigated by enriching the space of kinematic variables of the mixture with a set of structural descriptors, each of which is power-conjugate to generalised forces satisfying a balance law. Establishing the influence of the structural change of the porous medium on the transport

A. Grillo (✉) · C. Giverso
DISMA Dipartimento di Scienze Matematiche, Politecnico di Torino,
Corso Duca degli Abruzzi 24, I-10129, Torino, Italy
e-mail: alfio.grillo@polito.it

C. Giverso
e-mail: chiara.giverso@polito.it

M. Favino · R. Krause
Institute of Computational Science, USI Universitá
della Svizzera Italiana, Lugano, Switzerland
e-mail: marco.favino@usi.ch

R. Krause
e-mail: rolf.krause@usi.ch

M. Lampe · G. Wittum
G-CSC Goethe Center for Scientific Computing, Goethe Universität Frankfurt,
Kettenhofweg 139, D-60325, Frankfurt am Main, Germany
e-mail: lampe@gcsc.uni-frankfurt.de

G. Wittum
e-mail: wittum@gcsc.uni-frankfurt.de

J. M. P. Q. Delgado et al. (eds.), *Numerical Analysis of Heat and Mass Transfer in Porous Media*, Advanced Structured Materials 27,
DOI: 10.1007/978-3-642-30532-0_2,

properties of the mixture and, thus, on the quantities characterising fluid flow is the crux of our contribution.

Keywords Mass transport · Porous media · Inelastic deformations · Variable mass · Dissipation

1 Introduction

The scope of this contribution is to study the transport of fluid in a deformable porous medium whose mass and internal structure may vary in time. Our level of investigation is coarse enough, so that we can formulate our problem with the aid of Mixture Theory. Thus, we consider a biphasic mixture consisting of a solid and a fluid phase. The solid phase represents the porous medium, whereas the fluid phase consists of a fluid that completely saturates the pores of the solid and may move throughout it. A system of this type is perhaps the most essential model of those soft biological tissues whose main constituents are a porous solid matrix and an interstitial fluid. An example of such tissues is articular cartilage, in which the solid consists of cells (chondrocytes), collagen fibres and matrix of proteoglycans, and the fluid comprises water, ions and various chemical compounds. The latter ones can be either nutrients for the cells or byproducts of the cellular metabolic activity. From here on, we call "fluid phase" the system made of water and all chemicals dissolved in it, and we refer to these as "constituents of the fluid phase".

Among the factors that assess the health of a tissue, an important role is played by the constituents that supply nutriment to the cells and regulate their metabolism. Other relevant constituents are those that can either promote or hinder processes, which can damage the tissue (e.g., growth of tumours). In any of these cases, a mathematical model of a tissue should provide information about the concentration of constituents and the transport processes to which they are subjected. As long as continuum models are concerned (these models do not explicitly track molecular or sub-cellular processes), the evolution of constituents is put in the form of diffusion-advection-reaction equations. Therefore, if the validity of Fick's and Darcy's laws is accepted, it is essential to determine the diffusivity and permeability of the tissue in order to quantify its capability of transporting matter. This capability, on the other hand, depends on geometrical (i.e., geometry of the pore space) and mechanical factors (i.e., deformation and stress), and couplings between them. Indeed, when the tissue deforms, the geometry of the pore space changes and so do the transport properties (diffusivity and permeability). Moreover, different compositions of the fluid phase lead to different hydraulic conductivities. However, we shall neglect this effect in the rest of our study, and we shall focus on the much simpler case of a fluid phase comprising a single constituent only. We shall also assume that the fluid phase is incompressible.

The picture sketched above becomes more complex when the structural changes of the tissue are considered beside deformation. By "structural changes" we mean processes that contribute to modify the properties of the tissue (e.g., the stiffness, diffusivity and permeability) in response to growth, mass exchange between the fluid and the solid phase, and re-organisation of cells and fibrils. Even though these phenomena are all intermingled with each other, a conceptual classification is pointed out in the biomechanical literature (cf., for example, Fung [25] and Taber [57]), where growth and mass exchange are said to lead to the variation of mass of a body, while the re-organisation of cells and fibrils is referred to as remodelling.

From the point of view of Continuum Mechanics, the structural change of the solid phase of a tissue is modelled by means of a class of deformations that describe how the material particles are distributed in the tissue. The mathematical object used to define these deformations is a second-order tensor, which we denote by $\mathbf{F}_a$. With the terminology of [22], $\mathbf{F}_a$ measures the material inhomogeneities triggered by growth, mass exchange processes, and cellular re-organisation. There is, thus, a strong conceptual difference between the standard deformations and those described by $\mathbf{F}_a$: whereas the former ones are related to the gradient of displacement of the body, the latter deformations need not be the gradient of any vector field. Rather, they are primitive entities that define, together with displacements, the parameters that are necessary and sufficient for describing the kinematics of deformable bodies with variable internal structure.

In "classical" Continuum Mechanics[1], tensor $\mathbf{F}_a$ represents the anelastic part of the deformation gradient tensor, $\mathbf{F}_s$, which describes the overall change of shape of a solid. This anelastic deformation may be due to, for example, plastic deformations, thermal distortions, and damage [42]. Tensor $\mathbf{F}_a$ individuates an evolving relaxed configuration of body elements. The accommodating deformation, which determines the actual configuration of the body from the relaxed one, is denoted by $\mathbf{F}_e$, and obeys the multiplicative decomposition $\mathbf{F}_s = \mathbf{F}_e\mathbf{F}_a$ [36, 37, 39]. Usually, $\mathbf{F}_e$ is said to be the elastic part of $\mathbf{F}_s$.

Rodriguez et al. [53] used the decomposition of $\mathbf{F}_s$ to study growth mechanics, and identified $\mathbf{F}_a$ with the deformations due to growth. When $\mathbf{F}_a$ is not the gradient of any vector field, it is said to be incompatible. Physically, this represents the case in which grown material points lose their geometric compatibility (this situation usually leads to residual stresses). The rate of anelastic deformation, $\mathbf{L}_a = \dot{\mathbf{F}}_a(\mathbf{F}_a)^{-1}$, is related to the variation of body mass in such a way that the mass density of the body is constant when measured with respect to the relaxed configuration (cf., for example, [22]).

The kinematic entities $\mathbf{F}_s$ and $\mathbf{F}_a$ are the mathematical objects describing the physical processes that influence the transport properties of a tissue (i.e., diffusivity and permeability). To be more specific, $\mathbf{F}_a$ may be decomposed as the product of tensors, which distinguish the anelastic deformations associated with growth from those associated with the exchange of mass between the solid and the fluid phase.

[1] By "classical" we mean here the Continuum Mechanics that studies non-living matter.

The case in which both processes are modelled together has been studied in [29, 27]. In our present contribution, however, we do not consider mass exchange, so that $\mathbf{F}_a$ accounts for the anelastic deformations associated with growth only.

Under the hypothesis of negligible inertial terms and incompressible solid and fluid phases, the unknowns to be determined are given by the displacement field (whose material gradient is $\mathbf{F}_s$), pressure, and $\mathbf{F}_a$. We formulate a boundary value problem for the calculation of pressure and displacements, and discuss how to find an equation for $\mathbf{L}_a$ (tensor $\mathbf{F}_a$ is then found by solving the initial value problem $\dot{\mathbf{F}}_a = \mathbf{L}_a \mathbf{F}_a$, with $\mathbf{F}_a(t_0) = \mathbf{F}_{a0}$). Accepting Darcy's law amounts to say that fluid flow is determined by the hydraulic conductivity, $\mathbf{K}$, and pressure gradient. Therefore, the study of the transport of fluid in a deformable porous medium with variable mass and internal structure reduces to the determination of the influence of $\mathbf{F}_a$ on $\mathbf{K}$ and pressure. In particular, we show that different choices of the initial value $\mathbf{F}_{a0}$ lead to different pressure distributions and displacements. The latter ones, in turn, affect $\mathbf{K}$ and are thus able to influence the capability of the medium of conveying fluid. The alteration of pressure may be relevant for some biomechanical applications in which the health of the cells of a tissue depends on the pressure (for example, this is the case of chondrocytes in articular cartilage). Our study aims to put some of concepts presented in [30] in a more rigorous framework.

Our contribution is organised as follows. In Sect. 2, we review the kinematics of biphasic mixtures with consideration of the structural changes of the solid phase; we study the balance laws relevant for our purposes and discuss the dissipation inequality characterising the system under investigation. In Sect. 3, we formulate the mathematical model in weak form in order to supply the basis for finite element implementations. Since our equations are non-linear, we present the linearisation procedure and discuss the spatial discretization of the computational domain as well as some numerical issues related to the solution of the problem. In Sect. 4, we study a benchmark problem. Finally, we summarise our results in Sect. 5.

2 Theoretical Background

The equations governing the coarse-scale evolution of biphasic systems can be found by averaging the relations that describe the systems at a smaller scale, e.g., the pore scale. The averaging procedures are often based on volume- and mass-averages [33, 10]. These assume the existence of a representative volume element (RVE) $\Omega(\boldsymbol{x})$, which supplies information about the composition and structure of the mixture at the point $\boldsymbol{x} \in \mathbb{R}^3$, where $\mathbb{R}^3$ is the ambient space. The characteristic size of the RVE depends on the system under investigation. The subvolume $\Omega_\alpha(t, \boldsymbol{x})$, with $\alpha = \ell, s$, represents the subset of $\Omega(\boldsymbol{x})$ occupied by the αth phase at time t. The ratio $\phi_\alpha(t, \boldsymbol{x}) := |\Omega_\alpha(t, \boldsymbol{x})| / |\Omega(\boldsymbol{x})|$ is referred to as volumetric fraction.

Moreover, the abbreviations $\mathscr{P}_s$ and $\mathscr{P}_\ell$ will be sometimes used to denote the solid and the fluid phase, respectively.

If the void space of the porous medium is completely filled with the fluid, the medium is said to be "saturated", and the condition $\phi_s + \phi_\ell = 1$ applies at all times and all points. The distribution of mass of $\mathscr{P}_\alpha$ in Ω_α is the "true" mass density of $\mathscr{P}_\alpha$, and is denoted here by $\hat{\rho}_\alpha$. The product $\rho_\alpha = \phi_\alpha \hat{\rho}_\alpha$ measures the distribution of mass of $\mathscr{P}_\alpha$ in Ω, and is called "apparent" mass density of $\mathscr{P}_\alpha$.

2.1 Kinematics

We formulate the kinematic description of the mixture in two steps. We consider first the velocities $\boldsymbol{v}_s$ and $\boldsymbol{v}_\ell$, which characterise the standard motion of $\mathscr{P}_s$ and $\mathscr{P}_\ell$. Subsequently, we introduce the second-order tensors $\mathbf{F}_a$ and $\mathbf{L}_a$, which describe the anelastic deformation and the rate of anelastic deformation associated with the change of internal structure of the solid phase.

Among the various ways of describing the motion of the phases $\mathscr{P}_s$ and $\mathscr{P}_\ell$, we choose that based on the set of "standard" velocities

$$\mathscr{V}_{\text{st}} := \{\boldsymbol{v}_s, \boldsymbol{w}_{\ell s}\}, \tag{1}$$

where $\boldsymbol{w}_{\ell s} := \boldsymbol{v}_\ell - \boldsymbol{v}_s$ is the relative velocity of $\mathscr{P}_\ell$ with respect to $\mathscr{P}_s$. Another relevant kinematic quantity is the velocity of the centre of mass of the mixture, which is defined by

$$\boldsymbol{v} := \frac{\rho_s \boldsymbol{v}_s + \rho_\ell \boldsymbol{v}_\ell}{\rho}, \qquad \rho := \rho_\ell + \rho_s. \tag{2}$$

Here, ρ denotes the mass density of the mixture as a whole. The relative velocities $\tilde{\boldsymbol{v}}_\ell := \boldsymbol{v}_\ell - \boldsymbol{v}$ and $\tilde{\boldsymbol{v}}_s := \boldsymbol{v}_s - \boldsymbol{v}$ describe the relative motion of each phase with respect to the motion of the mixture as a whole. For each phase, we denote by $\boldsymbol{a}_\alpha$ (with $\alpha = \ell, s$) the acceleration of that phase, which is defined by the convective derivative of $\boldsymbol{v}_\alpha$ with respect to the motion of the phase $\mathscr{P}_\alpha$, i.e.

$$\boldsymbol{a}_\alpha := \partial_t \boldsymbol{v}_\alpha + \text{grad}(\boldsymbol{v}_\alpha)\boldsymbol{v}_\alpha, \qquad \alpha = \ell, s. \tag{3}$$

The acceleration of the mixture is indicated by $\boldsymbol{a}$ and is defined as the convective derivative of $\boldsymbol{v}$ with respect to the motion of the mixture as a whole, i.e.

$$\boldsymbol{a} := \partial_t \boldsymbol{v} + \text{grad}(\boldsymbol{v})\boldsymbol{v}. \tag{4}$$

The description of kinematics shown so far is based on the spatial, or Eulerian, formalism. In order to make it consistent with the Lagrangian description, we need to relate the velocities $\boldsymbol{v}_s$ and $\boldsymbol{v}_\ell$ with the motions of the solid and fluid phases. To this end, we present a formulation that follows, with some slight differences, the picture put forward by Quiligotti [48]. The starting point is that the biphasic mixture represents a porous medium whose void space is saturated by a fluid. If we ideally remove the fluid, we are left with a solid skeleton, which we embed in the three-dimensional Euclidean space $\mathbb{R}^3$. We call $\mathscr{B}_r$ the subset of $\mathbb{R}^3$ in which the solid skeleton is embedded, and denote by $\boldsymbol{X}$ the coordinates of the centroid of the RVE associated with $\mathscr{B}_r$. The set $\mathscr{B}_r$ is also said to be the "reference configuration" of the solid phase $\mathscr{P}_s$, and the coordinates $\boldsymbol{X}$ may be called "material coordinates". A smooth motion of $\mathscr{P}_s$, which is referred to as s-motion in [48], is a sequence of mappings $\chi_s(t,\cdot)$: $\mathscr{B}_r \mapsto \mathbb{R}^3$, such that $\boldsymbol{x} = \chi_s(t, \boldsymbol{X}) \in \mathbb{R}^3$. The material gradient of the map χ_s equals the deformation gradient $\mathbf{F}_s := \text{Grad}(\chi_s)$, i.e. $(F_s)_{iI} = (\chi_s)_{i,I}$ in Cartesian coordinates, where the comma means differentiation with respect to X_I. The time derivative of χ_s is defined by

$$\partial_t \chi_s(t, \boldsymbol{X}) = \boldsymbol{v}_s(t, \chi_s(t, \boldsymbol{X})) = \boldsymbol{v}_s(t, \boldsymbol{x}). \tag{5}$$

In order to describe the kinematics of the fluid phase $\mathscr{P}_\ell$, we introduce the material manifold $\mathscr{B}_\ell$, which consists of fluid particles. A fluid particle, labelled by X_ℓ, is placed in the Euclidean space by means of an embedding that locates the particle at $\boldsymbol{x}$. A smooth motion of $\mathscr{P}_\ell$, the ℓ-motion, is a sequence of mappings defined by $\chi_\ell(t,\cdot)$: $\mathscr{B}_\ell \mapsto \mathbb{R}^3$, such that $\boldsymbol{x} = \chi_\ell(t, X_\ell)$. The velocity of the particle labelled by X_ℓ satisfies the identity

$$\partial_t \chi_\ell(t, X_\ell) = \boldsymbol{v}_\ell(t, \chi_\ell(t, X_\ell)) = \boldsymbol{v}_\ell(t, \boldsymbol{x}). \tag{6}$$

The definitions of s- and ℓ-motion imply the chain of identities

$$\boldsymbol{x} = \chi_s(t, \boldsymbol{X}) = \chi_\ell(t, X_\ell), \tag{7}$$

which means that both the solid and the fluid phase co-exist at the same point $\boldsymbol{x}$ of the Euclidean space. A quantity associated with the fluid phase, e.g., the velocity $\boldsymbol{v}_\ell$, can be expressed in terms of the material coordinates by using the following composition of maps:

$$\boldsymbol{v}_\ell(t, \boldsymbol{x}) = \boldsymbol{v}_\ell(t, \chi_s(t, \boldsymbol{X})) = [\boldsymbol{v}_\ell(t,\cdot) \circ \chi_s(t,\cdot)](\boldsymbol{X}). \tag{8}$$

An analogous result holds true for $\boldsymbol{v}_s$. The portion of $\mathbb{R}^3$ occupied by the mixture at time t is given by the intersection $\mathscr{B}_t = \chi_s(t, \mathscr{B}_r) \cap \chi_\ell(t, \mathscr{B}_\ell)$.

The gradient of $\boldsymbol{v}_s$ is denoted by $\mathbf{L}_s := \text{grad}(\boldsymbol{v}_s)$. The deformation process of $\mathscr{P}_s$ is determined by a tensor field $\mathbf{F}_s$, which satisfies the condition $\dot{\mathbf{F}}_s = \mathbf{L}_s \mathbf{F}_s$. The determinant of $\mathbf{F}_s$, $J_s := \det(\mathbf{F}_s)$, accounts for the change of volume associated with the change of configuration. The definitions introduced so far can be found in several treatises about the classical Theory of Mixtures (cf., for example, [7, 13, 58, 51]).

In order to complete the kinematic analysis of the considered biphasic mixture, we have to introduce a non-standard descriptor in addition to the standard velocities collected in $\mathscr{V}_{\text{st}}$. This descriptor has to model the structural change of the solid phase, $\mathscr{P}_s$, in response to interactions that lead to the variation and redistribution of its mass. On the basis of the motivations reported, for example, in

[53, 22, 20, 41, 3, 4], these types of structural evolution are viewed as anelastic processes. Accordingly, the kinematic descriptor of these processes is the tensor of anelastic deformation, $\mathbf{F}_a$, which is related to the rate of anelastic deformation, $\mathbf{L}_a$, through the differential equation $\dot{\mathbf{F}}_a := \mathbf{L}_a\mathbf{F}_a$. If the solid phase exhibits elastic behaviour, the tensor $\mathbf{F}_s$ is decomposed as $\mathbf{F}_s = \mathbf{F}_e\mathbf{F}_a$, where $\mathbf{F}_e$ represents the elastic contribution to the overall deformation. The variations of volume of the solid phase due to the elastic and the anelastic deformations are denoted by $J_e = \det(\mathbf{F}_e)$ and $J_a = \det(\mathbf{F}_a)$, respectively. The multiplicative decomposition of $\mathbf{F}_s$ implies $J_s = J_e J_a$. The determinants J_s, J_e and J_a are strictly positive.

The multiplicative decomposition $\mathbf{F}_s = \mathbf{F}_e\mathbf{F}_a$ was introduced by Kröner [36, 37], Lee [39] and other scientists working in Continuum Mechanics and, in particular, in the Theory of Plasticity. In Biomechanics, it was firstly used by Rodriguez et al. [53]. The tensor $\mathbf{F}_a$ maps vectors attached to $\mathscr{B}_r$ into vectors attached to a relaxed configuration, which is often referred to as "natural configuration" [49] and denoted by $\mathscr{B}_n$.

In the following, the index "s" associated with $\mathbf{F}_s$ and J_s will be dropped for the sake of simpler notation.

2.2 Balance Laws

The Eulerian, local forms of the balance of mass of the solid and fluid phase read

$$D_s\rho_s + \rho_s \mathrm{div}(\mathbf{v}_s) = \rho_s\gamma_s, \tag{9}$$

$$D_s\rho_\ell + \rho_\ell \mathrm{div}(\mathbf{v}_s) + \mathrm{div}(\rho_\ell \mathbf{w}_{\ell s}) = 0, \tag{10}$$

where the operator $D_s\mathbf{A} = \partial_t\mathbf{A} + \mathrm{grad}(\mathbf{A}) \cdot \mathbf{v}_s$ is the convective derivative of the generic tensor field $\mathbf{A}$ with respect to the motion of the solid phase, and γ_s is the rate at which the mass of the solid phase is produced or depleted. Multiplying (9) and (10) by J, and passing to the material description lead to the following form of the mass balance laws of the constituents of the mixture:

$$\overline{(J\rho_s)}^{\boldsymbol{\cdot}} = J\rho_s\gamma_s, \tag{11}$$

$$\overline{(J\rho_\ell)}^{\boldsymbol{\cdot}} + \mathrm{Div}(J\rho_\ell \mathbf{F}^{-1}\mathbf{w}_{\ell s}) = 0. \tag{12}$$

The operators "Div" and "Grad" are the divergence and gradient operators computed with respect to the material coordinates. They are related to "div" and "grad" by the formulae $\mathrm{grad}(\boldsymbol{A}) = \mathrm{Grad}(\boldsymbol{A}_\chi)\mathbf{F}^{-1}$ and $\mathrm{div}(\boldsymbol{A}) = \mathrm{Grad}(\boldsymbol{A}_\chi)\colon \mathbf{F}^{-T}$, where $\boldsymbol{A}$ is a given vector field, the symbol ":" denotes the inner product between tensors, and the index "χ", which will be dropped for here on, means $\boldsymbol{A}_\chi(t,\cdot) = \boldsymbol{A}(t,\cdot) \circ \chi_s(t,\cdot)$.

The product $\rho_{sr} := J\rho_s$ in (11) defines the mass of $\mathscr{P}_s$ measured per unit volume of $\mathscr{B}_r$. By using the definition of apparent mass density, and the fact that $J = J_e J_a$, the quantity ρ_{sr} can be rewritten as

$$\rho_{sr} = J_e J_a \rho_s = J_a \rho_{sn}, \tag{13}$$

where $\rho_{sn} := J_e \rho_s$ indicates the mass density of $\mathscr{P}_s$ computed with respect to the natural configuration $\mathscr{B}_n$. Furthermore, substituting (13) into (11) yields

$$\mathrm{tr}(\mathbf{L}_a)\rho_{sn} + \dot{\rho}_{sn} = \rho_{sn}\gamma_s. \tag{14}$$

We enforce now the condition that the variation of body mass is compensated for by the rate $\mathrm{tr}(\mathbf{L}_a)$, which implies that the mass density ρ_{sn} is constant in time. Thus, we arrive at the results

$$\gamma_s = \mathrm{tr}(\mathbf{L}_a), \quad \rho_{sn} = \rho_{s0}, \tag{15}$$

where ρ_{s0} may be a function of material coordinates only. A consequence of (13)–(15) is that the solution to (11) is given by

$$\rho_s = \frac{\rho_{sn}}{J_e} = \rho_{sn}\frac{J_a}{J}. \tag{16}$$

This means that the apparent density of the solid phase, ρ_s, is determined if the constant mass distribution ρ_{sn} is assigned, and the volumetric deformations J and J_a are known.

A simplification may be obtained under the hypothesis that the true mass density $\hat{\rho}_s$ is a given constant, which implies that $\mathscr{P}_s$ is incompressible. This allows to reformulate (16) in terms of the stronger condition

$$\phi_s = \frac{\phi_{sn}}{J_e} = \phi_{sn}\frac{J_a}{J}, \tag{17}$$

which involves only different measures of the volumetric fraction of the solid phase. In (17), ϕ_{sn} denotes the volumetric fraction of $\mathscr{P}_s$ "seen" by $\mathscr{B}_n$. We remark that ϕ_{sn} is constant and should be regarded as a known quantity of the model.

Another simplification follows from requiring that the true mass density of the fluid phase, $\hat{\rho}_\ell$, is a given constant, so that $\mathscr{P}_\ell$ is incompressible too. Granted this condition and the saturation constraint, a consequence of (17) and $(15)_1$ is that (12) acquires the simpler form

$$\mathrm{Div}[J\mathbf{F}^{-1}\boldsymbol{q}_{\ell s}] = -\dot{J} + J_a\phi_{sn}\mathrm{tr}(\mathbf{L}_a), \tag{18}$$

with $\boldsymbol{q}_{\ell s} := \phi_\ell \boldsymbol{w}_{\ell s}$. In summary, (17) and (18) provide the balances of mass of the solid and the fluid phase, respectively.

Together with mass balance, also the balance of momentum of the phases $\mathscr{P}_s$ and $\mathscr{P}_\ell$ has to be studied. The Eulerian, local form of these balance laws is given by

$$\partial_t(\rho_s \boldsymbol{v}_s) + \operatorname{div}(\rho_s \boldsymbol{v}_s \otimes \boldsymbol{v}_s) - \operatorname{div}(\mathbf{T}_s) = \rho_s \boldsymbol{m}_s + (\rho_s \boldsymbol{p}_s + \rho_s \gamma_s \boldsymbol{v}_s), \tag{19}$$

$$\partial_t(\rho_\ell \boldsymbol{v}_\ell) + \operatorname{div}(\rho_\ell \boldsymbol{v}_\ell \otimes \boldsymbol{v}_\ell) - \operatorname{div}(\mathbf{T}_\ell) = \rho_\ell \boldsymbol{m}_\ell, \tag{20}$$

where $\mathbf{T}_s$ and $\mathbf{T}_\ell$ denote the Cauchy stress tensors of the solid and fluid phase, $\rho_s \boldsymbol{m}_s$ and $\rho_\ell \boldsymbol{m}_\ell$ are the rate of exchange of momentum between the two phases, and $\rho_s \boldsymbol{p}_s$ is rate of change of momentum due to growth. The forces $\rho_s \boldsymbol{m}_s$ and $\rho_\ell \boldsymbol{m}_\ell$ satisfy the condition

$$\rho_s \boldsymbol{m}_s + \rho_\ell \boldsymbol{m}_\ell = \mathbf{0}, \tag{21}$$

which states that, in the absence of growth (i.e., when $\gamma_s = 0$ and $\boldsymbol{p}_s = \mathbf{0}$), the mixture is closed with respect to momentum. An explanation of the physical meaning of (21) in terms of pore scale considerations can be found, for example, in [33]. The use of (21), and the definition of the relative velocities $\tilde{\boldsymbol{v}}_\ell$ and $\tilde{\boldsymbol{v}}_s$ as well as of the accelerations $\boldsymbol{a}$ and $\boldsymbol{a}_\ell$ allow for reformulating the balance laws (19) and (20) in the following way

$$\rho \boldsymbol{a} - \operatorname{div}(\mathbf{T}) = \rho_s \boldsymbol{p}_s + \rho_s \gamma_s \tilde{\boldsymbol{v}}_s, \tag{22}$$

$$\rho_\ell \boldsymbol{a}_\ell - \operatorname{div}(\mathbf{T}_\ell) = \rho_\ell \boldsymbol{m}_\ell, \tag{23}$$

where $\mathbf{T}$, which denotes the Cauchy stress tensor of the mixture, is defined by

$$\mathbf{T} := \mathbf{T}_s + \mathbf{T}_\ell - \left[\rho_s \tilde{\boldsymbol{v}}_s \otimes \tilde{\boldsymbol{v}}_s + \rho_\ell \tilde{\boldsymbol{v}}_\ell \otimes \tilde{\boldsymbol{v}}_\ell\right]. \tag{24}$$

Equation (22) represents the balance of momentum of the mixture as a whole, and is obtained by adding together (19) and (20) and applying the definitions (2)–(4) to the result. Furthermore, substituting the identity

$$\rho \boldsymbol{a} := \textstyle\sum_{\alpha=\ell,s} \rho_\alpha \boldsymbol{a}_\alpha - \sum_{\alpha=\ell,s} \operatorname{div}(\rho_\alpha \tilde{\boldsymbol{v}}_\alpha \otimes \tilde{\boldsymbol{v}}_\alpha) + \rho_s \gamma_s \tilde{\boldsymbol{v}}_s \tag{25}$$

into (22) yields

$$\rho_s \boldsymbol{a}_s + \rho_\ell \boldsymbol{a}_\ell - \operatorname{div}(\mathbf{T}_s + \mathbf{T}_\ell) = \rho_s \boldsymbol{p}_s, \tag{26}$$

$$\rho_\ell \boldsymbol{a}_\ell - \operatorname{div}(\mathbf{T}_\ell) = \rho_\ell \boldsymbol{m}_\ell. \tag{27}$$

Finally, neglecting the inertial forces of both phases, the balances of momentum (19) and (20) become

$$\operatorname{div}(\mathbf{T}_s + \mathbf{T}_\ell) + \rho_s \boldsymbol{p}_s = \mathbf{0}, \tag{28}$$

$$\operatorname{div}(\mathbf{T}_\ell) + \rho_\ell \boldsymbol{m}_\ell = \mathbf{0}. \tag{29}$$

By means of the Piola transformations of (28) and (29), the momentum balance laws of the mixture can be written with respect to the reference placement $\mathscr{B}_r$, i.e.

$$\operatorname{Div}(\mathbf{P}_s + \mathbf{P}_\ell) + J \rho_s \boldsymbol{p}_s = \mathbf{0}, \tag{30}$$

$$\mathrm{Div}(\mathbf{P}_\ell) + J\rho_\ell \boldsymbol{m}_\ell = \mathbf{0}, \tag{31}$$

where

$$\mathbf{P}_s := J\mathbf{T}_s\mathbf{F}^{-T}, \quad \mathbf{P}_\ell := J\mathbf{T}_\ell\mathbf{F}^{-T} \tag{32}$$

denote, respectively, the first Piola–Kirchhoff stress tensors of the solid and fluid phase. In order to close the mathematical problem resulting from (18), (30) and (31), it is necessary to provide information about the stresses $\mathbf{P}_s$ and $\mathbf{P}_\ell$, and the force densities $\rho_\ell \boldsymbol{m}_\ell$ and $\rho_s \boldsymbol{p}_s$.

The change of internal structure of the solid phase is a process whose kinematics are described by the tensor map $\mathbf{F}_a$ and the generalised velocity $\mathbf{L}_a = \dot{\mathbf{F}}_a(\mathbf{F}_a)^{-1}$. The set of generalised forces that perform working on $\mathbf{L}_a = \dot{\mathbf{F}}_a(\mathbf{F}_a)^{-1}$ comprises an internal force, $\mathbf{Z}_n$, which drives the structural evolution, and an external force, $\mathbf{Y}_n$, which models the interaction of the system with its surrounding environment. Both forces are second-order tensors. It is postulated that they obey the balance law

$$\mathbf{Z}_n = \mathbf{Y}_n. \tag{33}$$

The index "n" means that $\mathbf{Z}_n$ and $\mathbf{Y}_n$ are conceived as forces acting on the natural configuration $\mathscr{B}_n$, although they can be also written with respect to $\mathscr{B}_r$ and $\mathscr{B}_t$ by performing proper transformations. The reader is referred to [20] for an exhaustive explanation of the physical concepts leading to (33). Here, we simply state that, in analogy with the balance laws (28) and (29) (which describe a balance of forces that perform working on the set of standard velocities), also the forces that perform working on the non-standard descriptor $\mathbf{L}_a$ should satisfy a balance law. Some extensions of the results presented in [20] can be found, for example, in [1, 26, 2, 5, 29, 28, 27].

The internal force-like variables $\rho_\ell \boldsymbol{m}_\ell$ and $\mathbf{Z}_n$ are responsible for dissipation, and should thus comply with the dissipation inequality that characterises the system under investigation.

2.3 Study of the Residual Dissipation

We introduce the total internal working associated with $\mathscr{P}_t \subset \mathscr{B}_t$

$$\mathfrak{W}^{in}(\mathscr{P}_t) = \mathfrak{W}^{in}_{\mathrm{st}}(\mathscr{P}_t) + \mathfrak{W}^{in}_{\mathrm{n-st}}(\mathscr{P}_t), \tag{34}$$

where $\mathfrak{W}^{in}_{\mathrm{st}}(\mathscr{P}_t)$ and $\mathfrak{W}^{in}_{\mathrm{n-st}}(\mathscr{P}_t)$ describe, respectively, the working performed by the standard and non-standard forces acting on the system. These two contributions are defined by the following expressions

$$\mathfrak{W}^{in}_{\text{n-st}}(\mathscr{P}_t) := \int_{\mathscr{P}_t} (J_e)^{-1} \mathbf{Z}_n : \mathbf{L}_a, \tag{35}$$

$$\mathfrak{W}^{in}_{\text{st}}(\mathscr{P}_t) := \int_{\mathscr{P}_t} \{ -\rho_\ell \boldsymbol{m}_\ell \cdot \boldsymbol{w}_{\ell s} + \mathbf{T}_\ell : \text{grad}(\boldsymbol{w}_{\ell s}) + (\mathbf{T}_s + \mathbf{T}_\ell) : \text{grad}(\boldsymbol{v}_s) \}. \tag{36}$$

In a purely mechanical context, it can be proven that dissipation is given by

$$\int_{\mathscr{P}_t} \mathscr{D} = -\int_{\mathscr{P}_t} [\rho_s D_s \Psi_s + \rho_\ell D_\ell \Psi_\ell] + \mathfrak{W}^{in}_{\text{st}}(\mathscr{P}_t) + \mathfrak{W}^{in}_{\text{n-st}}(\mathscr{P}_t) \geq 0, \tag{37}$$

which is assumed to be non-negative. In terms of the overall Helmholtz free energy density of the system, $\rho\Psi$, the first term on the RHS of (37) can be written as

$$-\int_{\mathscr{P}_t} [\rho_s D_s \Psi_s + \rho_\ell D_\ell \Psi_\ell] = -\text{d}_t \int_{\mathscr{P}_t} \rho\Psi + \int_{\partial\mathscr{P}_t} \boldsymbol{q}_\Psi \cdot \boldsymbol{n} + \int_{\mathscr{P}_t} \rho_s \gamma_s \Psi_s, \tag{38}$$

where

$$\rho\Psi := \rho_s \Psi_s + \rho_\ell \Psi_\ell, \tag{39}$$

$$\boldsymbol{q}_\Psi := -[\rho_s \Psi_s \tilde{\boldsymbol{v}}_s + \rho_\ell \Psi_\ell \tilde{\boldsymbol{v}}_\ell]. \tag{40}$$

Under the hypotheses of hyperelastic solid phase and macroscopically inviscid fluid, the study of the inequality (37) yields the following results for the Cauchy stresses $\mathbf{T}_s$ and $\mathbf{T}_\ell$:

$$\mathbf{T}_s = -\phi_s \pi \mathbf{g}^{-1} + \mathbf{g}^{-1} \left(\rho_s \frac{\partial \Psi_s}{\partial \mathbf{F}_e} \right) (\mathbf{F}_a)^{-T} \mathbf{F}^T, \tag{41}$$

$$\mathbf{T}_\ell = -\phi_\ell \pi \mathbf{g}^{-1}, \tag{42}$$

where $\mathbf{g}$ is the metric tensor associated with $\mathscr{B}_t$

Requiring the invariance of constitutive laws under superimposed rigid motions places further restrictions on the results (41) and (42). If a rigid motion is impressed, the points $\boldsymbol{x} \in \mathscr{B}_t$ transform as $\boldsymbol{x} \mapsto \bar{\boldsymbol{x}} = \mathbf{R}\,\boldsymbol{x} + \boldsymbol{c}$, where $\mathbf{R}$ is a proper orthogonal tensor defining a pure rotation, and $\boldsymbol{c}$ is a vector defining a pure rigid translation [44, 32]. Consequently, $\mathbf{F}$, $\mathbf{F}_e$ and $\mathbf{F}_a$ transform as follows

$$\mathbf{F} \mapsto \bar{\mathbf{F}} = \mathbf{R}\mathbf{F}, \quad \mathbf{F}_e \mapsto \bar{\mathbf{F}}_e = \mathbf{R}\mathbf{F}_e, \quad \mathbf{F}_a \mapsto \bar{\mathbf{F}}_a = \mathbf{F}_a. \tag{43}$$

However, the Helmholtz free energy density Ψ_s has to remain invariant under these transformations. Therefore, Ψ_s may depend on $\mathbf{F}_e$ only through the Cauchy stretch tensor $\mathbf{C}_e = (\mathbf{F}_e)^T \mathbf{F}_e$, which is independent on $\mathbf{R}$. This yields the relation

$$\mathbf{g}^{-1} \left(\rho_s \frac{\partial \Psi_s}{\partial \mathbf{F}_e} \right) = \mathbf{F}_e \left(2\rho_s \frac{\partial \mathscr{W}_s}{\partial \mathbf{C}_e} \right), \tag{44}$$

where $\mathscr{W}_s$ is the Helmholtz free energy density of the solid phase written as function of $\mathbf{C}_e$. On the other hand, the Cauchy stresses $\mathbf{T}_s$ and $\mathbf{T}_\ell$ transform as $\bar{\mathbf{T}}_\alpha = \mathbf{R}\mathbf{T}_\alpha\mathbf{R}^T$, with $\alpha = s, \ell$.

Furthermore, by invoking the incompressibility of the solid phase and using the definitions (17) and (32), the first Piola–Kirchhoff stress tensors become

$$\mathbf{P}_s = -J_a\phi_{sn}\pi\mathbf{g}^{-1}\mathbf{F}^{-T} + \mathbf{F}\left[J_a(\mathbf{F}_a)^{-1}\left(2\frac{\partial\mathscr{W}_{sn}}{\partial\mathbf{C}_e}\right)(\mathbf{F}_a)^{-T}\right], \tag{45}$$

$$\mathbf{P}_\ell = -(J - J_a\phi_{sn})\pi\mathbf{g}^{-1}\mathbf{F}^{-T}, \tag{46}$$

where $\mathscr{W}_{sn} = \phi_{sn}\hat{\rho}_s\mathscr{W}_s$. Since $\mathbf{F}^{-T} = (\mathbf{F}_e)^{-T}(\mathbf{F}_a)^{-T}$, the stress $\mathbf{P}_s$ can be rewritten as

$$\mathbf{P}_s = J_a\mathbf{P}_{sn}(\mathbf{F}_a)^{-T}, \quad \mathbf{P}_{sn} := -\phi_{sn}\pi\mathbf{g}^{-1}(\mathbf{F}_e)^{-T} + \mathbf{F}_e\left(2\frac{\partial\mathscr{W}_{sn}}{\partial\mathbf{C}_e}\right). \tag{47}$$

Finally, we call Mandel stress the tensor

$$\mathbf{M}_{sn} := (\mathbf{F}_e)^t\mathbf{P}_{sn} := -\phi_{sn}\pi(\mathbf{g}_n)^{-1} + \mathbf{C}_e\left(2\frac{\partial\mathscr{W}_{sn}}{\partial\mathbf{C}_e}\right), \tag{48}$$

with $(\mathbf{F}_e)^t = (\mathbf{g}_n)^{-1}(\mathbf{F}_e)^T\mathbf{g}$.

The constitutive results (45) and (46) allow for a simplification of the expression of dissipation. After localisation, we obtain

$$\mathscr{D} = -\{\rho_\ell\boldsymbol{m}_\ell - \pi\mathrm{grad}(\phi_\ell)\}\cdot\boldsymbol{w}_{\ell s} + (J_e)^{-1}\{\mathbf{M}_{sn} + \mathbf{Z}_n\} : \mathbf{L}_a \geq 0 \tag{49}$$

We introduce now the dissipative part of the force-like variables $\boldsymbol{m}_\ell$ and $\mathbf{Z}_n$, which are given by

$$\rho_\ell\check{\boldsymbol{m}}_\ell := \rho_\ell\boldsymbol{m}_\ell - \pi\mathrm{grad}(\phi_\ell), \quad \check{\mathbf{Z}}_n := \mathbf{M}_{sn} + \mathbf{Z}_n, \tag{50}$$

and are constrained to satisfy the inequality

$$\mathscr{D} = -\rho_\ell\check{\boldsymbol{m}}_\ell\cdot\boldsymbol{w}_{\ell s} + (J_e)^{-1}\check{\mathbf{Z}}_n : \mathbf{L}_a \geq 0. \tag{51}$$

Dissipation has to be zero when $\boldsymbol{w}_{\ell s}$ and $\mathbf{L}_a$ vanish. Substitution of (50) into the balance laws (29) and (33), and use of the constitutive result (42) yield

$$\rho_\ell\check{\boldsymbol{m}}_\ell = \phi_\ell\mathrm{grad}(\pi), \tag{52}$$

$$\check{\mathbf{Z}}_n = \mathbf{M}_{sn} + \mathbf{Y}_n. \tag{53}$$

Equations (52) and (53) represent the new form of the force balances that define, respectively, the dynamics of the fluid phase and the internal structure of the solid phase (the dynamics of the mixture as a whole are defined by (28) or (30)). Before proceeding with the determination of the forces $-\rho_\ell\check{\boldsymbol{m}}_\ell$ and $(J_e)^{-1}\check{\mathbf{Z}}_n$, a discussion about the study of the dissipation inequality (51), and some of its implications, is mandatory.

Let us set $\mathbf{L}_a = \mathbf{0}$ and focus on the pair $(-\rho_\ell \check{\boldsymbol{m}}_\ell, \boldsymbol{w}_{\ell s})$. In Biomechanics, it is often assumed that fluid flow obeys Darcy's law (cf., for example, [6]). Darcy's model of flow can be retrieved consistently with the study of dissipation (cf., for example, [33]) by expressing $-\rho_\ell \check{\boldsymbol{m}}_\ell$ as a constitutive function of $\boldsymbol{w}_{\ell s}$ that vanishes when $\boldsymbol{w}_{\ell s} = 0$. This function is then expanded in Taylor series in a neighbourhood of $\boldsymbol{w}_{\ell s} = 0$ and, for small velocities, only the first-order term of the expansion is maintained. Therefore, one obtains

$$-\rho_\ell \check{\boldsymbol{m}}_\ell = \mathbf{A}.\boldsymbol{q}_{\ell s}, \tag{54}$$

where $\boldsymbol{q}_{\ell s} := \phi_\ell \boldsymbol{w}_{\ell s}$ is called "specific discharge" and $\mathbf{A}$ is a positive-definite second-order tensor that represents the resistivity of the medium (in this discussion, ϕ_ℓ is assumed to be strictly different from zero). Substitution of this result into (52), and inverting $\mathbf{A}$ yield

$$\boldsymbol{q}_{\ell s} = -\mathbf{K}.\mathrm{grad}(\pi), \tag{55}$$

where $\mathbf{K} = \phi_\ell \mathbf{A}^{-1}$ is said to be the hydraulic conductivity of the medium. Equation (55) is a simplified form of Darcy's law, in which the buoyancy term is neglected. According to (55), $\boldsymbol{q}_{\ell s}$ vanishes in a permeable medium when the pressure gradient is zero. It is known, however, that Darcy's law may cease to be valid if, for example, the flow starts only when $-\rho_\ell \check{\boldsymbol{m}}_\ell$ exceeds a certain threshold, which is given by the pressure gradient [9]. In this case, the force $-\rho_\ell \check{\boldsymbol{m}}_\ell$ need not be smooth at $\boldsymbol{w}_{\ell s} = 0$.

Let us put now $\boldsymbol{w}_{\ell s} = 0$ and study the pair $((J_e)^{-1}\check{\mathbf{Z}}_n, \mathbf{L}_a)$. In some models of growth mechanics, constitutive laws of the type $\check{\mathbf{Z}}_n = \mathbb{H}_n \colon \mathbf{L}_a$ have been proposed (cf., for example, [1, 40, 2, 28, 5, 56, 27]), with $\mathbb{H}_n$ being a diagonally symmetric, positive-definite fourth-order tensor. The rate of anelastic deformation $\mathbf{L}_a$ was thus presented in the form $\mathbf{L}_a = \mathbb{G}_n \colon \check{\mathbf{Z}}_n$, with $\mathbb{G}_n = (\mathbb{H}_n)^{-1}$. Substitution of (53) in this relation yields

$$\mathbf{L}_a = \mathbb{G}_n \colon (\mathbf{M}_{sn} + \mathbf{Y}_n). \tag{56}$$

Equation (56) follows from the hypothesis that $\check{\mathbf{Z}}_n$ can be assigned as a constitutive function of $\mathbf{L}_a$ that vanishes when $\mathbf{L}_a = \mathbf{0}$. This function is then assumed to be smooth and linearised in a neighbourhood of $\mathbf{L}_a = \mathbf{0}$. For a positive-definite $\mathbb{G}_n$, the formula (56) admits the following interpretation: the rate of anelastic deformation, $\mathbf{L}_a$, becomes zero when the external force $\mathbf{Y}_n$ can be tuned in such a way that the sum $(\mathbf{M}_{sn} + \mathbf{Y}_n)$ vanishes. This situation implies that $\mathbf{F}_a$ (which always satisfies the kinematic relation $\dot{\mathbf{F}}_a = \mathbf{L}_a \mathbf{F}_a$) either ceases to evolve in time or remains equal to its initial value. In some biomechanical applications, $\mathbf{Y}_n$ is thought of as the "target stress" that regulates the process with which it is associated (when the target stress is reached, the process ceases). For example, in the model of arterial growth proposed in [45], $\mathbf{Y}_n$ is related to the homeostatic stress.

On the other hand, if the tensor $\mathbf{Y}_n$ is zero (or negligibly small), the equality $\mathbf{L}_a = \mathbf{0}$ cannot be recovered in general, since the Mandel stress, $\mathbf{M}_{sn}$, is not compensated by any external force. Furthermore, the relation (56) could be too restrictive for some applications. In fact, one may relax (56) and postulate an evolution law of the type [42]

$$\dot{\mathbf{V}}_a = \langle f \rangle \sum_{h=0}^{2} \sum_{k=0}^{2} \beta_{hk}(\mathbf{V}_a, \mathbf{S}_{sn}) \left[(\mathbf{V}_a)^h (\mathbf{S}_{sn})^k + (\mathbf{S}_{sn})^k (\mathbf{V}_a)^h \right]. \tag{57}$$

Here, $\mathbf{V}_a$ is the symmetric part of $\mathbf{F}_a$ (cf. the decomposition $\mathbf{F}_a = \mathbf{V}_a \mathbf{R}_a$ [44], where $\mathbf{R}_a$ –the tensor of anelastic rotation– is set equal to the identity), $\mathbf{S}_{sn} := (\mathbf{F}_e)^{-1} \mathbf{P}_{sn}$ is the second Piola–Kirchhoff stress tensor, β_{hk} is a given constitutive function of the invariants of $\mathbf{V}_a$, $\mathbf{S}_{sn}$ and compositions of these tensors, f is said to be dynamic yield function, and the symbol $\langle f \rangle$ equals unity when the anelastic deformation changes with time and equals zero otherwise. Constitutive restrictions on $\mathbf{L}_a$, $\mathbf{S}_{sn}$ and the dynamic yield function are then found by using (57) in the computation of the extrema of the anelastic working.

Another method for determining evolution laws is given in [17], where rate-independent plasticity is investigated. The dissipation is defined as a function of $\mathbf{L}_a$ and is assumed to be continuous, but generally non-differentiable, at $\mathbf{L}_a = \mathbf{0}$, while the tensor $\check{\mathbf{Z}}_n$ is constitutively indeterminate at $\mathbf{L}_a = \mathbf{0}$. Within this framework, a maximum-dissipation criterion is formulated and it is proven that the dissipation function is everywhere sub-differentiable and, thus, convex with respect to $\mathbf{L}_a$, and that $\check{\mathbf{Z}}_n$ must belong to the sub-differential of the dissipation function. The evolution of $\mathbf{L}_a$ is determined by introducing a scalar yield criterion through the yield function f and showing that $\mathbf{L}_a$ has to be an element of the sub-differential of f. In the case of a smooth yield function f, it is found that $\mathbf{L}_a$ follows the "normality rule"

$$\mathbf{L}_a = \lambda \frac{\partial f}{\partial \check{\mathbf{Z}}_n}, \tag{58}$$

where λ is a Kuhn-Tucker multiplier satisfying the conditions $\lambda \geq 0, f \leq 0, \lambda f = 0$, and determined by the consistency requirement $\lambda \dot{f} = 0$. Many of the mathematical tools for presenting this theory can be found in [52]. In the context of biological materials, an evolution law of the type (58) was proposed in [46] for modelling the reorganisation of cells (an anelastic deformation) of a tissue in the presence of growth (e.g., a tumour).

We remark that (56–58) are all plausible ways to determine the evolution of $\mathbf{L}_a$. They are, however, different from each other since they are conceived for modelling different physical situations. On the other hand, the common feature of all these models is that the relations linking the generalised forces $-\rho_\ell \check{\boldsymbol{m}}_\ell$ and $\check{\mathbf{Z}}_n$ with their power-conjugate generalised velocities $\boldsymbol{q}_{\ell s}$ and $\mathbf{L}_a$ satisfy a maximum-dissipation principle (cf. [50]).

2.4 Summary of the Model

Our purpose is to study how the structural change of the solid phase influences fluid flow through the modulation of the transport properties of the mixture. We accept the validity of Darcy's law, so that the fluid flow depends on hydraulic conductivity and pressure gradient. Therefore, to accomplish our task, we have to show how, for a given type of problem and assigned boundary conditions, different tensors $\mathbf{F}_a$ modulate the hydraulic conductivity of the medium and the pressure field inside it.

We remark that the medium is assumed to be isotropic with respect to both its elastic properties and permeability.

By substituting (55) into (18), using the constitutive results (45) and (46), neglecting $\boldsymbol{p}_s$ in (28), and writing the pressure gradient in material coordinates, i.e. $\text{grad}(\pi) = \mathbf{F}^{-T}\text{Grad}(\pi)$, the equations to solve are

$$-\text{Div}\big[-J\pi\mathbf{F}^{-T} + \mathbf{P}\big] = \mathbf{0}, \tag{59}$$

$$\text{Div}\big[\mathbf{K}_r.\text{Grad}(\pi)\big] = \dot{J} - J_a\phi_{sn}\text{tr}(\mathbf{L}_a), \tag{60}$$

$$\dot{\mathbf{F}}_a = \mathbf{L}_a\mathbf{F}_a \tag{61}$$

where $\mathbf{P} = J_a\mathbf{F}\mathbf{F}_a^{-1}\mathbf{S}_n\mathbf{F}_a^{-T}$ denotes from here on the constitutive part of $\mathbf{P}_s$ (cf. (45)), $\mathbf{S}_n$ is the second Piola–Kirchhoff stress tensor measured with respect to $\mathscr{B}_n$, and $\mathbf{K}_r$ is the material form of the tensor of hydraulic conductivity, i.e., $\mathbf{K}_r = J\mathbf{F}^{-1}\mathbf{K}\mathbf{F}^{-T}$. The material is assumed to be hyperelastic and isotropic, and is modelled by the Neo-Hookean elastic energy given below [12], which leads to the following expressions of $\mathbf{S}_n$ and elasticity tensor $\mathbb{C}_n$:

$$\mathscr{W}_{sn}(\mathbf{C}_e) = \frac{\mu_n}{2}\left[\text{tr}(\mathbf{C}_e) - 3\right] - \mu_n\ln(J_e) + \frac{\lambda_n}{2}\left[\ln(J_e)\right]^2, \tag{62}$$

$$\mathbf{S}_n = 2\frac{\partial\mathscr{W}_{sn}}{\partial\mathbf{C}_e} = \mu_n[(\mathbf{g}_n)^{-1} - (\mathbf{C}_e)^{-1}] + \lambda_n[\ln(J_e)](\mathbf{C}_e)^{-1}, \tag{63}$$

$$\mathbb{C}_n = 4\frac{\partial^2\mathscr{W}_{sn}}{\partial\mathbf{C}_e\otimes\partial\mathbf{C}_e} = \lambda_n(\mathbf{C}_e)^{-1}\otimes(\mathbf{C}_e)^{-1} + 2[\mu_n - \lambda_n\ln(J_e)]\mathbb{I}_n, \tag{64}$$

where $\mathbb{I}_n$ is defined in the Appendix. Tensor $\mathbf{K}$ is taken from [34] and adapted to our framework, i.e.

$$\mathbf{K} = k_0\left[\frac{\phi_{s0}}{1-\phi_{s0}}\frac{J - \phi_{sn}J_a}{J_a\phi_{sn}}\right]^{m_0}\exp\left\{\frac{m_1}{2}\left[\frac{J^2 - J_a^2}{J_a^2}\right]\right\}\mathbf{g}^{-1}. \tag{65}$$

The numbers m_0 and m_1 featuring in (65) are material parameters. To close the problem, $\mathbf{L}_a$ should be supplied by one of the formulae (56)–(58). The formulae of the elasticity tensor and hydraulic conductivity given in (64) and (65), respectively, define isotropic tensors. However, other forms of hydraulic conductivity, which account for tissue anisotropy, have been recently proposed in [6, 24].

In the absence of growth, the field equations (59) and (60) were studied in [21] in the case of a linear viscoelastic biphasic model for soft tissues.

Equations (59) and (60), which hold in the internal points of $\mathscr{B}_r$, are completed with conditions prescribed on the boundary $\partial\mathscr{B}_r$. The unknowns of the problem are displacements, $\boldsymbol{u}(t,\boldsymbol{X}) := \chi(t,\boldsymbol{X}) - \boldsymbol{X}$, and pressure, π. For each unknown, the boundary $\partial\mathscr{B}_r$ is split into a Dirichlet- and a Neumann-type subset. This means that $\partial\mathscr{B}_r$ admits the representations $\partial\mathscr{B}_r = \Gamma_N^{\boldsymbol{u}} \cup \Gamma_D^{\boldsymbol{u}}$ and $\partial\mathscr{B}_r = \Gamma_N^{\pi} \cup \Gamma_D^{\pi}$, where $\Gamma_N^{\boldsymbol{u}}$ and Γ_N^{π} are the subsets of $\partial\mathscr{B}_r$ on which Neumann boundary conditions for the displacements and pressure are prescribed, while $\Gamma_D^{\boldsymbol{u}}$ and Γ_D^{π} are the subsets on which Dirichlet conditions are supplied. Formally, boundary conditions are written as

$$\begin{cases} \boldsymbol{u} = \boldsymbol{u}_b, & \text{on } \Gamma_D^{\boldsymbol{u}}, \\ \left[- J\pi\mathbf{F}^{-T} + \mathbf{P} \right]\boldsymbol{N} = \boldsymbol{f}_{rb}, & \text{on } \Gamma_N^{\boldsymbol{u}}, \\ \pi = \pi_b, & \text{on } \Gamma_D^{\pi}, \\ \left[- \mathbf{K}_r \mathrm{Grad}(\pi) \right] \cdot \boldsymbol{N} = Q_{rb}, & \text{on } \Gamma_N^{\pi}, \end{cases} \tag{66}$$

where $\boldsymbol{N}$ is the unit vector normal to $\partial\mathscr{B}_r$. The surface force $\boldsymbol{f}_{rb}$ is defined per unit area of the reference boundary $\Gamma_N^{\boldsymbol{u}}$ and is, thus, generally different from the force $\boldsymbol{f}_b$ associated with the actual configuration $\mathscr{B}_t$. An analogous argument holds true for the quantities Q_{rb} and Q_b, the latter being the flux prescribed per unit are of the boundary $\partial\mathscr{B}_t$ of the actual configuration. The pairs $\boldsymbol{f}_{rb} - \boldsymbol{f}_b$ and $Q_{rb} - Q_b$ are related to each other by the formulae [12]

$$\boldsymbol{f}_{rb} = \boldsymbol{f}_b J\sqrt{\boldsymbol{N} \otimes \boldsymbol{N} \colon \mathbf{C}^{-1}}, \quad Q_{rb} = Q_b J\sqrt{\boldsymbol{N} \otimes \boldsymbol{N} \colon \mathbf{C}^{-1}}. \tag{67}$$

In order to reduce the number of equations to solve numerically, we consider the very particular case in which the tensor of anelastic deformation is kept constant (i.e. $\mathbf{F}_a$ is constant and known from the outset), so that no anelastic evolution occurs. This implies that $\mathbf{L}_a$ is zero. In other words, we assume that anelastic deformations have already taken place, which means that the tissue has already grown and remodelled. Physically, this can be rephrased by saying that the tissue grows and remodels over a time scale much larger than the scale over which fluid flows and elastic deformations take place.

3 Weak Formulation of the Field Equations

Throughout this section, we adopt Cartesian coordinates. To obtain the weak form of the field equations (59) and (60), we multiply (59) by the virtual velocity $\boldsymbol{v}_v$, and (60) by the virtual pressure π_v, and apply Leibniz's rule of differentiation. By integrating the resulting expressions over the reference configuration $\mathscr{B}_r$, and using Gauss' Theorem, we obtain

$$-\int_{\mathscr{B}_r} J\pi\mathbf{F}^{-T}\colon \dot{\mathbf{F}}_v \, \mathrm{d}\Omega + \int_{\mathscr{B}_r} \mathbf{P}\colon \dot{\mathbf{F}}_v \, \mathrm{d}\Omega - \int_{\Gamma_N^{\boldsymbol{u}}} \boldsymbol{v}_v \cdot \boldsymbol{f}_{rb} \, \mathrm{d}A = 0, \tag{68}$$

$$-\int_{\mathscr{B}_r} \mathrm{Grad}(\pi_v) \cdot \mathbf{K}_r \mathrm{Grad}(\pi) \, \mathrm{d}\Omega - \int_{\Gamma_N^{\pi}} \pi_v Q_{rb} \mathrm{d}A - \int_{\mathscr{B}_r} \pi_v \dot{J} \, \mathrm{d}\Omega = 0, \tag{69}$$

where $\dot{\mathbf{F}}_v = \mathrm{Grad}(\boldsymbol{v}_v)$. In the jargon of Numerical Analysis, the virtual fields $\boldsymbol{v}_v$ and π_v are referred to as test functions. These functions are chosen in such a way that they vanish identically on $\Gamma_D^{\boldsymbol{u}}$ and Γ_D^{π}, respectively.

Apart from boundary conditions, which might be time-dependent in general, the only place in which time features explicitly in (68) and (69) is in the time derivative $\dot{J}$. We approximate this derivative with a finite difference, and use an implicit Euler method [47]. In order to do that, we consider the time interval (t_i, t_f) and discretize it in N sub-intervals (t^{m-1}, t^m). Thus, we write $\dot{J} \approx (J^m - J^{m-1})/(\Delta t)^m$, where $(\Delta t)^m$ is the size of the mth time step, and reformulate the system of equations (68) and (69) in the following way [12]

$$-\int_{\mathscr{B}_r} J^m \pi^m (\mathbf{C}^m)^{-1}\colon \dot{\mathbf{E}}_v^m \, \mathrm{d}\Omega + \int_{\mathscr{B}_r} \mathbf{S}^m\colon \dot{\mathbf{E}}_v^m \, \mathrm{d}\Omega - \int_{\Gamma_N^{\boldsymbol{u}}} \boldsymbol{v}_v \cdot \boldsymbol{f}_{rb}^m \, \mathrm{d}A = 0, \tag{70}$$

$$\begin{aligned} &-\int_{\mathscr{B}_r} \mathrm{Grad}(\pi_v) \cdot \mathbf{K}_r^m \mathrm{Grad}(\pi^m) \, \mathrm{d}\Omega - \int_{\Gamma_N^{\pi}} \pi_v Q_{rb}^m \mathrm{d}A \\ &-\int_{\mathscr{B}_r} \pi_v \frac{J^m}{(\Delta t)^m} \, \mathrm{d}\Omega = -\int_{\mathscr{B}_r} \pi_v \frac{J^{m-1}}{(\Delta t)^m} \, \mathrm{d}\Omega, \end{aligned} \tag{71}$$

where $\mathbf{C}^m = (\mathbf{F}^m)^T \mathbf{F}^m$, $\mathbf{S}^m := (\mathbf{F}^m)^{-1} \mathbf{P}^m$, and $\dot{\mathbf{E}}_v^m$ is given by

$$\dot{\mathbf{E}}_v^m := \frac{1}{2}\left[(\mathbf{F}^m)^T \dot{\mathbf{F}}_v + (\dot{\mathbf{F}}_v)^T \mathbf{F}^m\right]. \tag{72}$$

In this notation, the Green-Lagrange strain tensor at time t^m reads $\mathbf{E}^m = \frac{1}{2}(\mathbf{C}^m - \mathbf{I}_r)$, with $\mathbf{I}_r$ being the identity tensor in the reference configuration. Given a generic function $\mathbf{A}$, the symbol $\mathbf{A}^m$ means $\mathbf{A}^m = \mathbf{A}(t^m, \cdot)$.

3.1 Linearisation

Formally, equations (70) and (71) can be rewritten in compact form as

$$\mathfrak{F}_{\boldsymbol{u}}(\boldsymbol{u}^m, \pi^m, \boldsymbol{v}_v) = 0, \tag{73}$$

$$\mathfrak{F}_{\pi}(\boldsymbol{u}^m, \pi^m, \pi_v) = 0, \tag{74}$$

where $\mathfrak{F}_{\boldsymbol{u}}$ and $\mathfrak{F}_{\pi}$ are functionals of displacement and pressure, and their virtual counterparts $\boldsymbol{v}_{\nu}$ and π_{ν}. In the sequel, we drop the dependence of these functionals on $\boldsymbol{v}_{\nu}$ and π_{ν} for the sake of a simpler notation. The non-linear system of equations (73) and (74) is solved by Newton's method [12, 47]. At the kth iteration, the linearised system reads

$$\mathfrak{F}_{\boldsymbol{u}}(\boldsymbol{u}^{m,k-1}, \pi^{m,k-1}) + \mathfrak{D}\mathfrak{F}_{\boldsymbol{u}}(\boldsymbol{u}^{m,k-1}, \pi^{m,k-1})[\boldsymbol{h}^{m,k}, \theta^{m,k}] = \mathbf{0}, \tag{75}$$

$$\mathfrak{F}_{\pi}(\boldsymbol{u}^{m,k-1}, \pi^{m,k-1}) + \mathfrak{D}\mathfrak{F}_{\pi}(\boldsymbol{u}^{m,k-1}, \pi^{m,k-1})[\boldsymbol{h}^{m,k}, \theta^{m,k}] = 0. \tag{76}$$

where $\mathfrak{D}\mathfrak{G}(\boldsymbol{u}^{m,k-1}, \pi^{m,k-1})[\boldsymbol{h}^{m,k}, \theta^{m,k}]$ denotes the Gateaux-derivative of a generic functional $\mathfrak{G}$, evaluated at $(\boldsymbol{u}^{m,k-1}, \pi^{m,k-1})$, and along the direction $[\boldsymbol{h}^{m,k}, \theta^{m,k}]$.

Given the pair $(\boldsymbol{u}^{m,k-1}, \pi^{m,k-1})$, in the neighbourhood of which the functionals (73) and (74) are linearised, the increments $\boldsymbol{h}^{m,k}$ and $\theta^{m,k}$ are computed by solving (75) and (76), and the updated pair of fields $(\boldsymbol{u}^{m,k}, \pi^{m,k})$ is determined as follows

$$\boldsymbol{u}^{m,k} = \boldsymbol{u}^{m,k-1} + \boldsymbol{h}^{m,k}, \tag{77}$$

$$\pi^{m,k} = \pi^{m,k-1} + \theta^{m,k}. \tag{78}$$

We notice that $\dot{\mathbf{E}}^{m}_{\nu}$, as defined in (72), is a bilinear functional of $\boldsymbol{u}^{m}$ and $\boldsymbol{v}_{\nu}$. Thus, by introducing the notation

$$\dot{\mathbf{E}}^{m}_{\nu} = \dot{\mathbf{E}}_{\nu}(\boldsymbol{u}^{m}) = \frac{1}{2}\left[(\mathbf{F}^{m})^{T}\dot{\mathbf{F}}_{\nu} + (\dot{\mathbf{F}}_{\nu})^{T}\mathbf{F}^{m}\right], \tag{79}$$

replacing $\boldsymbol{u}^{m}$ with $\boldsymbol{u}^{m,k}$, and invoking (77), we conclude that

$$\dot{\mathbf{E}}^{m,k}_{\nu} = \dot{\mathbf{E}}_{\nu}(\boldsymbol{u}^{m,k}) = \dot{\mathbf{E}}^{m,k-1}_{\nu} + \mathrm{sym}\left[(\mathbf{H}^{m,k})^{T}\dot{\mathbf{F}}_{\nu}\right], \tag{80}$$

with $\mathbf{H}^{m,k} := \mathrm{Grad}(\boldsymbol{h}^{m,k})$, and that the following identity holds true:

$$\dot{\mathbf{E}}^{m,k-1}_{\nu} := \frac{1}{2}\left[(\mathbf{F}^{m,k-1})^{T}\dot{\mathbf{F}}_{\nu} + (\dot{\mathbf{F}}_{\nu})^{T}\mathbf{F}^{m,k-1}\right] = \mathfrak{D}\mathbf{E}(\boldsymbol{u}^{m,k-1})[\boldsymbol{v}_{\nu}]. \tag{81}$$

Furthermore, we recall that [12]

$$\mathfrak{D}\mathbf{F}(\boldsymbol{u}^{m,k-1})[\boldsymbol{h}^{m,k}] = \mathrm{Grad}(\boldsymbol{h}^{m,k}) = \mathbf{H}^{m,k}. \tag{82}$$

With the formalism of this section, the boundary terms f^{m}_{rb} and Q^{m}_{rb} should be written as

$$f^{m}_{rb} = f_{rb}(t^{m}, \boldsymbol{u}^{m}), \quad Q^{m}_{rb} = Q_{rb}(t^{m}, \boldsymbol{u}^{m}). \tag{83}$$

Since f^{m}_{rb} and Q^{m}_{rb} depend on displacements in a non-linear way, they should be linearised too. In the computational praxis, however, this linearisation is sometimes avoided, and the approximations $f^{m}_{rb} \approx f^{m}_{rb}(t^{m}, \boldsymbol{u}^{m,k-1})$ and $Q^{m}_{rb} \approx Q^{m}_{rb}(t^{m}, \boldsymbol{u}^{m,k-1})$ are made. This is done because it is often too difficult to connect the boundary finite element over which f^{m}_{rb} and Q^{m}_{rb} are defined with the volume element to which it belongs. This approximation does not usually destroy the convergence of Newton's method.

When $\mathbf{L}_a$ is different from zero (which means that anelastic deformations evolve in time), the rate of anelastic deformation could be either given by one of (56) and (58) or computed as $\mathbf{L}_a = \dot{\mathbf{V}}_a(\mathbf{V}_a)^{-1}$, with $\dot{\mathbf{V}}_a$ specified in (57). In any case, the resulting expression is a function of deformation and should, thus, be linearised. If a plasticity-like flow rule is taken (cf. (58)), dedicated numerical procedures (e.g., the Return Mapping Algorithm [54] or the Linearised Projection Algorithm [59]) have to be implemented for the computation of the admissible stresses. However, the use of such numerical schemes may rise stability issues. In the case of associative multiplicative elasto-plasticity, these stability problems were studied, for example, by Miehe et al. [43]. On the other hand, if $\mathbf{L}_a$ is determined by (57), the linearisation procedure may become too demanding. Finally, also in the case in which the validity of (56) is assumed, a stable numerical procedure should be tested in order to obtain reliable solutions of (70), (71) and the evolution equation $\dot{\mathbf{F}}_a = \mathbf{L}_a\mathbf{F}_a$.

3.2 Spatial Discretization

The variational problem associated with (56) and (58) can be written in abstract form as:

$$\boxed{\begin{array}{l}\text{Find } \boldsymbol{h}^{m,k} \text{ and } \theta^{m,k} \in Q, \text{ such that :}\\ \qquad\begin{cases} a(\boldsymbol{h}^{m,k}, \boldsymbol{v}_v) - b(\boldsymbol{v}_v, \theta^{m,k}) = f(\boldsymbol{v}_v) & \forall \boldsymbol{v}_v \in V \\ -c(\boldsymbol{h}^{m,k}, \pi_v) - d(\theta^{m,k}, \pi_v) = g(\pi_v) & \forall \pi_v \in Q \end{cases}\end{array}} \tag{84}$$

where $V := (H^1_{\Gamma^u_D}(\mathscr{B}_r))^3$ and $Q := H^1_{\Gamma^\pi_D}(\mathscr{B}_r)$ are Hilbert spaces. The bilinear forms $a(\cdot,\cdot)$, $b(\cdot,\cdot)$, $c(\cdot,\cdot)$, $d(\cdot,\cdot)$ are defined as follows

$$\begin{aligned} a(\boldsymbol{h}^{m,k}, \boldsymbol{v}_v) := & \int_{\mathscr{B}_r} \left\{ \mathfrak{D}\mathbf{E}^{m,k-1}[\boldsymbol{v}_v] \colon \left[\mathbb{C}_c^{m,k-1} + \mathbb{C}_r^{m,k-1}\right] \colon \mathfrak{D}\mathbf{E}^{m,k-1}[\boldsymbol{h}^{m,k}] \right\} \mathrm{d}\Omega \\ & + \int_{\mathscr{B}_r} \left[-J^{m,k-1}\pi^{m,k-1}(\mathbf{C}^{m,k-1})^{-1} + \mathbf{S}^{m,k-1} \right] \colon \mathfrak{D}\dot{\mathbf{E}}_v^{m,k-1}[\boldsymbol{h}^{m,k}] \mathrm{d}\Omega, \end{aligned} \tag{85}$$

$$b(\boldsymbol{v}_v, \theta^{m,k}) := \int_{\mathscr{B}_r} J^{m,k-1}\theta^{m,k}(\mathbf{C}^{m,k-1})^{-1} \colon \mathfrak{D}\mathbf{E}^{m,k-1}[\boldsymbol{v}_v] \mathrm{d}\Omega, \tag{86}$$

$$c(\boldsymbol{h}^{m,k}, \pi_v) := \frac{b(\boldsymbol{h}^{m,k}, \pi_v)}{(\Delta t)^m} + \int_{\mathscr{B}_r} \mathrm{Grad}(\pi_v) \cdot \mathfrak{D}\mathbf{K}_r^{m,k-1}[\boldsymbol{h}^{m,k}].\mathrm{Grad}(\pi^{m,k-1})\, \mathrm{d}\Omega, \tag{87}$$

$$d(\theta^{m,k}, \pi_v) := \int_{\mathscr{B}_r} \mathrm{Grad}(\pi_v) \cdot \mathbf{K}_r^{m,k-1} \mathrm{Grad}(\theta^{m,k}), \mathrm{d}\Omega, \tag{88}$$

and the right-hand sides $f(\cdot)$, $g(\cdot)$ are given by

$$f(\boldsymbol{v}_v) := -\mathfrak{F}_{\boldsymbol{u}}(\boldsymbol{u}^{m,k-1}, \pi^{m,k-1}, \boldsymbol{v}_v), \tag{89}$$

$$g(\pi_v) := -\mathfrak{F}_{\pi}(\boldsymbol{u}^{m,k-1}, \pi^{m,k-1}, \pi_v). \tag{90}$$

We notice that two "elasticity" tensors feature in (85). They are defined by:

$$\mathbb{C}_r^{m,k-1} := J_a(\mathbf{F}_a)^{-1} \underline{\otimes} (\mathbf{F}_a)^{-1} \colon \mathbb{C}_n^{m,k-1} \colon (\mathbf{F}_a)^{-T} \underline{\otimes} (\mathbf{F}_a)^{-T}, \tag{91}$$

$$\mathbb{C}_c^{m,k-1} := 2J^{m,k-1}\pi^{m,k-1}\left[\mathbb{I}^{m,k-1} - \frac{1}{2}(\mathbf{C}^{m,k-1})^{-1} \otimes (\mathbf{C}^{m,k-1})^{-1}\right]. \tag{92}$$

The tensor $\mathbb{C}_r^{m,k-1}$ is the true elasticity tensor, which can be found by the constitutive prescriptions. The tensor $\mathbb{C}_c^{m,k-1}$, on the other hand, is due to the incompressibility constraint, which, however, does not imply the restriction $J(\mathbf{C}) = 1$ to the motion of the solid phase of a biphasic material.

All quantities with indices "$(m, k-1)$" (e.g., $J^{m,k-1}$) are computed at the "point" $(\boldsymbol{u}^{m,k-1}, \pi^{m,k-1})$, i.e. $J^{m,k-1} \equiv J(\boldsymbol{u}^{m,k-1})$. Moreover, for sake of shorter notation, the Gauteax derivatives have been rewritten as

$$\mathfrak{D}\mathbf{E}(\boldsymbol{u}^{m,k-1})[\boldsymbol{v}_v] \equiv \mathfrak{D}\mathbf{E}^{m,k-1}[\boldsymbol{v}_v], \tag{93}$$

$$\mathfrak{D}\mathbf{E}(\boldsymbol{u}^{m,k-1})[\boldsymbol{h}^{m,k}] \equiv \mathfrak{D}\mathbf{E}^{m,k-1}[\boldsymbol{h}^{m,k}], \tag{94}$$

$$\mathfrak{D}\dot{\mathbf{E}}_v(\boldsymbol{u}^{m,k-1}, \boldsymbol{v}_v)[\boldsymbol{h}^{m,k}] \equiv \mathfrak{D}\dot{\mathbf{E}}_v^{m,k-1}[\boldsymbol{h}^{m,k}], \tag{95}$$

$$\mathfrak{D}\mathbf{K}_r(\boldsymbol{u}^{m,k-1})[\boldsymbol{h}^{m,k}] \equiv \mathfrak{D}\mathbf{K}_r^{m,k-1}[\boldsymbol{h}^{m,k}]. \tag{96}$$

These results (85)–(96) can be obtained by adopting the formulae in Appendix, and using (64) and (65).

In order to write the finite element formulation of the problem, we introduce a conforming shape-regular mesh $\mathscr{T}_h$ consisting of N_h non-overlapping elements $\{K_i\}_{i=1}^{N_h}$. Moreover, we define the following finite dimensional spaces:

$$\begin{aligned} V_h^l &= \left\{\boldsymbol{v}_h \in V\colon \ \boldsymbol{v}_h|_{K_i} \in (\mathbb{P}_l)^3, \text{ for } i = 1, \ldots N_h\right\} \\ Q_h^\sigma &= \left\{\pi_h \in Q\colon \ \pi_h|_{K_i} \in \mathbb{P}_\sigma, \text{ for } i = 1, \ldots N_h\right\}, \end{aligned} \tag{97}$$

where $(\mathbb{P}_l)^3$ and $\mathbb{P}_\sigma$ are spaces of polynomials of degree l and σ, respectively. In particular, the notation $(\mathbb{P}_l)^3$ means that all components of the three-dimensional vector $\boldsymbol{v}_h|_{K_i}$ are polynomials of degree l. Denoting by $\{\boldsymbol{\varphi}_q\}_{q=1}^M$ and $\{\phi_s\}_{s=1}^N$, with $M = \dim(V_h^l)$ and $N = \dim(Q_h^\sigma)$, the Lagrangian basis functions of V_h^l and Q_h^σ, respectively, the discrete solutions can be written as

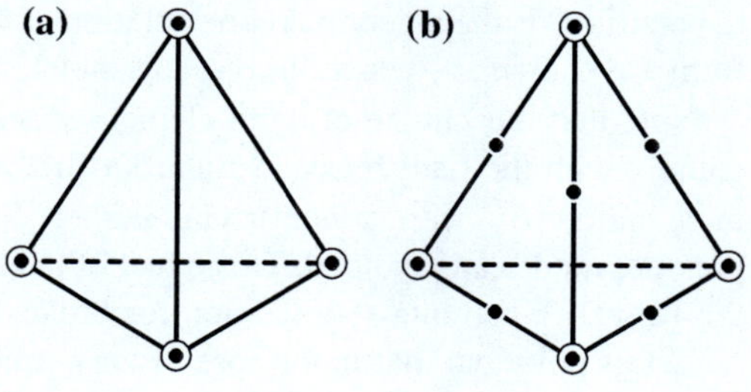

Fig. 1 DOFs for the employed P1-P1 couple (**a**) and for Taylor-Hood P2-P1 elements (**b**). The full circles indicate DOFs for displacement, while the empty circles indicate the DOFs for pressure

$$\boldsymbol{h}_h^{m,k} = \sum_{q=1}^{M} h_q^{m,k} \boldsymbol{\varphi}_q, \qquad \theta_h^{m,k} = \sum_{s=1}^{N} \theta_s^{m,k} \phi_s, \tag{98}$$

Employing these definitions, the finite element formulation of the problem reads

$$\boxed{\begin{array}{l}\text{Find } \boldsymbol{h}_h^{m,k} \in V_h^l \text{ and } \theta_h^{m,k} \in Q_h^\sigma, \text{ such that :}\\ \quad \left\{\begin{array}{ll} a(\boldsymbol{h}_h^{m,k}, \boldsymbol{\varphi}_p) - b(\boldsymbol{\varphi}_p, \theta_h^{m,k}) = f(\boldsymbol{\varphi}_p), & p = 1 \ldots M,\\ -c(\boldsymbol{h}_h^{m,k}, \phi_r) - d(\theta_h^{m,k}, \phi_r) = g(\phi_r), & r = 1 \ldots N,\end{array}\right.\end{array}} \tag{99}$$

or in an algebraic form:

$$\begin{pmatrix} A & -B^T \\ -C & -D \end{pmatrix} \begin{pmatrix} \boldsymbol{h} \\ \boldsymbol{\theta} \end{pmatrix} = \begin{pmatrix} \boldsymbol{f} \\ \boldsymbol{g} \end{pmatrix}. \tag{100}$$

The matrices $A \in \mathbb{R}^{M \times M}$, $B, C \in \mathbb{R}^{M \times N}$, and $D \in \mathbb{R}^{N \times N}$ are the algebraic representations of the weak forms $a(\cdot,\cdot)$, $b(\cdot,\cdot)$, $c(\cdot,\cdot)$, and $d(\cdot,\cdot)$, respectively, and their entries are given by

$$\begin{array}{ll} A_{pq} = a(\boldsymbol{\varphi_q}, \boldsymbol{\varphi_p}) & B_{rq} = b(\boldsymbol{\varphi_q}, \boldsymbol{\phi_r}) \\ C_{rq} = c(\boldsymbol{\varphi_q}, \boldsymbol{\phi_r}) & D_{rs} = d(\phi_s, \phi_r). \end{array} \tag{101}$$

The entries of vectors $\boldsymbol{f}$ and $\boldsymbol{g}$ are given by

$$f_p = f(\boldsymbol{\varphi_p}), \quad g_r = g(\boldsymbol{\phi_r}), \tag{102}$$

while the vectors $\boldsymbol{h}$ and $\boldsymbol{\theta}$ are defined such that $\boldsymbol{h}_q = h_q$ and $\boldsymbol{\theta}_s = \theta_s$.

The discrete problem (100) has the same form as the saddle-point problem arising from the discretization of the incompressible Navier-Stokes equations or the almost incompressible elasticity equations. However, there are some important differences. On the one hand, the matrix A is symmetric but not positive definite since, in general, ellipticity does not hold for the terms defining the bilinear form $a(\cdot,\cdot)$ (cf. (85)). On the other hand, the second row of the system does not arise from the constraint $\operatorname{div}(\boldsymbol{v}) = 0$, rather it comes from the discretization of (69). In fact, we have $C \neq B$ and the matrix D is the algebraic representation of a diffusion operator. Problems of this form are known as "generalised saddle-point problems" [11].

Note that our choice of finite element spaces V_h^1 and Q_h^1 (see Fig. 1a) does not comply with the Ladyženskaja-Babuška-Brezzi (LBB) condition [15]. This results in a matrix B^T with a non-trivial kernel. Still, since D is non-zero, spurious pressure modes are avoided. This effect of the Laplacian is the same as obtained by the Brezzi-Pitktäranta stabilization for Stokes and Navier-Stokes problems [16].

Let us point out that in the case of very small permeability the stabilizing effect of the diffusion matrix D becomes less effective, which can be the source of numerical instabilities. In this case, the choice of discrete spaces that satisfy the LBB condition is necessary, e.g. Taylor-Hood elements (Fig. 1b).

In our simulation, no occurrence of volume locking was observed. However, despite the formulation of the problem in a saddle-point form, volume locking can occur in the case of large bulk moduli. To overcome this issue, different variational formulations involving a mixed problem [55, 14] or second order finite elements can be applied.

3.3 Numerical Solution

The discretization method described above has been implemented in the software toolbox UG/Obslib++ [8, 31].

For small number of degrees of freedom, direct solvers are advantageous due to their robustness and fast solution phase. In our numerical experiments (see Sect. 3) we used UMFPACK [19].

For large system sizes, direct solvers are prohibitively expensive in terms of computing time and memory consumption. However, standard efficient solution strategies like domain-decomposition and multigrid methods are not readily applicable to the type of *coupled* problems discussed here. For example, in the case of an additive Schwarz preconditioned, it can be shown that the iterations of a GMRES solver grows proportional to the square root of the number of the processors [35]. In the case of multigrid methods, the challenge is to find suitable smoothers for the indefinite global matrix.

An alternative solution method is based on the Schur complement of the system (100). This method is based on the elimination of the displacement degrees of freedom which results in a reduced system of size N:

$$(CA^{-1}B^T + D)\boldsymbol{\theta} = -CA^{-1}\boldsymbol{f} - \boldsymbol{g}. \tag{103}$$

Note that if $\boldsymbol{f} = \boldsymbol{0}$, forming the right-hand side does not require any solution of a linear system. A well-known solution method for the system (103) is the Uzawa algorithm. This method is a two-step method and reads:

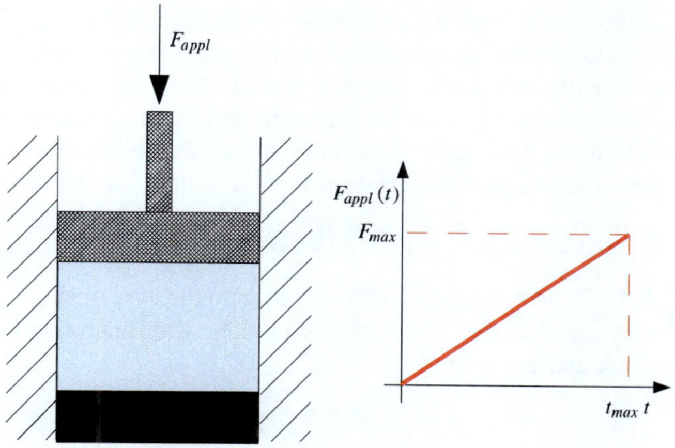

Fig. 2 Schematic representation of the parallel plate apparatus used in the benchmark problem: the lower plate is impermeable, whereas liquid can flow through the upper plate. Results are reported in the case of an applied load linearly increasing in time

Given the initial guess $\boldsymbol{\theta_0}$, at every step k solve :

$$A\boldsymbol{h}_k = B^T\boldsymbol{\theta}_k + \boldsymbol{f},$$

and update pressure :

$$\boldsymbol{\theta}_{k+1} = \boldsymbol{\theta}_k + \omega(-\boldsymbol{g} - C\boldsymbol{h}_k - D\boldsymbol{\theta}_k).$$

The index k used in this box should not be confused with that denoting the kth Newton step in the procedure sketched above.

This method is equivalent to the Richardson method for the Schur complement system and requires the solution of one linear system $A\boldsymbol{h_k} = \boldsymbol{B^T}\boldsymbol{\theta_k} + \boldsymbol{f}$ at every step. The matrix A obtained from the discretization of a non-linear elasticity problem can be efficiently solved with a multigrid method. An efficient preconditioning strategy for the Schur complement arising in the linear problem has been presented in [38, 23].

4 Solution of a Benchmark Problem

We apply the model presented in Sect. 2.4 to describe a confined compression test under given loading conditions. We consider the case in which the biphasic material is positioned inside a rigid cylinder and left free to grow. The cylindrical sample is then compressed between two plates: the lower plate is impermeable, whereas the upper plate allows fluid exudation, so that the liquid embedded in the material can escape from the specimen due to compression (see Fig. 2).

The formulation of the confined compression is based on the assumption that the matrix representation of the deformation gradient is given a very simple diagonal form. Indeed, since the cylindrical wall of the parallel-plate-apparatus is supposed to be rigid and impervious, it is reasonable to assume deformations and velocities of all constituents to be along the Z-axis. Therefore, using a cylindrical coordinate system, the deformation generated by a uniaxial force applied along the Z-axis is

$$\chi^r(t,\boldsymbol{X}) = R, \qquad \chi^\vartheta(t,\boldsymbol{X}) = \Theta, \qquad \chi^z(t,\boldsymbol{X}) = z, \tag{104}$$

where $\boldsymbol{X} = (R, \Theta, Z)$. We restrict our investigation to the case in which χ^z depends on Z and t only, so that the matrix representation of the deformation gradient tensor is diagonal and given by

$$\mathsf{F} = \mathrm{diag}\{1, 1, \partial_Z \chi^z\}. \tag{105}$$

We remark that, due to the particular form of F, the identity $J = \partial_Z \chi^z$ holds true. Then, we assume that $\mathbf{F}_a$, the tensor of anelastic deformation that maps the tangent space of the reference configuration, $T\mathscr{B}_r$, onto the tangent space of the natural configuration, $T\mathscr{B}_n$, has the diagonal form

$$\mathsf{F}_a = \mathrm{diag}\{g_1, g_1, g_3\}. \tag{106}$$

We choose a non-spherical growth in order to study the influence of non-isotropic growth on the distribution of load and pressure throughout compression. For the problem under investigation, g_1 and g_3 are assumed to be constant in time and given from the outset. From (105) and (106), it follows that

$$\mathsf{F}_e = \mathrm{diag}\left\{\frac{1}{g_1}, \frac{1}{g_1}, \frac{J}{g_3}\right\}. \tag{107}$$

The next step is to re-write (59) and (60) in cylindrical coordinates. For ease of notation, we introduce the symbol

$$\boldsymbol{Q} := -\mathbf{K}_r.\mathrm{Grad}(\pi). \tag{108}$$

It follows from (60) that

$$\dot{J} = -\mathrm{Div}(\boldsymbol{Q}) = -\frac{1}{R}\frac{\partial}{\partial R}(RQ^R) - \frac{1}{R}\frac{\partial Q^\Theta}{\partial \Theta} - \frac{\partial Q^Z}{\partial Z} = -\frac{\partial Q^Z}{\partial Z}, \tag{109}$$

where the last identity holds true by requiring that the derivatives with respect to both the radial and tangential directions vanish identically, and that Q^R is zero. The latter condition amounts to say that there is no fluid flow along the radial direction. It is thus sufficient to determine Q^Z, which is given by

$$Q^Z = -(K_r)^{ZZ}\frac{\partial \pi}{\partial Z} = -JK\frac{1}{J^2}\frac{\partial \pi}{\partial Z} = -\frac{K}{J}\frac{\partial \pi}{\partial Z}. \tag{110}$$

By substituting (110) into (109), we obtain

$$\dot{J} = \frac{\partial}{\partial Z}\left(\frac{K}{J}\frac{\partial \pi}{\partial Z}\right). \tag{111}$$

Taking into account that the Piola–Kirchhoff stress tensor $\mathbf{P}$ is diagonal, i.e.

$$\begin{aligned}\mathbf{P} =& \mathrm{diag}\{P^{rR}, P^{\vartheta\Theta}, P^{zZ}\} \\ =& \mu_n J_a \mathrm{diag}\left\{\frac{1}{g_1^2}, \frac{1}{g_1^2}, \frac{J}{g_3^2}\right\} + \left[\lambda_n \log\left(\frac{J}{J_a}\right) - \mu_n\right] \mathrm{diag}\left\{J_a, J_a, \frac{J_a}{J}\right\},\end{aligned} \tag{112}$$

and that the liquid and the solid phases move only along the z-direction, the balance of momentum (30) reduces to

$$\frac{\partial \pi}{\partial Z} = \frac{\partial P^{zZ}}{\partial Z}. \tag{113}$$

By coupling (113) with (111), we obtain

$$\begin{aligned}\dot{J} = \frac{\partial}{\partial Z}\left(\frac{K}{J}\frac{\partial P^{zZ}}{\partial Z}\right) = \\ \frac{\partial}{\partial Z}\left[\frac{K}{J}\frac{\partial}{\partial Z}\left(\mu_n \frac{g_1^2}{g_3} J + \left(\lambda_n \log\left(\frac{J}{g_1^2 g_3}\right) - \mu_n\right)\frac{g_1^2 g_3}{J}\right)\right].\end{aligned} \tag{114}$$

It can be proven that the partial derivative of P^{zZ} with respect to the axial direction reads

$$\frac{\partial P^{zZ}}{\partial Z} = \left[\frac{J_a \mu_n}{g_3^2} + \frac{J_a \mu_n}{J^2} + \frac{J_a \lambda_n}{J^2} - \frac{J_a \lambda_n}{J^2} \ln\left(\frac{J}{J_a}\right)\right]\frac{\partial J}{\partial Z}. \tag{115}$$

Therefore, the mass balance law acquires the form of a nonlinear diffusion equation, in which the "transported" quantity is the volumetric deformation J. Indeed, by substituting (115) into (114), we obtain

$$\dot{J} = \frac{\partial}{\partial Z}\left[D(J)\frac{\partial J}{\partial Z}\right] \tag{116}$$

where $D(J)$ represents a fictitious diffusion coefficient defined by

$$D(J) := K\left[\frac{J_a \mu_n}{J g_3^2} + \frac{J_a \mu_n}{J^3} + \frac{J_a \lambda_n}{J^3} - \frac{J_a \lambda_n}{J^3} \ln\left(\frac{J}{J_a}\right)\right]. \tag{117}$$

It is important to notice that $D(J)$ is always positive, being $J < J(0, Z) = J_a$ because of compression, which leads to $J_e < 1$. Therefore, all the terms on the right-hand-side of (117) are positive. This is consistent with the diffusive nature of the problem and it will be important also in defining boundary conditions.

Finally, the diffusion equation (116) has to be solved together with the auxiliary equations

$$\frac{\partial \chi^z}{\partial Z} = J, \tag{118}$$

$$\frac{\partial \pi}{\partial Z} = \left[\frac{J_a \mu_n}{g_3^2} + \frac{J_a \mu_n}{J^2} + \frac{J_a \lambda_n}{J^2} - \frac{J_a \lambda_n}{J^2} \ln\left(\frac{J}{J_a}\right) \right] \frac{\partial J}{\partial Z}. \tag{119}$$

We remark, however, that (118) and (119) are decoupled from (116), and can thus be solved *a posteriori* once J is known, provided proper boundary conditions are supplied.

4.1 Boundary and Initial Conditions

In order to solve (114), (118) and (119), we have to supply boundary conditions (BCs) and an initial condition (IC). In particular, the mass balance (114) requires two BCs and one IC, whereas both (118) and (119) require one BC only. Boundary conditions have to be provided at the boundary points $Z = 0$ and $Z = L$, which identify the lower and upper boundary of the specimen, respectively.

The boundary conditions have to be consistent with the following requirements: (i) the axial stress at the upper boundary of the specimen has to be equal to the applied load, $P_{appl}(t)$; (ii) the velocity of the fluid and of the solid phase have to be zero at the bottom because the lower plate is impermeable and fixed; and (iii) the pressure π has to be zero at $Z = L$ since the liquid is in equilibrium with the atmosphere. These observations are translated in the following set of boundary conditions

$$\chi^z(t, 0) = 0, \tag{120}$$

$$Q^Z(t, 0) = 0 \quad \Rightarrow \quad \frac{\partial \pi}{\partial Z}(t, 0) = 0, \tag{121}$$

$$\tilde{P}^{zZ}(t, L) + P^{zZ}(t, L) = P_{appl}(t), \tag{122}$$

$$\pi(t, L) = 0, \tag{123}$$

where $\tilde{\mathbf{P}} = -J\pi \mathbf{g}^{-1}\mathbf{F}^{-T}$. We remark that, by virtue of the identity (113), we may rephrase (121) as follows

$$\frac{\partial P^{zZ}}{\partial Z}(t, 0) = \left[\frac{J_a \mu_n}{g_3^2} + \frac{J_a \mu_n}{J^2} + \frac{J_a \lambda_n}{J^2} - \frac{J_a \lambda_n}{J^2} \ln\left(\frac{J}{J_a}\right) \right]_{(t,0)} \frac{\partial J}{\partial Z}(t, 0) = 0. \tag{124}$$

We recall that the argument in the square brackets is always positive. Therefore, this condition leads to a zero-Neumann BC for J at the lower boundary:

$$\frac{\partial J}{\partial Z}(t, 0) = 0. \tag{125}$$

On the other hand, (122) leads to a Dirichlet condition on J at the upper boundary:

$$\mu_n \frac{J_a J(t,L)}{g_3^2} - \mu_n \frac{J_a}{J(t,L)} + \lambda_n \frac{J_a}{J(t,L)} \ln\left(\frac{J(t,L)}{J_a}\right) = P_{appl}(t). \tag{126}$$

Since this equation is nonlinear with respect to $J(t,L)$, solutions can be found by applying Newton's method or other techniques.

As initial condition, we take $J(0,Z) = J_a(0,Z)J_e(0,Z) = J_a$. Indeed, at the initial time, there is no elastic deformation, although the anelastic deformation has already occurred. We remark that, for consistency, the condition $\mathbf{F}_e(0,X) = \boldsymbol{\delta}$ entails that $g_1 = 1$.

4.2 Discretization

Equation (114) can be solved using central differences for space derivatives and then a proper ODE solver to obtain the temporal evolution. In the following we depict the main steps of this procedure.

The 1D-domain, represented by the interval $[0,L]$, is divided into $N-1$ subintervals of the same width ΔZ through the introduction of N equispaced nodes

$$0 = Z_1 < Z_2 = Z_1 + \Delta Z < \ldots < Z_j < \ldots < Z_{N-1} < Z_N = L.$$

Spaces derivatives are then approximated by finite differences, so that the following system of $N-2$ equations is obtained:

$$\dot{J}_j = \frac{1}{(\Delta Z)^2}\left(\frac{K_{j+1}}{J_{j+1}} P_{j+1}^{zZ} - \left(\frac{K_{j+1}}{J_{j+1}} + \frac{K_j}{J_j}\right) P_j^{zZ} + \frac{K_j}{J_j} P_{j-1}^{zZ}\right). \tag{127}$$

Here, j enumerates the nodes of the grid, i.e. $J_j = J(t,Z_j)$, $K_j = K(t,Z_j)$ and $P_j^{zZ} = P^{zZ}(t,Z_j)$, with $j = 2, \ldots, N-1$. The boundary values J_1 and J_N are given by (125) and (126). A special treatment is performed for the initial node: in order to preserve the second-order-accuracy of the discretization method, a fictitious node Z_0 is introduced, and the Neumann boundary condition (125) is approximated by the central difference

$$\frac{J_2 - J_0}{2(\Delta Z)} = 0,$$

which implies $J_0 = J_2$. This allows to prolong the validity of the discretization used in (127) to the node $j = 1$.

At the upper boundary, we solve (126) numerically in order to determine J_N. For this purpose, we implement a standard Newton-Raphson method. According to this procedure, the initial partial differential equation (114) is approximated by a system of ordinary differential equations that can be integrated by choosing a stable ODE solver, with the initial condition $J_j(0) = J(0,Z_j) = g_1^2 g_3$, for $j = 1, \ldots, N$.

Table 1 Parameters of the benchmark problem

Parameter	Description	Value
L	Height of the specimen	10 mm
$2R$	Diameter of the specimen	5 mm
$F_{\max}$	Maximum applied force	$-0.2 \cdot 9.81$N
$t_{\max}$	Time of load application	30 s
k_0	Hydraulic conductivity	$3.6454 \cdot 10^{-12}$ m$^4/(\mathrm{N} \cdot \mathrm{s})$
m_0	Material parameter	0.0848
m_1	Material parameter	4.638
ϕ_{s0}	Referential value of solidity.	0.6
ϕ_{sn}	Solidity in the relaxed configuration	0.6
λ_n	Lamé's first modulus	0.3137 MPa
μ_n	Shear modulus	0.3566 MPa

After computing J, the function χ^z is calculated by invoking (118) coupled with (120), and using a standard forward Euler method. The variation of pressure inside the specimen can be calculated a posteriori, once P^{zZ} is known. Indeed, integrating the balance of momentum (113), with the boundary condition $\pi(t, L) = 0$ MPa, yields

$$\pi(t, Z) = P^{zZ}(t, Z) - P^{zZ}(t, L). \tag{128}$$

4.3 Results

The model presented in the previous sections is applied to describe the compression of a cylindric specimen of soft biological tissue, which is positioned in a chamber delimited by a rigid cylindric wall and two parallel plates. The wall and the lower plate are impermeable to liquid, whereas the upper plate allows for fluid exudation. An external compressive force is applied at the upper plate, parallel to the symmetry axis of the specimen. The experimental apparatus is schematically shown in Fig. 2.

We restrict our analysis to the case in which the external force increases linearly in time until $t_{\max} = 30$ s when the maximum force, $F_{\max} = -0.2 \cdot 9.81$ N, is reached.

In (106), we consider

$$\mathsf{F}_a = \mathrm{diag}\{g_1, g_1, g_3\} = \mathrm{diag}\{g, g, g + \epsilon\}, \tag{129}$$

where ϵ measures the deviation of $\mathbf{F}_a$ from a spherical anelastic deformation. The numerical results shown in this section are obtained by implementing in Matlab® the procedure described in Sect. 4.2. All the parameters are listed in Table 1.

We recall that, for the considered problem, the only possible value of g is unity. This implies that the deviations of $\mathbf{F}_a$ from a referential spherical tensor are actually the deviations from the identity tensor. We start the simulations with $\epsilon = 0.1$.

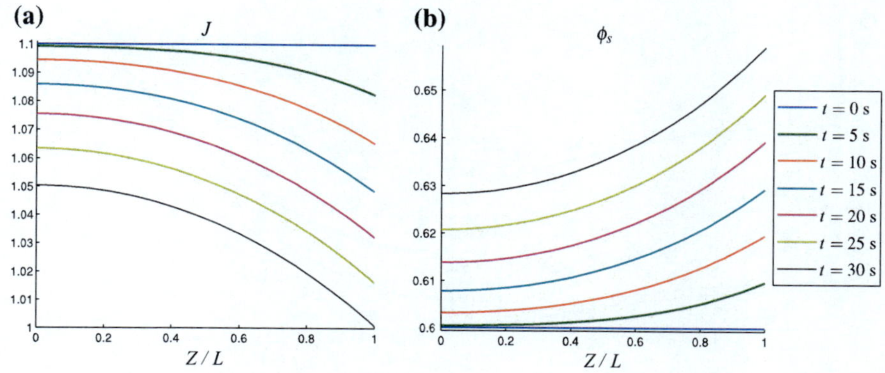

Fig. 3 Evolution in time and space of $J(t,Z)$ (**a**) and solid volumetric fraction (**b**), when $\epsilon = 0.1$, starting from the initial values $J(0,Z) = J_a$ and $\phi_s = \phi_{sn}$. Solutions are reported every 5 s for 30 s, which correspond to the maximum time of load application. All the parameters used in the simulation are listed in Table 1

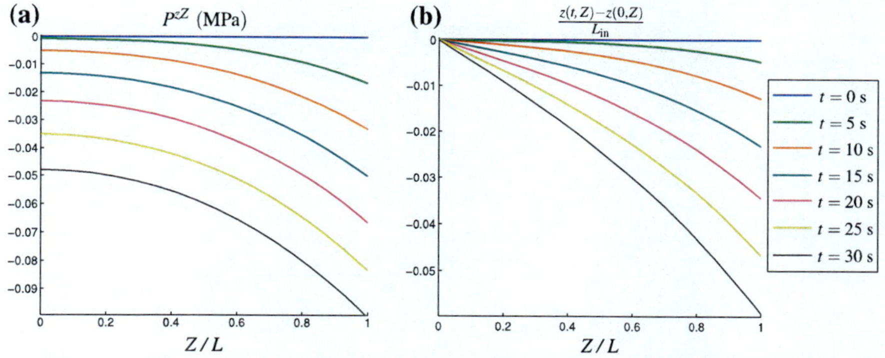

Fig. 4 Evolution in time and space of $P^{zZ}(t,Z)$ (**a**) and relative displacement, $[z(t,Z) - z(0,Z)]/L_{\text{in}}$ (**b**), with $P^{zZ}(0,Z) = 0$ MPa and $z(0,Z) = J_a Z$. Results are plotted every 5 s up to 30 s, in the case of the parameters listed in Table 1 and $\epsilon = 0.1$

Equation (114), integrated with the initial condition $J(0,Z) = J_a$ and BCs (125) and (126), gives origin to the curves in Fig. 3a, which represent $J(t,Z)$ plotted over space, at different instants of time (every 5 s for 30 s). The corresponding volumetric fraction of the solid, $\phi_s(t,Z) = \phi_{sn} J_a [J(t,Z)]^{-1}$, is reported in Fig. 3b. The characteristic time of the diffusive process described in (114) is defined by $t_d := L^2[D(J)]^{-1}$, which is a function of time and material coordinates through J. For the considered load, $D(J)$ is an increasing function of J. Thus, the maximum characteristic time, t_d^M, corresponds to the minimum value of J, which is reached at the end of the simulation in $Z = L$ (cf. Fig. 3a). On the other hand, the minimum characteristic time, t_d^m, is reached at the beginning of the simulation, for $J(0,Z) = J_a$ and $D(J(0,Z)) = k_0(\mu_n g_3^{-2} + \mu_n J_a^{-2} + \lambda_n J_a^{-2}) = k_0(2\mu_n + \lambda_n)J_a^{-2}$. We remark

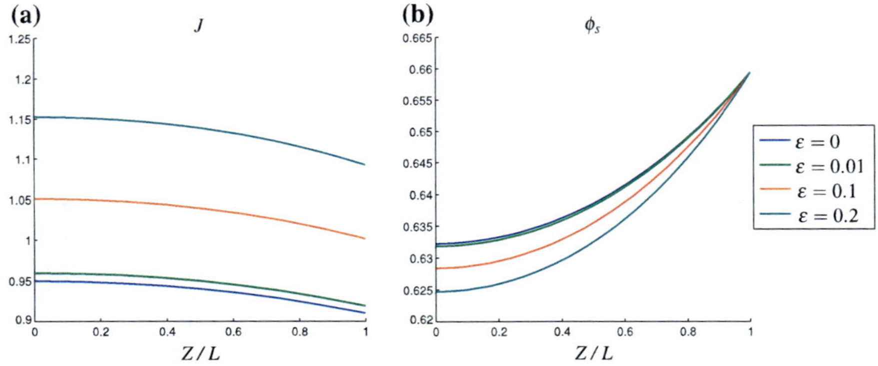

Fig. 5 Distribution of J (**a**) and solid volumetric fraction (**b**) at the final time of compression $t_{\max} = 30$ s, for different values of ϵ

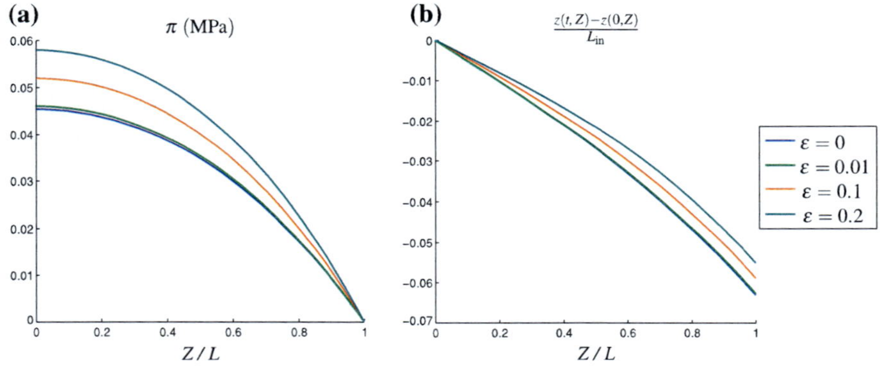

Fig. 6 Evolution of the pressure over space (**a**) and relative displacement (**b**) at the final time of compression $t_{\max} = 30$ s, for different values of ϵ

that the factor $(2\mu_n + \lambda_n)$ coincides with the P-wave modulus of the material. For the parameters listed in Table 1, we find $t_d^M \approx 38.2625$ s and $t_d^m \approx 32.3206$ s. We notice that t_d^M and t_d^m are of the same order as $t_{\max}$.

The axial component of the constitutive part of the first Piola–Kirchhoff stress tensor, P^{zZ}, and the relative displacement, $[z(t,Z) - z(0,Z)]/L_{\text{in}}$ are plotted in Fig. 4a and b, respectively. Here, L_{in} denotes the length of the specimen at time $t = 0$ s, which is defined by $L_{\text{in}} := \int_0^L \partial_Z \chi^z(0,Z)\mathrm{d}Z = J_a L$ consistently with (118).

The value of P^{zZ} at the upper boundary is given by $P^{zZ}(t,L) = \big(F_{appl}(t,L)\big)/S$, where $F_{appl}(t,L) = F_{\max}[t/t_{\max}]$ and $S = \pi R^2$, the area of the surface over which the applied load is distributed, coincides with the cross section of the specimen in the reference configuration. The amplitude of the displacement increases in time with the applied load (cf. Fig. 4b). This behaviour is qualitatively the same also for

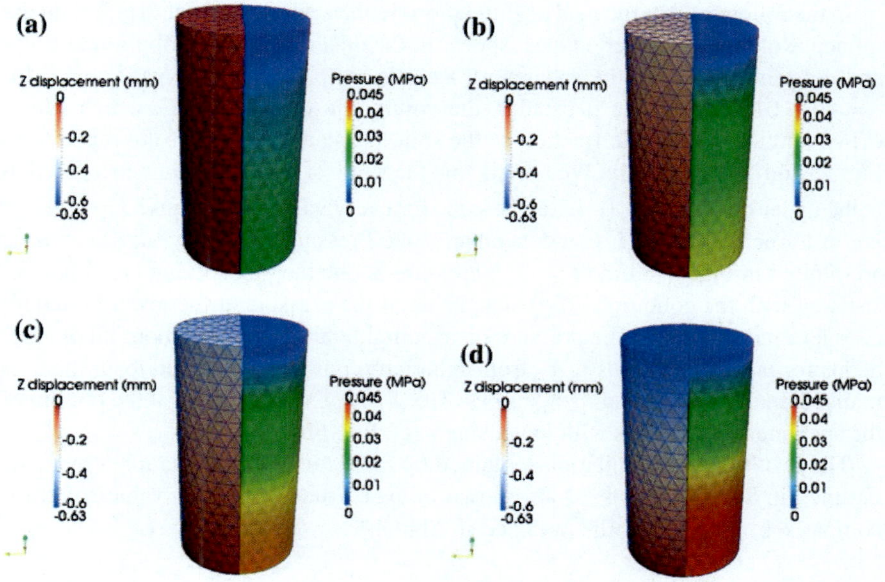

Fig. 7 Time evolution of the displacement and the pressure without growth ($\epsilon = 0$)

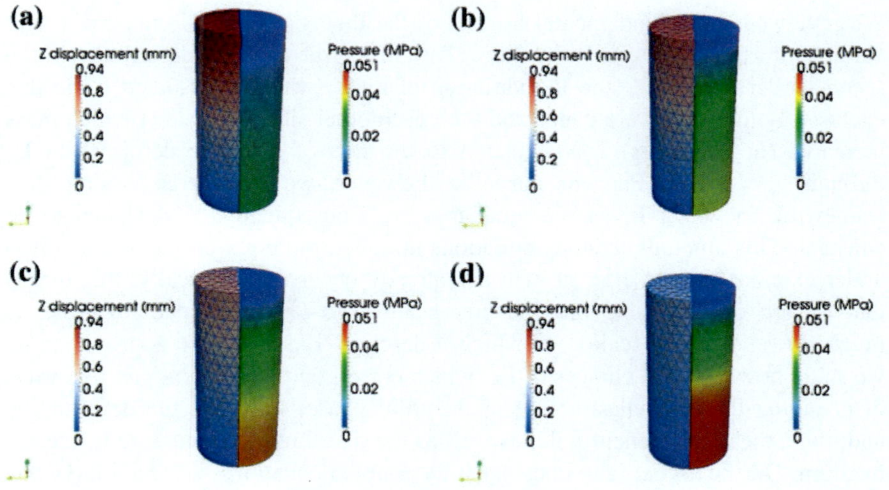

Fig. 8 Time evolution of the displacement and the pressure with $\epsilon = 0.1$

the other values of ϵ considered in the following. However, the diffusive process tends to become slower as ϵ gets bigger.

In the following, we run a set of simulations with varying ϵ in order to highlight the influence of this parameter on the response of the material (e.g., distribution of stress and deformation inside the specimen). Results are presented in Figs. 5 and 6 for $\epsilon = \{0, 0.01, 0.1, 0.2\}$. In particular, the volumetric deformation J, which solves (114), and the volumetric fraction of the solid phase at $t_{\max} = 30$ s are reported in Fig. 5a and b, respectively. We remark that the value of the solid volumetric fraction at the upper boundary, $\phi_s(t, L)$, is the same for every value of ϵ because $J_a[J(t, L)]^{-1}$ is constrained to satisfy (126) independently of ϵ. Pressure and relative displacements are plotted in Fig. 6 at time $t = t_{\max}$. Pressure is obtained by solving (128) in consistency with the condition (123). The value of the pressure at the lower boundary rises as ϵ increases, and the pressure distribution tends to become more inhomogeneous for larger deviations of $\mathbf{F}_a$ from sphericity. For the considered load, the normalised final displacement, $[z(t_{\max}, Z) - z(0, Z)]/L_{\text{in}}$, which is zero at the bottom of the specimen, diminishes with increasing ϵ (cf. Fig. 6b).

The results of the simulations obtained by means of the computational methods outlined in Sects. 3.2 and 3.3 are shown in Figs. 7 and 8, for two values of ϵ, for comparison with the results obtained in Matlab®.

5 Conclusions and Outlook

We reviewed some fundamental aspects of the theory of biphasic materials with variable mass and internal structure. The structural change, described by the second-order tensor $\mathbf{F}_a$, and the variation of mass (which is assumed to be due exclusively to growth) are connected with each other since the rate at which mass increases (or decreases), γ_s, is related to the rate of anelastic deformation $\mathbf{L}_a$ through $\gamma_s = \text{tr}(\mathbf{L}_a)$. For our purposes, however, we considered a simplified framework in which $\mathbf{L}_a$ is set equal to zero. Consequently, $\mathbf{F}_a$ is taken to be constant. This amounts to study situations in which the biphasic medium evolves under external actions after growth has already occurred. Physically, this means that we are hypothesising that the time scale over which the medium grows is much slower than the scale over which it deforms. Based on this approximation, we study how different choices of $\mathbf{F}_a$ (which correspond to different possible ways of changing the internal structure of the solid phase) influence the deformation and, thus, the displacement field, as well as the distribution of pressure inside the medium. Our next goal is to consider fully coupled equations, in which the value of $\mathbf{F}_a$ changes in time according one of the evolution laws given in (56), (57), or (58). Furthermore, on the basis of a model presented in [27], we would like to study the influence of the structural change of the porous medium on the transport mechanisms governing the evolution of chemical agents.

Acknowledgments This work was supported by the Politecnico di Torino (Turin, Italy), Universitá della Svizzera Italiana (Lugano, Switzerland), and Goethe-Universität Frankfurt am Main, Germany. We thank Prof. Luigi Preziosi (Polytechnic of Turin, Italy) for useful discussions.

A.1 Appendix

The following relations are used in the computations above:

$$\frac{\partial J}{\partial \mathbf{C}} = \frac{1}{2} J \mathbf{C}^{-1}, \quad \frac{\partial \mathbf{C}}{\partial \mathbf{F}} = \delta \,\overline{\otimes}\, \mathbf{F}^T + \mathbf{F}^T \underline{\otimes}\, \delta, \tag{130}$$

$$\frac{\partial \mathbf{C}^{-1}}{\partial \mathbf{C}} = -\mathbb{I} = -\frac{1}{2}\left[\mathbf{C}^{-1} \underline{\otimes}\, \mathbf{C}^{-1} + \mathbf{C}^{-1} \overline{\otimes}\, \mathbf{C}^{-1}\right]. \tag{131}$$

The tensor $\mathbb{I}_n$ is obtained from (131) by substituting $\mathbf{C}$ with $\mathbf{C}_e$.

Given two second-order tensors $\mathbf{A}$ and $\mathbf{B}$, the products $\mathbf{A} \,\overline{\otimes}\, \mathbf{B}$ and $\mathbf{A} \underline{\otimes}\, \mathbf{B}$ have the following index representation [18]

$$[\mathbf{A} \,\overline{\otimes}\, \mathbf{B}]_{IJMN} = A_{IN} B_{JM}, \quad [\mathbf{A} \underline{\otimes}\, \mathbf{B}]_{IJMN} = A_{IM} B_{JN}. \tag{132}$$

To perform the linearisation procedure, we set $\boldsymbol{u} = \boldsymbol{u}_0 + \boldsymbol{h}$, and use the following Gateaux-derivatives:

$$\begin{aligned} \mathfrak{D}(J\pi\mathbf{C}^{-1})(\boldsymbol{u}_0, \pi_0)[\boldsymbol{h}, \theta] = {} & (J_0(\mathbf{C}_0)^{-T} \colon \mathfrak{D}\mathbf{E}(\boldsymbol{u}_0)[\boldsymbol{h}])\pi_0(\mathbf{C}_0)^{-1} \\ & + J_0\theta(\mathbf{C}_0)^{-1} - J_0\pi_0 2(\mathbf{C}_0)^{-1}\{\mathfrak{D}\mathbf{E}(\boldsymbol{u}_0)[\boldsymbol{h}]\}(\mathbf{C}_0)^{-1}, \end{aligned} \tag{133}$$

$$\mathfrak{D}\dot{\mathbf{E}}_v(\boldsymbol{u}_0, \boldsymbol{v}_v)[\boldsymbol{h}] = \frac{1}{2}\left[(\mathbf{H})^T \dot{\mathbf{F}}_v + (\dot{\mathbf{F}}_v)^T \mathbf{H}\right], \tag{134}$$

$$\mathfrak{D}\mathbf{S}(\boldsymbol{u}_0)[\boldsymbol{h}] = \mathbb{C}_r(\boldsymbol{u}_0) \colon \mathfrak{D}\mathbf{E}(\boldsymbol{u}_0)[\boldsymbol{h}], \tag{135}$$

$$\mathbb{C}_r = J_a(\mathbf{F}_a)^{-1} \underline{\otimes} (\mathbf{F}_a)^{-1} \colon \mathbb{C}_n \colon (\mathbf{F}_a)^{-T} \underline{\otimes} (\mathbf{F}_a)^{-T}. \tag{136}$$

The formulae in Sect. 3.1 are retrieved by setting $\boldsymbol{u} \equiv \boldsymbol{u}^{m,k}$, $\boldsymbol{u}_0 \equiv \boldsymbol{u}^{m,k-1}$ and $\boldsymbol{h} \equiv \boldsymbol{h}^{m,k}$.

References

1. Ambrosi, D., Guana, F.: Stress-modulated growth. Math. Mech. Solids **12**, 319–342 (2007)
2. Ambrosi, D., Guillou, A., Di Martino, E.: Stress-modulated remodeling of a non-homogeneous body. Biomechan. Model. Mechanobiol **7**, 63–76 (2008)
3. Ambrosi, D., Mollica, F.: On the mechanics of a growing tumor. Int. J. Eng. Sci. **40**, 1297–1316 (2002)
4. Ambrosi, D., Mollica, F.: The role of stress in the growth of a multicell spheroid. J. Math. Biol. **48**, 477–499 (2004)
5. Ambrosi, D., Preziosi, L., Vitale, G.: The insight of mixtures theory for growth and remodeling. Z. Angew. Math. Phys. **61**, 177–191 (2010)
6. Ateshian, G.A., Weiss, J.A.: Anistropic hydraulic permeability under finite deformation. J. Biomech. Eng. **132**, 111,004–1 (2010)
7. Atkin, R., Craine, R.: Continuum theories of mixtures: basic theory and historical development. Quart. J. Mech. Appl. Math. **29**, 209–244 (1976)

8. Bastian, P., Birken, K., Johannsen, K., Lang, S., Neuß, N., Rentz-Reichert, H., Wieners, C.: UG – a flexible software toolbox for solving partial differential equations. Comput. Vis. Sci. **1**, 27–40 (1997)
9. Bear, J.: Dynamics of fluid in Porous Media. Dover Publications, INC, New York (1972)
10. Bennethum, L.S., Murad, M.A., Cushman, J.H.: Macroscale thermodynamics and the chemical potential for swelling porous media. Transport in Porous Media **39**, 187–225 (2000)
11. Benzi, M., Golub, G.H., Liesen, J.: Numerical solution of saddle point problems. Acta Numerica **14**, 1–137 (2005). DOI:10.1017/S0962492904000212. http://dx.doi.org/10.1017/S0962492904000212
12. Bonet, J., Wood, R.D.: Nonlinear Continuum Mechanics for Finite Element Analysis. Cambridge University Press, Cambridge, New York, Melbourne (2008)
13. Bowen, R.: (1976) Theory of Mixtures. vol. III, pages 1–127. Academic Press, Inc., New York
14. Braess, D.: Finite Elements : Theory, Fast Solvers, and Applications in Solid Mechanics. Cambridge University Press (2009). http://www.worldcat.org/isbn/9780521588348
15. Brezzi, F.: On existence, uniqueness, and approximation of saddle-point problems arising from Lagrangian multipliers. RAIRO Model. Math. Anal. Numer. **21**, 129–151 (1974)
16. Brezzi, F., Pitkäranta, J.: (1984) On the stabilization of finite element approximations of the Stokes equations. In: Hackbusch W. (ed.) Efficient Solutions of Elliptic Systems, pp. 11–19. Vieweg
17. Cermelli, P., Fried, E., Sellers, S.: Configurational stress, yield and flow in rate-independent plasticity. Proc. R. Soc. Lond. A **457**, 1447–1467 (2001)
18. Curnier, A., He, Q.C., Zysset, P.: Conewise linear elastic materials. J. Elast. **37**, 1–38 (1995)
19. Davis, T.A.: Algorithm 832: Umfpack v4.3—an unsymmetric-pattern multifrontal method. ACM Trans. Math. Softw. **30**, 196–199 (2004). http://doi.acm.org/10.1145/992200.99220.. http://doi.acm.org/10.1145/992200.992206
20. DiCarlo, A., Quiligotti, S.: Growth and balance. Mechanics Research Communications **29** (2002)
21. Ehlers, W., Markert, B.: A linear viscoelastic biphasic model for soft tissues based on the theory of porous media. J. Biomech. Eng. **123**, 418–424 (2001)
22. Epstein, M., Maugin, G.A.: Thermomechanics of volumetric growth in uniform bodies. Int. J. Plast. **16**, 951–978 (2000)
23. Favino, M., Krause, R., Johannes, S.: An efficient preconditioning strategy for schur complements arising from biphasic models. In: Biomedical Engineering / 765: Telehealth / 766: Assistive Technologies (2012)
24. Federico, S., Grillo, A.: Elasticity and permeability of porous fibre-reinforced materials under large deformations. Mech. Mater. **44**, 58–71 (2012)
25. Fung, Y.C.: Biomechanics: Motion, Flow, Stress, and Growth. Springer-Verlag Inc., New York (1990)
26. Fusi, L., Farina, A., Ambrosi, D.: Mathematical modelling of solid liquid mixture with mass exchange between constituents. Mathematics and Mechanics of Solids pp. 575–595 (11)
27. Grillo, A., Federico, S., Wittum, G.: Growth, mass transfer, and remodeling in fiber-reinforced, multi-constituent materials. International Journal of Non-Linear Mechanics (doi:10.1016/j.ijnonlinmec.2011.09.026) (2011)
28. Grillo, A., Federico, S., Wittum, G., Imatani, S., Giaquinta, G., Mićunović, M.V.: Evolution of a fibre-reinforced growing mixture. Il Nuovo Cimento **32C**(1), 97–119 (2009)
29. Grillo, A., Wittum, G., Giaquinta, G., Mićunović, M.V.: A multiscale analysis of growth and diffusion dynamics in biological materials. Int. J. Eng. Sci. **47**, 261–283 (2009)
30. Grillo, A., Zingali, G., Borrello, D., Giaquinta, G.: Transport phenomena in living systems and continuum physics. La Rivista del Nuovo Cimento **31**(11), 485–562 (2007)
31. Groß, C., Krause, R.: Import of geometries and extended informations into obslib++ using the exodus ii and exodus parameter file formats. Tech. Rep. 712, Institute for Numerical Simulation, University of Bonn, Germany (2008)
32. Gurtin, M.E., Fried, E., Anand, L.: The Mechanics and Thermodynamics of Continu. Cambridge University Press, Cambridge, New York, Melbourne (2010)

33. Hassanizadeh, S.M., Gray, W.G.: Derivation of conditions describing transport across zones of reduced dynamics within multiphase systems. Water Resour. Res. **25**, 529–539 (1989)
34. Holmes, M.H., Mow, V.C.: The nonlinear characteristics of soft gels and hydrated connective tissues in ultrafiltration. J. Biomech. **23**(11), 1145–1156 (1990)
35. Hwang, F., Cai, X.: Parallel fully coupled schwarz preconditioners for saddle point problems. Eelectronic transactions on numerical analysis (2006)
36. Kröner, E.: Die inneren Spannungen und die Inkompatibilitätstensor in der Elastizitätstheorie. Zeitschrift für Angewandte Mathematik und Physik **7**, 249–257 (1959)
37. Kröner, E.: Kontinuumstheorie der versetzung und eigenspannungen. Arch Ration. Mech. Anal. **4**, 273–334 (1960)
38. Lang, S., Wieners, C., Wittum, G.: The application of adaptive parallel multigrid methods to problems in nonlinear solid mechanics, stein, e. (ed): adaptive methods in solid mechanics edn. J. Wiley (2002)
39. Lee, E.H.: Elastic-plastic deformation at finite strains. ASME Transaction on Journal of Applied Mechanics **36**, 1–6 (1969)
40. Loret, B., Simões, F.M.F.: A framework for deformation, generalized diffusion, mass transfer and growth in multi-species multi-phase biological tissues. Eur. J. Mech. A/Solids**24**, 757–781 (2005)
41. Lubarda, V.A., Hoger, A.: On the mechanics of solids with a growing mass. Int. J. Solids Struct. **39**, 4627–4664 (2002)
42. Mićunović, M.V.: Thermodynamics of Viscoplasticity, Advances in Mathematics for Applied Sciences, vol. 20. Springer Verlag, Heidelberg (2009)
43. Miehe, C., Stein, E., Wagner, W.: Associative multiplicative elasto-plasticity. formulation and aspects of the numerical implementation including stability analysis. Comput. Struct. **52**, 969–978 (1994)
44. Ogden, R.W.: Non-Linear Elastic Deformations. Dover Publications, INC, Mineola (1984)
45. Olsson, T., Klarbring, A.: Residual stresses in soft tissue as a consequence of growth and remodeling: application to an arterial geometry. Eur. J. Mech. A/Solids **27**, 959–974 (2008)
46. Preziosi, L., Vitale, G.: A multiphase model of tumor and tissue growth including cell adhesion and plastic reorganization. Math. Models Methods Appl. Sci. **21**(9), 1901–1932 (2011)
47. Quarteroni, A., Sacco, R., Saleri, F.: Numerical Mathematics, Texts in Applied Mathematics, vol. 37. Springer-Verlag Inc., New York (2000)
48. Quiligotti, S.: On bulk growth mechanics of solid-fluid mixtures: kinematics and invariance requirements. Theor. Appl. Mech. **28**(−29), 277–288 (2002)
49. Rajagopal, K.R.: Multiple natural configurations in continuum mechanics. Rep. Inst. Comput. Appl. Mech. **6**, (1995)
50. Rajagopal, K.R., Srinivasa, A.R.: On thermodynamical restrictions of continua. Proc. R. Soc. Lond. A **460**, 631–651 (2004)
51. Rajagopal, K.R., Tao, L.: Mechanics of Mixtures, Advances in Mathematics for Applied Sciences, vol. 35. World Scientific, Singapore (1995)
52. Rockafellar, R.T.: Convex Analysis. Princeton University Press, Princeton, NJ (1970)
53. Rodriguez, E.K., Hoger, A., D., M.A.: Stress-dependent finite growth in soft elastic tissues. Journal of Biomechanics **27**, 455–467 (1994)
54. Simo, J.C., Hughes, T.J.R.: Compuational Inelasticity. Springer Verlag, (1998)
55. Simo, J.C., Rifai, M.S.: A class of mixed assumed strain methods and the method of incompatible modes. International Journal for Numerical Methods in Engineering **29**(8), 1595–1638 (1990). doi:DOI:10.1002/nme.1620290802. http://dx.doi.org/10.1002/nme.1620290802
56. T.E. Simos, G.P., Tsitouras, C. (eds.): Growth and Mass Transfer in Multi-Constituent Biological Materials, vol. 1 (2010)
57. Taber, L.A.: Biomechanics of growth, remodeling and morphogenesis. ASME Appl. Mech. Rev. **48**, 487–545 (1995)
58. Truesdell, C.A.: Sulle basi della termomeccanica. Lincei. Rend. Sc. fis. mat. nat. **22**, 33–38 (1957)
59. Wieners, C.: Nonlinear solution methods for infinitesimal perfect plasticity. Z. Angew. Math. Mech. **87**(8–9), 643–660 (2007)

Study of Diffusion in a One-Dimensional Lattice-Gas Model of Zeolites: The Analytical Approach and Kinetic Monte Carlo Simulations

Alexander Tarasenko and Lubomir Jastrabík

Abstract The diffusion of molecules adsorbed in a one-dimensional channel with side pockets is investigated in the framework of a one-dimensional lattice-gas model. The model can describe the molecules migration in some type of zeolites. We obtained the exact expression for the free energy of this model. Using the local equilibrium approximation we derived the analytical expressions for the diffusion coefficients. The concentration dependencies of the center-of-mass and Fickian diffusion coefficients are calculated for some representative values of the lateral interactions between molecules. The theoretical dependencies are compared with the numerical data obtained by the kinetic Monte Carlo simulations. The data obtained by the two completely different methods coincide amazingly well in the whole concentration and wide interaction regions.

Keywords Zeolites · Lattice-gas model · Monte Carlo

1 Introduction

Zeolites are nanoporous crystalline solids with pore sizes comparable to the molecular size (an established term for these substances is molecular sieves to define porous solid materials that act as sieves on a molecular scale).

A. Tarasenko (✉)
Institute of Physics, v.v.i., AV CR, Na Slovance 2,
182 21 Prague 8, Czech Republic
e-mail: taras@fzu.cz

L. Jastrabík
Palacký University, RCPTM, Joint Laboratory of Optics,
17. listopadu 12, 771 46 Olomouc, Czech Republic
e-mail: jastrab@fzu.cz

J. M. P. Q. Delgado et al. (eds.), *Numerical Analysis of Heat and Mass Transfer in Porous Media*, Advanced Structured Materials 27,
DOI: 10.1007/978-3-642-30532-0_3,

Structurally, zeolites are complex inorganic polymers based on an infinitely extending framework of SiO_4 and AlO_4 tetrahedra linked to each other by the sharing of oxygen ions (hence another name-aluminosilicates). Si or Al atoms (also possible other atoms with relatively low electronegativity P, Ge, Zn, etc.) are in the center of the tetrahedra, and oxygen anions are in the corners. There are several types of building units in zeolite frameworks. The tetrahedral units, which form rings of different sizes, are the primary ones. A ring containing n tetrahedra is called an n-ring. The most common rings contain 4, 5, 6, 8, 10, or 12 tetrahedra. The larger units are polyhedra constructed by n-rings. They are classified according to their form and sizes. Cages are polyhedra formed by small 6-rings. Cavities are larger polyhedral units with wide windows that allow the passage of molecules in and out of the cavities. A channel (void) is a pore that is infinitely extended in one dimension with a minimum aperture n-ring that allows guest molecules to diffuse along the pore. In many zeolites the channels intersect forming the two- and three-dimensional channel systems. The typical zeolite pore size ranges are 0.3–0.45 nm (small pore zeolites with 8-ring pores), 0.45–0.6 nm (medium pore zeolites with 10-ring pores), and 0.6–0.8 nm (large pore zeolites with 12-ring pores).

Zeolites have a rich variety of interesting properties and applications. They are extensively used in industrial applications due to the special characteristics like a large free volume, low density, and void spaces in the form of cages and channels. Catalysis, adsorption, ion exchange, molecular sieving to name but a few are now the common fields of application for these materials [5]. The mechanisms and rates of guest molecules migration are of great scientific and practical importance as many of these processes are diffusion controlled. The correct assessment of their efficiency depends on the accurate determinations of the dependence of diffusion on the topological and textural properties. Diffusion and transport phenomena in porous media like zeolites have been the subject of an intense research for several decades. The pore architecture, i.e. pore size, pore size distribution, pore volume, and pore topology all have a large influence on adsorption and transport phenomena (see, for example, [20]).

Zeolites are ideal systems to study the effect of confinement on the migration of adsorbed molecules. The diffusion in zeolites differs from the ordinary diffusion considerably. The guest molecules have to move through the channels of molecular dimensions. When the molecules are sufficiently large and cannot pass each other, their diffusion along one-dimensional channels proceeds by a rather specific mode. This type of diffusion is known as a single-file diffusion [17–19, 21].

It is rather difficult to develop the reliable approach that can account for all the factors influencing the diffusion in these materials. It is often hard even to establish the character of the interactions between atoms and molecules. Usually, the interactions are modelled by the universal pair wise Lennard-Jones potentials:

$$V_{LJ}(r_{ij}) = 4\varepsilon_{ij}\left[\left(\frac{\sigma_{ij}}{r_{ij}}\right)^{12} - \left(\frac{\sigma_{ij}}{r_{ij}}\right)^{6}\right]. \tag{1}$$

The interaction parameters ε_{ij} and σ_{ij} depend on the properties of the interacting atoms. In principle, they should be calculated using ab initio the quantum–mechanical approach. But really, they are the fitting parameters the values of which are chosen to mimic some experimental data of the investigated systems. It means that for the description of the equilibrium and transport processes, in principle, different sets of the Lennard-Jones parameters (for the same pairs of interacting atoms) are required.

The overwhelming number of theoretical studies of the diffusion in zeolites have been carried out using numerical techniques and powerful computation facilities. There are two distinctly different methods for calculations of the diffusion coefficients: the kinetic Monte Carlo (kMC) simulations and molecular dynamics (MD). They are rather reliable tools for the investigations of the adsorptive and diffusive phenomena in zeolites and other microporous materials. The main limitations of the molecular approach, especially for the investigation of the molecule migration, are the relatively short time and length scales that are accessible even with the modern computational power. The kMC techniques use the stochastic description of the diffusion. The migration of molecules proceeds via random jumps between the adsorption sites on some lattice. It is a reasonable assumption for the guest molecules with high activation energies. The stochastic algorithms use jumps as the fundamental acts of migration. The kMC simulations are capable of modeling the system evolution over much longer time scales than the molecular dynamics. Lattice models are very suitable for the diffusion simulations.

We model the diffusion in zeolites by replacing the zeolite framework with a one-dimensional (1D) chain of binding sites with side arms. Although this model is far too simple to make any quantitative predictions about the behavior of real adsorption systems, it includes the most important components that are expected to influence their properties, particularly the transport in such structures. It should be noted that despite its simplicity the model can be applied for the investigations of real zeolites. The extremely rich variety of the known zeolite frameworks (see, for example [4]) permits to find some possible candidates for this model. It can be the pure silica zeolite SSZ-73 with the framework of Si-ITE type. It has a very large micropore volume of 0.25 cm^3/g, due to its sizable cages, and, also, a high surface area of about 585 m^2/g. The two-dimensional network of channels connected by the 8-ring SiO_4 windows of different opening sizes generates four large cavities per unit cell. The marked difference of the window openings allows the diffusion of short chain hydrocarbons (methane, ethane, propane, and propene) along the only one crystallographic axis. The diffusion has the one-dimensional character. The temperature dependencies of the diffusion of this material was investigated by the MD simulations [11, 12]. The other candidates are the MOR-type zeolites, which are important catalysts in the petrochemical industry. The framework of the MOR-type zeolite (named after the mineral mordenite) consists of main channels formed by the large 12-rings with an elliptical shape. Small side pockets are connected to these main channels by 8-rings. Since mordenite is, in practice, a 1D large-pore zeolite, transport of molecules within the zeolite occurs only along the c axis.

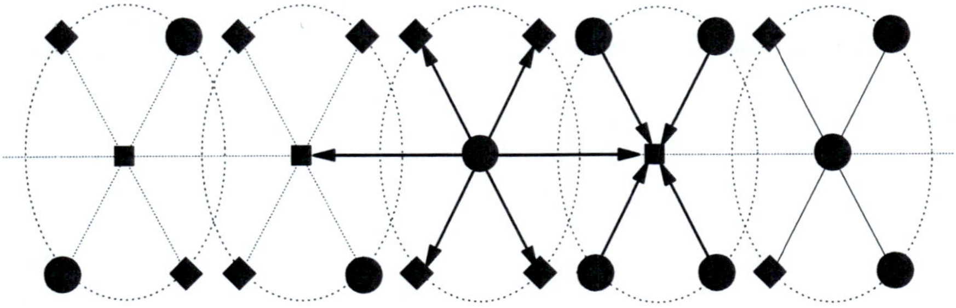

Fig. 1 The schematic view of the lattice. The *squares* and *diamonds* denote the channel and pocket adsorption sites, correspondingly. Guest molecules shown as *circles*. The *arrows* denote the possible jumps of the molecules

We have obtained the exact expression for the free energy of the lattice-gas model. Using the theoretical approach, developed in [14], we have derived the analytical expressions for the diffusion coefficients. The kMC simulations were used to check the applicability of the derived expressions. The coincidence between the analytical dependencies and numerical data is amazingly good.

The Chapter is organized as follows. In Sects. 2 and 3 we describe the lattice-gas model and give its exact solution. The definitions of the diffusion coefficients are presented in Sect. 4. In Sect. 5 we analyse the molecule migration in the heterogeneous lattice and describe the derivation of the approximate analytical expressions for the diffusion coefficients. In Sect. 6 we briefly summarize some technical details of our kMC simulations. The main results are presented and discussed in Sect. 7. Finally, the concluding remarks are offered in Sect. 8.

2 The Lattice-Gas Model and Hamiltonian

We consider a heterogeneous 1D lattice composed of the two different types of sites. There is a linear chain of sites (channel) evenly spaced at the distance a apart. These sites are analogous to the window sites in a zeolite framework. Each channel site is surrounded by a 'cavity' with p pocket sites placed symmetrically on the planes transverse to the channel (p characterizes the cavity volume). The lattice with the bonds connecting the sites is depicted in Fig. 1. A guest molecule adsorbed in a channel cite can jump to any pocket site of the cavity and to the nearest neighbor (NN) channel sites along the chain. It has $p + 2$ NN sites. The molecules adsorbed in the pocket sites jump only to the NN channel site. They have only one NN site. The lattice looks like a hair brush. The model resembles the comb model which was put forward in [43, 44]. But there is a crucial difference: in the comb model all branches have an infinite length, which results in an anomalous behavior of the migration [3, 27, 28], whereas in our case there are branches with a finite length only.

The system of adsorbed molecules is described by a set of occupation numbers $\{n_i\}$ with $n_i = 1/0$ if the ith site is occupied/empty. We introduce the concentration of the molecules $\theta = N_a/N(p+1)$, the occupancy of the pocket sites $\theta_p = N_{ap}/pN$, and the occupancy of the channel sites $\theta_c = N_{ac}/N$, where N_{ac}, N_{ap}, and $N_a = N_{ac} + N_{ap}$ are the numbers of molecules adsorbed on the channel and pocket sites, and the total number of molecules, respectively. There is a simple relation between the concentrations

$$\theta = (\theta_c + p\theta_p)/(p+1). \tag{2}$$

If the adsorption energies ε_c, ε_p are large relative to the thermal energy $k_B T$, the molecules will populate the sites jumping occasionally to the empty NN sites. There are two lateral interactions: φ between the molecules adsorbed in the NN channel sites, and ϕ between the molecules adsorbed in the channel site, and any molecule occupying the pocket sites of the cavity. We neglected the interactions between the molecules adsorbed in the pocket sites (it can be a reasonable approximation only for the small values $p \leq 4$). Now the Hamiltonian H of the molecule system reads

$$H = \sum_{i=1}^{N} \left[n_i^c \left(\varphi n_{i+1}^c - \varepsilon_c \right) + \left(\phi n_i^c - \varepsilon_p \right) \sum_{j=1}^{p} n_j^p \right]. \tag{3}$$

Here sums being over the channel sites and pocket sites, respectively. The upper index c, p indicates the type of the site where the molecule is adsorbed. The grand partition function Q and the free energy F are written

$$Q = \sum_{\{n_i\}} \exp(\mu N_a - H), \quad F = \frac{\ln Q}{N(p+1)}, \tag{4}$$

where the sum being over all possible $2^{N(p+1)}$ configurations of the molecules, $k_B T = 1$, and μ is the chemical potential.

3 Exact Solution of the Model

The unique property of the 1D case is the possibility to get an exact solution for any value of the chemical potential μ. The grand partition function of the 1D lattice-gas system can be written as a trace of the product of the N transfer matrices $\hat{P}^{(i,i+1)}$ of the order 2^{p+1}

$$Q = Tr(\hat{P}^N), \tag{5}$$

where

$$\hat{P} = \begin{pmatrix} P_{11}^{(i,i+1)} & P_{12}^{(i,i+1)} & \cdots \\ P_{12}^{(i,i+1)} & P_{22}^{(i,i+1)} & \cdots \\ \vdots & \vdots & \ddots \end{pmatrix}$$

is the transfer matrix with the elements

$$P_{nm}^{(i,i+1)} = exp\left\{\frac{1}{2}\left[E_n^{(i)} + E_m^{(i+1)}\right] + V_{nm}^{(i,i+1)}\right\}. \tag{6}$$

Here $E_n^{(i)}$ is the energy of the nth configuration of molecules in the ith cavity. There are 2^{p+1} molecule configurations in total for every cavity, but the energy depends only on the number of the molecules occupying the pocket sites. If k molecules are adsorbed in a cavity with p sites, there are C_k^p configurations with the same energy $E_n^{(i)}$ as

$$E_n^{(i)} = (\mu + \varepsilon_p)k \quad \textit{if the central channel site is empty, or}$$

$$E_n^{(i)} = (\mu + \varepsilon_p - \phi)k + \mu + \varepsilon_c \quad \textit{if the channel site is occupied.}$$

Here $C_k^p = p!/k!(p-k)!$ is the binomial coefficient. The term $V_{nm}^{(i,i+1)} = -\varphi n_i^c n_{i+1}^c$ describes the interaction between the molecules adsorbed in the ith and $(i+1)$th channel sites.

In the representation diagonalizing the matrix $\hat{P}$ the trace is easily calculated. The grand partition function Q and the free energy F are expressed via the largest eigenvalue Λ of the transfer matrix

$$Q = \Lambda^N \textit{ and } F = (p+1)^{-1} ln\Lambda, \textit{ as } N \to \infty \tag{7}$$

where

$$\Lambda = \frac{1}{2}\left(B + \sqrt{B^2 + 4C}\right),$$

$$B = e^{\mu+\varepsilon_c-\varphi}\left(1 + e^{\mu+\varepsilon_p-\phi}\right)^p + \left(1 + e^{\mu+\varepsilon_p}\right)^p,$$

$$C = e^{\mu+\varepsilon_c}\left(1 - e^{-\varphi}\right)\left(1 + e^{\mu+\varepsilon_p-\phi}\right)^p\left(1 + e^{\mu+\varepsilon_p}\right)^p,$$

$$p = 0, 1, 2, \ldots \tag{8}$$

For the analytical calculations of the diffusion coefficients we need some thermodynamic quantities which can be obtained from the exact expression Eq. 7. The molecule concentrations (absorption isotherms) are the first derivatives of the free energy over the chemical potential and depths of the adsorption sites

$$\theta \equiv \frac{\partial F}{\partial \mu} = \frac{1}{(p+1)\Lambda}\frac{\partial \Lambda}{\partial \mu},$$

$$\theta_c \equiv (p+1)\frac{\partial F}{\partial \varepsilon_c} = \frac{1}{\Lambda}\frac{\partial \Lambda}{\partial \varepsilon_c},$$

$$\theta_p \equiv \frac{p+1}{p}\frac{\partial F}{\partial \varepsilon_p} = \frac{1}{p\Lambda}\frac{\partial \Lambda}{\partial \varepsilon_p}. \tag{9}$$

The isothermal susceptibility is the second derivative of the free energy over the chemical potential. It equals the mean square concentration fluctuations

$$\chi_T \equiv \frac{\partial^2 F}{\partial \mu^2} = \frac{1}{(p+1)\Lambda}\frac{\partial^2 \Lambda}{\partial \mu^2} - (p+1)\theta^2. \tag{10}$$

The correlation function P_{00}^{cc} is the probability to find a pair of the NN empty channel sites. It is a combination of the first derivatives of the free energy as

$$P_{00}^{cc} = 1 - 2\theta_c - \frac{1}{\Lambda}\frac{\partial \Lambda}{\partial \varphi}. \tag{11}$$

4 Diffusion Coefficients

The guest molecule migration is described by some diffusion coefficients. It should be noted that there are no well-established terms for these quantities. The readers may encounter many diffusion coefficients, diffusivities and diffusion constants. Necessary to mention them here, at least.

Conceptually, the simplest diffusion coefficient is a single molecule diffusion coefficient D_0. It describes the mean square displacement of a single molecule over an empty lattice. If a molecule performs random jumps in some medium, its mean square displacement $\langle \vec{R}^2(t) \rangle$ grows linearly with the time t as

$$\langle \vec{R}^2(t) \rangle = c\nu\lambda^2 t \equiv D_0 t, \tag{12}$$

where ν is the characteristic jump rate and λ is the jump length; $c \sim 1$ is a coefficient depending on the dimensionality (and symmetry) of this medium.

The self-diffusivity or tracer diffusion coefficient D_t (usually, the diffusion of the impurities is followed using the isotopic radioactive tracers, hence the name) describes the mean square displacement of the individual tagged molecules randomly walking over the lattice. In the 1D case one has the following expression for this coefficient

$$D_t = \lim_{t\to\infty} \frac{1}{2tN_a} \sum_{k=1}^{N_a} \langle x_k^2(t) \rangle. \tag{13}$$

Here $x_k(t)$ is the displacement of the kth molecule after the time t; the brackets $\langle \cdots \rangle$ denote the average over the initial molecule configurations.

The next diffusion coefficient, D_{cm}, describes the asymptotic behavior of the center of mass of the molecule system. D_{cm} is the diffusion coefficient for a fictitious molecule located at the center of mass of the molecule system normalized by the total number of molecules

$$D_{cm} = \lim_{t\to\infty} \frac{1}{2tN_a} \left\langle \left[\sum_{k=1}^{N_a} x_k(t) \right]^2 \right\rangle. \tag{14}$$

There is an obvious name for this diffusivity—the center-of-mass (CM) diffusion coefficient. The other names for this coefficient are the following: jump, limiting, corrected, Stefan-Maxwell, Darken. The coefficients D_t and D_{cm} are very suitable for the kMC simulations as easily expressed in terms of directly accessible quantities.

The Fickian (also chemical, collective, transport) diffusion coefficient D_c, describing the mass transfer over the lattice, is determined by the Fick's first law which constitutes the relationship between the flux of molecules $J(x,t)$ and the gradient of their concentration

$$J(x,t) = -D_c\left(\frac{\partial\theta}{\partial x}\right). \tag{15}$$

The CM and Fickian diffusion coefficients are related by the Kubo–Green equation [16]

$$D_c = \theta D_{cm}/\chi_T, \tag{16}$$

The coefficients mentioned above are different, but they are related, of course, as describe the same process—random walks of the molecules. All diffusion coefficients are equal in a zero-concentration limit:

$$D_t = D_{cm} = D_c = D_0, \quad \theta \to 0.$$

In the completely occupied lattice ($\theta \to 1$) the tracer and CM diffusion coefficients are equal zero: $D_{cm} = D_t = 0$, but their ratio depends on the dimensionality and symmetry of the lattice

$$\lim_{\theta\to 1} D_t/D_{cm} < 1.$$

It means that the concentration dependencies of the coefficients are different. The reason is the following: while the total change of the lattice-gas state is a Markov process, the migration of a tagged molecule is not Markovian due to the backward correlation. Any molecule left a vacancy—the empty site behind itself after every jump. If the other NN sites are empty, there is no effect, but if some of them are occupied, the probability for the backward jump is greater than for the forward or sideward jumps. The backward correlation is taken into account by a correlation factor $f(\theta)$ as follows:

$$D_t(\theta) = f(\theta)\, D_{cm}(\theta).$$

This correlation is rather weak in the 3D and 2D homogeneous lattices. The correlation factor $f(\theta)$ is a slowly varying, slightly nonlinear function. But in the 1D case f equals zero. The diffusion of tracers disappears: $D_t = 0$. Instead one has a slow single-file migration: $\langle x^2(t)\rangle \sim t^{1/2}$. In our model the migration is one-dimensional, but the molecules can pass each other performing jumps from the channel sites to the pocket sites and vice versa, therefore, the diffusion has a normal character. Nevertheless, the effect of the backward correlation is strong. The correlation factor drops down very fast almost to zero as $\theta \to 1$ and $D_t \ll D_{cm}$. It should be noted that the model permits to investigate the crossover to the single-file diffusion. There are two possibilities. One can reduce the number of the pocket sites, i.e. consider the lattice with the pocket sites attached to every second, third, etc. channel sites, and/or decrease the jump rates to the pocket sites.

Necessary to note that we are going to calculate the diffusion coefficients by the two different and independent methods using the analytical expressions and kMC simulations, and compare the obtained results. There is an analytical expression for the Fickian diffusion coefficient [14]

$$D_c = \frac{\nu a^2 exp(\mu) P_{00}^{cc}}{(p+1)\chi_T}, \tag{17}$$

Then, D_c can easily be calculated using the exact expressions Eqs. 9, 10 and 11. The CM diffusion coefficient is calculated via the Kubo-Green equation (16). Also we obtain the tracer and CM diffusion coefficients by the kMC simulations using Eqs. 13 and 14. The Fickian diffusion coefficient is calculated via Eq. 16. The two sets of data for the CM and Fickian diffusion coefficients will be compared to check the correctness of the approach developed for the description of the molecule diffusion in the inhomogeneous lattices.

5 Analysis of the Molecule Diffusion

As the site depths are different, one has deep d and shallow s adsorption sites. We consider the case when the channel sites are shallow and the pocket sites are deep (wide windows and deep cavities). The strong inhomogeneity $\exp(\varepsilon_p) \gg \exp(\varepsilon_c)$ results in the striking difference between the s and d sites. The molecules tend to occupy d sites. In the low concentration region $\theta < p/(p+1)$ one has almost empty s sites and partially filled d sites. In the high concentration region $\theta > p/(p+1)$ the d sites are almost completely occupied, and the channel sites are partially occupied. The inhomogeneity also implies much higher jump rates for the "fast" $s \rightarrow d$ and $s \rightarrow s$ jumps than for the "slow" $d \rightarrow s$ jumps (here 'fast' and 'slow' mean short and long waiting times, not the duration of the jumps themselves). It means that the average time of a molecule sojourn in the d sites is considerably longer than the corresponding time in the s sites. The stoichiometric coverage $\theta = p/(p+1)$ separates the regions where the molecules migration has qualitatively different character. We consider these cases separately in the following subsections.

5.1 Low Concentration

In the low concentration region $\theta < p/(p+1)$ almost all molecules occupy the pocket d sites. The channel s sites are empty: $\theta_c \ll 1$. Any migration act starts with a slow jump from an initial filled d site. After the jump the activated molecule does not stay long in the final s site. Almost immediately it jumps to some NN empty site (there is, at least, one empty d and 2 empty s NN sites). The molecule can jump to a deep site. Such jump does not change the molecule distribution and does not give any contribution to the total molecule displacement. But jumps to the NN s sites are effective. The molecule moves to the NN sites and can occupy a deep site in another cavity, or continue its traveling along the channel. These jumps give a contribution to the total molecule displacement.

As slow and fast jumps take place on the largely different time scales, the jumps collect into successions: any slow jump is followed by some fast jumps. The molecule diffusion proceeds by the jump sequences: the first slow $(d \rightarrow s)$ jump + $1, 2, \ldots$ fast $(s \rightarrow s)$ jumps + the final fast $(s \rightarrow d)$ jump. The total time spent by the molecule to perform such jump sequence is almost the same as in the case of a single slow $d \rightarrow s$ jump. We suppose the fast jumps are almost instant events giving the negligibly small contribution to the total time of the jump sequence. These jump successions control the molecule migration in the low concentration region. The probability to find an empty d site decreases as $\theta \rightarrow p/(p+1)$. The molecules should perform more and more fast jumps to find empty d sites. The jump sequences become longer, and the effective length of such sequences L increases considerably. The expression for the effective length L is derived in Appendix.

The specific features of the molecule diffusion in the low concentration region are the following:

- Almost all molecules are trapped in the d sites. The number of agitated molecules is exponentially small. Only a few molecules jump simultaneously over the s sites.
- The molecules perform long sequences of jumps. These sequences give the main contribution to the total molecule displacement.
- The effective length of the jump sequences equals L, and the effective jump rate of the sequences is determined by the slow jump rate v_d.
- The backward correlation is absent, and the tracer and CM diffusion coefficients are equal $D_t = D_{cm}$. The correlation factor is close to one $(f \rightarrow 1)$.

Then, the diffusion coefficient is the ordinary single-molecule diffusion coefficient for the molecule jumping over the empty lattice of d sites with the lattice constant L

$$D_{cm}, D_t = v_d L^2 = D_0 (1 - \theta_p)^{-1}. \tag{18}$$

Here $D_0 = v\exp(-\varepsilon_p)a^2/p$. If the concentration θ grows from zero to $p/(p+1)$, the occupancy of the deep sites θ_p approaches 1. The diffusion coefficients grow considerably. But it is necessary to remember that in our simple analysis of the molecule migration we implicitly set the diffusion coefficient over the shallow sites infinite, $D_s = v\exp(-\varepsilon_c)a^2 \rightarrow \infty$ (instant fast jumps). Then, the above expression Eq. 18 is valid until $D \leq D_s$. This expression can also be obtained from the exact expression Eq. 17. In the low concentration region almost all s sites are empty. The molecules adsorbed in the d sites do not interact with each other. Therefore, one can write the following approximate expressions for the thermodynamic quantities

$$P_{00}^{cc} = \langle (1 - n_i^c)(1 - n_{i+1}^c) \rangle \approx 1,$$

$$exp(\mu + \varepsilon_p) \approx \theta_p/(1 - \theta_p),$$

$$\chi_T \approx \frac{p}{(p+1)} \theta_p (1 - \theta_p). \tag{19}$$

Combining Eqs. 17, 16 and 19 one has the following approximate expressions for the CM and Fickian diffusion coefficients

$$D_{cm} = D_0(1 - \theta_p)^{-1}, \quad D_c = D_0(1 - \theta_p)^{-2}. \tag{20}$$

5.2 High Concentration

In the high concentration region $\theta > p/(p+1)$ the d sites are completely filled, $\theta_p \rightarrow 1$ and the s sites are partially occupied. The pN molecules are trapped, and the excess $N_a - pN$ molecules migrate over the s sites. The number of the slow jumps is exponentially small, so we can neglect them. As all pocket sites are occupied, the molecules migrate over the channel sites. They perform ordinary fast jumps. These mobile molecules give the main contribution to the total displacement. The diffusion over the channel s sites is described by the analytical expression derived for the homogeneous 1D lattice [14]. For D_{cm} and D_c the 1D approximation works rather well. But D_t is noticeably suppressed. The tracer diffusion coefficient does not equal zero but is a fast decaying function of the molecule concentration as the backward correlation is strong $f \ll 1$. For D_t we do not have any simple analytical expression. The analytical expressions for the CM and Fickian diffusion coefficients should be written with account of the following factors:

- The actual concentration of the 1D homogeneous lattice is $\theta_c \approx \theta - p(1 - \theta)$.
- The lateral interaction with the molecules adsorbed in the pocket sites changes the jump rates. We can account for the effect introducing multiplier $exp(p\phi)$ as $\theta_p \rightarrow 1$.
- The diffusion coefficients are reduced by the multiplier $\theta_c/(p+1)\theta$ due to the fact that the common mean square displacement of the mobile molecules should be shared equally with the trapped ones.

Then, we obtain the following approximate expressions for the CM and Fickian diffusion coefficients:

$$D_{cm}(\theta) \approx D_j^h(\theta_c) exp(p\phi) \theta_c/(p+1)\theta,$$

$$D_c(\theta) \approx D_c^h(\theta_c) exp(p\phi) \theta_c/(p+1)\theta. \tag{21}$$

Here $D_{cm,c}^h(\theta_c)$ are the diffusion coefficients calculated for the 1D homogeneous chain with the site depth ε_c and lateral interaction parameter φ.

The analysis of the molecule migration helps to understand the peculiarities of the molecule migration in the whole concentration region. We have the exact analytical expression Eqs. 17 for the description of the diffusion, but this simple analysis reveals an interesting relation between the diffusion coefficients for the homogeneous and inhomogeneous lattice-gas systems. The approximate expression Eq. 18 has really universal character. It describes diffusion in many other inhomogeneous systems like triangular, square, anisotropic inhomogeneous lattices [33, 34, 36].

6 The kMC Simulations

The migration proceeds by the jumps of the activated molecules over the lattice sites. To simulate the kinetic process one must know the detailed mechanism of the molecule transitions between the adsorption sites, i.e. the kMC transition algorithm. An activated molecule should surmount the barrier separating the initial ith and final fth sites. The activation energy is given by the difference between the saddle point energy ($= 0$) and the adsorption energy ε_i. The energy is affected by the NN molecules. But the interaction between the activated molecule at the saddle point with its NNs is usually neglected. Different algorithms can be used for the simulation of the molecule migration. Some of them are described in [1, 15]. We select the well established expression for the jump rate derived in the transition state theory [9, 10]

$$\nu_i = \nu \exp\left[-\varepsilon_c + \phi \sum_{k=1}^{p} n_k^p + \varphi \sum_{j=1}^{2} n_j^c\right], \quad i,j \in c,$$

$$\nu_i = \nu \exp(-\varepsilon_p), \quad i \in p. \tag{22}$$

The jump rate ν_i is the same for all possible jumps from the ith site. The repulsive interaction increases and the attractive one decreases the molecule energy and the jump frequency. The expressions should be appropriate for the short-range lateral interactions.

The system of molecules represented by the Hamiltonian Eq. 3 is realized by the 1D array of the adsorption sites with the periodic boundary conditions ($N = 4096$). The basic events are jumps of the adsorbed molecules with the associated jump probabilities given by Eqs. 22. Initially, a certain number of molecules, corresponding to the molecule concentration θ, is placed on the randomly chosen p and c sites.

The standard MC step corresponding to the Metropolis importance sampling [23] is the sequence of the following steps:

- Pick at random an initial ith site out of all eligible filled sites.
- Select an adjacent final fth site randomly.
- If the destination is vacant, a jump can occur with a probability $\sim \nu_i$.

Although the bookkeeping overhead is small, this MC algorithm suffers from the small transition probability from d sites, where the fraction of rejected jumps is high. Then, a large part of the computing time is spent in attempting jumps which are rejected. For chosen value of the difference $\varepsilon_p - \varepsilon_c = ln\ 100$ only 1 % of MC steps will be successful. This inefficiency can be overcome using the Bortz-Kalos-Lebowitz algorithm [7] (the detailed description of the BKL algorithm see, for example, in Ref. [25] and references therein). The BKL algorithm operates in a space of possible events, instead of real space as the ordinary Metropolis algorithm does. In contrast to the standard technique, where we consider fixed time intervals at which a jump may occur with a certain probability, the BKL algorithm randomly selects a jump out of all possible events and subsequently associates the number of clock ticks that must be passed for the event to take place. So, the transition probability does not result in rejected jump but to a different amount of the elapsed time, making the algorithm particularly efficient at low temperatures and for the systems with the complex potential relief. For the successful performance of the BKL algorithm it is necessary to classify all possible events into groups or classes of equal transition probabilities. In our case there are $3p$ different classes of events, in general.

The kMC step for the BKL algorithm is the sequence of the following steps:

- By drawing a random number, an event class is selected. At this step the binary-tree search method is used to speed up the kMC simulations.
- Another random number picks a specific event from the selected class.
- A third random number produces the number of clock ticks.

However, the bookkeeping overhead for updating the lists of classes after every jump is substantial. For the BKL algorithm the accumulated time during the kMC cycle can be used as a termination criterion of the cycle. Choosing of such concentration-dependent criterion offers a possibility for very substantial savings of the computer time without affecting the quality of results.

We used the kMC simulations for calculations of the tracer and CM diffusion coefficients. Also we determine the adsorption isotherms μ, and the isothermal susceptibility χ_T as functions of the concentration θ for different values of the interaction parameter φ. The Fickian diffusion coefficient has been determined via the Kubo–Green equation. The isothermal susceptibility χ_T is the mean-square fluctuations of the concentration. They are measured in a probe region which consists of the N_f sites. The average over the initial configurations squared fluctuations of the number of molecules in this region $\left\langle (\delta N)^2 \right\rangle$ is related to the isothermal susceptibility as [29]

$$\chi_T = \frac{N}{N_f(N - N_f)} \left\langle (\delta N)^2 \right\rangle. \tag{23}$$

To obtain adsorption isotherms we use the method of local states (see, for example, Ref. [2]). In this method one counts the frequencies of occurrence (ensemble average populations) σ_α of the local states α. The probability to find any molecule configuration α is proportional to the Boltzmann factor $exp(-E_\alpha)$ where $E_\alpha = \varepsilon_p + k\phi$ with

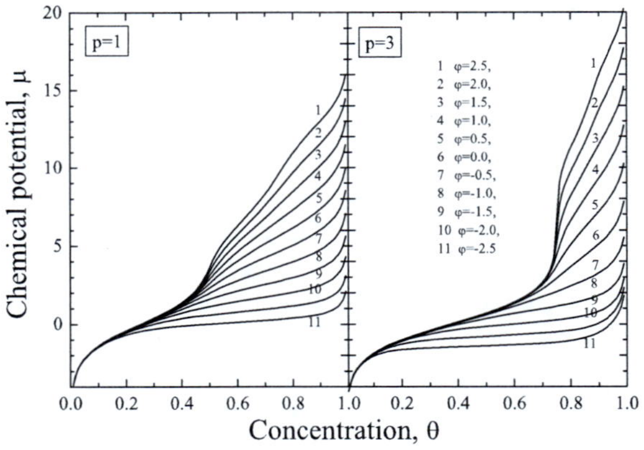

Fig. 2 The adsorption isotherms $\mu(\theta)$ plotted for $p = 1$ and $p = 3$ as indicated. The interaction parameter φ varies in the range ± 2.5. The *solid lines* are the analytical dependencies Eq. 9

$k = 0, 1$ and $\varepsilon_c + k\phi + l\varphi$ with $k = 0, 1, 2, \ldots, p$ and $l = 0, 1, 2$ are the energies of these configurations. The energies define the local states of the molecules. The set of conjugate states α^* is determined for empty sites. All conjugate states have the same zero energy: $E_\alpha^* = 0$. The frequencies of occurrence ensemble average populations of the α th local state and its conjugation are denoted as σ_α and σ_α^*. The condition of the detailed balance between the states requires

$$\mu = ln(\sigma_\alpha / \sigma_\alpha^*) + E_\alpha. \tag{24}$$

To smooth out the fluctuations it is advisable to average μ over the most probable local states and exclude rare configurations.

In order to obtain the good quality of the kMC data runs of up to 1,00,000 MC steps for up to 50,000 different initial configurations were carried out.

7 Results and Discussion

The values of the energetic parameters are the same for all dependencies: $\varepsilon_p - \varepsilon_c = ln\ (100)$; $\varphi = \phi = -2.5, -2, \ldots, 2.5$ and $p = 1, 3$.

The adsorption isotherms $\mu(\theta)$ are presented in Fig. 2. The strong lateral interaction produces rather specific peculiarities on the dependencies. The lateral attraction results in the almost horizontal plateau, which means jump of the system concentration at some value of the chemical potential. The behavior resembles the first order phase transition, but the dependencies $\mu(\theta)$ remain analytical and continuous. The strong lateral repulsion produces another peculiarity. The dependencies have very steep slope at the stoichiometric concentrations $p/(p + 1)$. The system is in the ordered state: all molecules occupy the deep pocket sites and all shallow channel sites are empty. The adsorption of extra molecules requires an additional energy, therefore, this state is stable in a wide region of the loading.

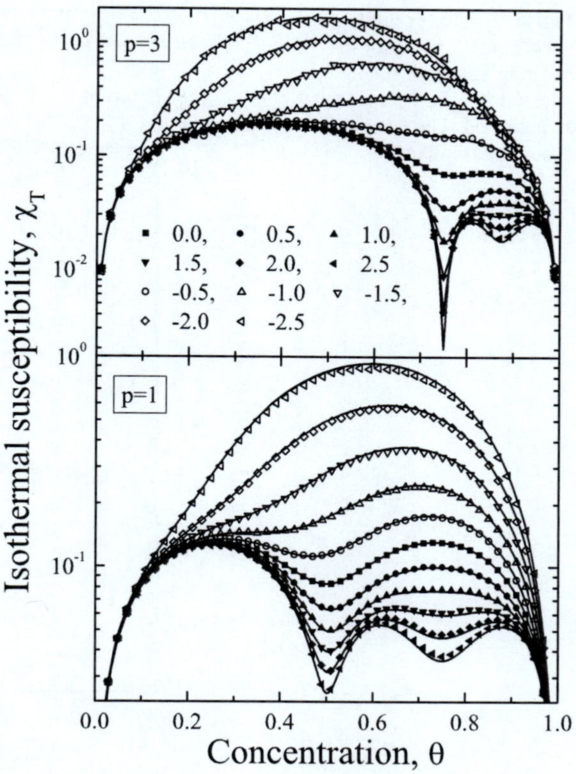

Fig. 3 The concentration dependencies of the isothermal susceptibility χ_T for the different values of the interaction parameter φ as indicated. The symbols denote the kMC data and *solid lines* are the analytical dependencies Eq. 10

It should be noted that there is no necessity to plot the kMC data as the coincidence with the analytical dependencies is perfect.

The concentration dependencies of the isothermal susceptibility $\chi_T(\theta)$ are plotted in Fig. 3. The repulsive interaction decreases the concentration fluctuations keeping molecules apart from each other. The fluctuations are strongly suppressed at the stoichiometric concentrations $\theta = p/(p+1)$. The dependencies $\chi_T(\theta)$ have a pronounced minimum at these concentrations. Strong repulsion results in the ordering of the molecules at the concentrations $\theta = (2p+1)/(2p+2)$ when all d sites are occupied, and a half of the s sites are occupied in the alternating order. The dependencies have a weak minimum at these concentrations. The attractive interaction decreases the relaxation of the concentration fluctuations and inhibits the molecule migration. The fluctuations grow considerably. The dependencies $\chi_T(\theta)$, obtained for the attractive interaction, are very similar for the different lattices and do not have any specific features related to the lattice inhomogeneity. In fact, one obtains almost the same dependencies for the homogeneous lattices too.

The concentration dependencies for the tracer diffusion coefficient $D_t(\theta)$ are plotted in Fig. 4. As mentioned above the tracer diffusion coefficient D_t describes the behavior of the tagged molecules. For the homogeneous systems D_t is a monotone decreasing convex function of the concentration. For the inhomogeneous lattice the

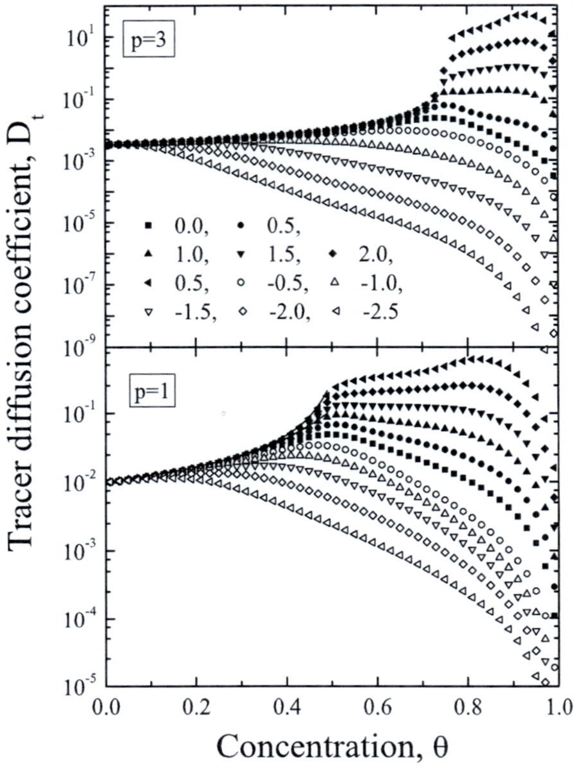

Fig. 4 The tracer diffusion coefficient D_t vs. θ for the different values of the interaction parameter $\varphi > 0$ as indicated. The symbols denote the kMC data

dependencies have a noticeable maximum at the stoichiometric concentrations $\theta = p/(p+1)$ if the lateral interaction is not very strong $|\varphi| \leq 1$. Such maximum on the dependencies $D_t(\theta)$ is a universal feature for the inhomogeneous lattices. In the low concentration region the migration proceeds by the long jump successions. The coefficient grows with the concentration $\sim (1-\theta_p)^{-1}$. At the stoichiometric concentrations the deep sites are completely occupied and do not inhibit the migration. The diffusion coefficient reaches its maximum. In the high concentration region $\theta > p/(p+1)$ the excess molecules migrate over the homogeneous lattice. There is an ordinary diffusion, and the tracer diffusion coefficient drops down to zero as the concentration approaches its maximum value.

It should be noted that the maxima on the concentration dependencies of the diffusion coefficients were observed in some experimental and theoretical investigations. For example, Gubbins and co-workers studied the diffusion of Lennard-Jones fluids in several models of microporous carbons using the microcanonical molecular dynamics simulations and found a maximum in the self diffusion coefficient as a function of loading [24, 26]. Also, experimentally, maxima of the diffusion coefficients are known for molecular fluids migrating in zeolites [8]. Using different experimental techniques the authors measured the diffusion of methanol in NaX zeolite crystals and obtained: '*...the somewhat unusual maximum in the trend of self-*

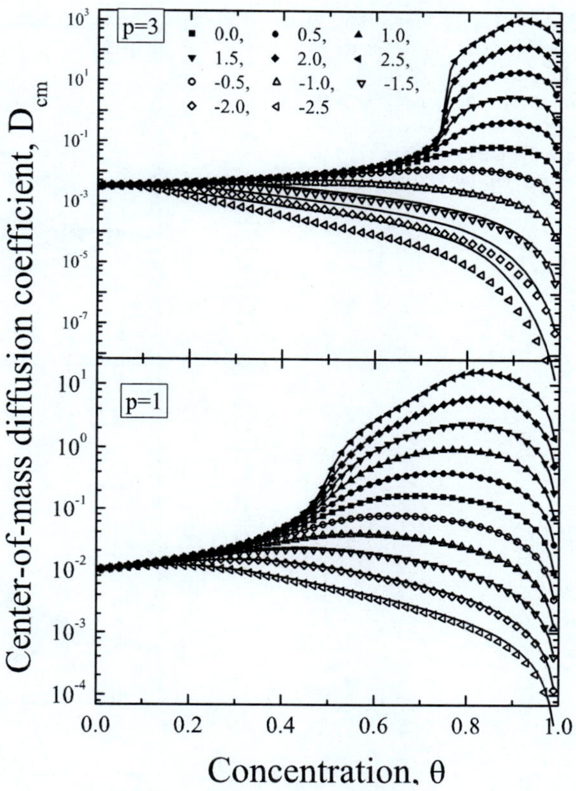

Fig. 5 The CM diffusion coefficient D_j vs. θ for the different values of the interaction parameter $\varphi > 0$ as indicated. Notations are the same as in Fig. 4. The *solid lines* are dependencies calculated by Eqs. 17 and 16

diffusivity with loading.' In another experiment they measured the diffusion of water in activated carbons [22]. There was a maximum of the tracer diffusion coefficient at $\theta = 1/2$. The experimental dependence plotted in Fig. 3b, Ref. [22], is qualitatively the same as plotted in Fig. 4 for the case $\varphi = 0$.

The investigated systems are different, but, what is essential, they are heterogeneous. This results in the specific mode of the molecule migration: long sequences of jumps between the empty d sites. The diffusion coefficients have the universal concentration dependence Eq. 21, which is really independent on the symmetry and dimensionality of the system.

The concentration dependencies for the CM diffusion coefficient D_{cm} are plotted in Fig. 5. The dependencies have qualitatively the same behavior as the corresponding dependencies for the tracer diffusion coefficient D_t. At low concentrations the dependencies D_{cm} and D_t are almost the same as the number of mobile molecules is very small and the backward correlation is absent $f \cong 1$. In the high concentration region the backward correlation is strong $f \ll 1$ and $D_t \ll D_{cm}$. The CM diffusion coefficient has a wide maximum in this region. The origin of this maximum is the same as in the case of the tracer diffusion coefficient. The repulsion considerably increases the molecule migration and CM diffusion coefficient, the attractive interaction inhibits the diffusion and suppress D_{cm}.

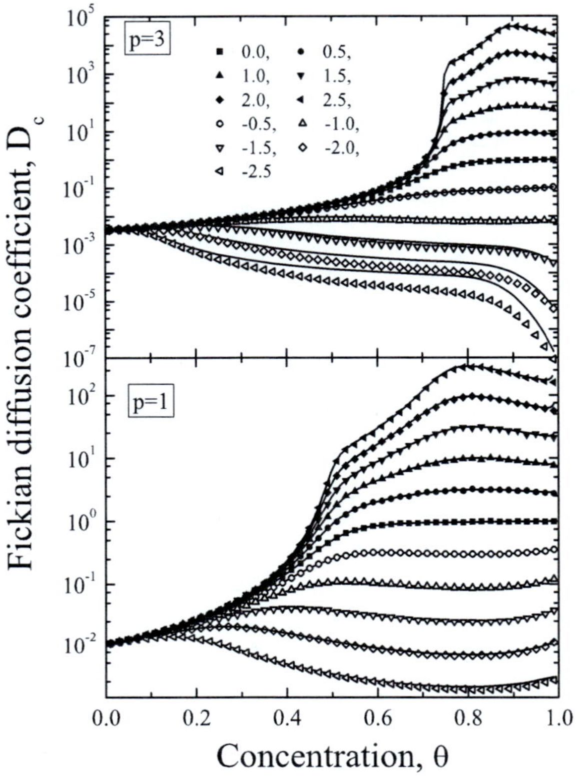

Fig. 6 The Fickian diffusion coefficient D_c vs. θ for the different values of the interaction parameter $\varphi > 0$ as indicated. The *solid lines* are the dependencies calculated by Eq. 17

There is a perfect coincidence of the kMC data with the analytical dependencies Eqs. 17 and 18. It should be noted that this coincidence is obtained without any fitting parameters. The expressions give the quantitatively correct description of the dependencies in the whole concentration region and in a wide region of the lateral interaction. The approach based on the local equilibrium approximation [14] gave the excellent results in the homogeneous lattices of different symmetries and dimensionality [37–41]. It occurs that it works very well in the 1D inhomogeneous lattice too.

It should also be noted that some authors, basing on the experimental studies of diffusion in zeolites, suggest the five different types of dependencies diffusivity vs. concentration [6, 20, 42]. The dependencies plotted in Fig. 4 are of type V having a quasi-parabolic increase of the tracer (and CM) diffusion coefficient with concentration. It is easy to see that the analytic expression describing the concentration behavior of the tracer and CM diffusion coefficients Eq. 18 for $p \gg 1$ (large cavities) can be approximated as follows

$$D_{t,cm} = D_0(1 - \theta_p)^{-1} \approx D_0(1 + \theta_p + \theta_p^2 + \ldots) \tag{25}$$

The concentration dependencies of the Fickian diffusion coefficient $D_c(\theta)$ are plotted in Fig. 6. Qualitatively, the concentration dependencies of the chemical diffusion coefficient resemble the corresponding curves obtained for the 2D and 3D heterogeneous lattices [30–36]. The coincidence between the numerical and analytical data is very good, like in the previous case for D_{cm}. The noticeable discrepancies arise for the strong attractive interaction $\varphi \leq -1.5$. The strong attraction causes clustering in the system of molecules, i.e. formation of islands of the dense phase with $\theta \to 1$ surrounded by a rarefied gas phase with $\theta \to 0$.

8 Summary

We have investigated the diffusion of molecules in the 1D inhomogeneous lattice with the two kinds of adsorption sites. Such systems have the specific features qualitatively influencing the molecule diffusion. The concentration dependencies of the tracer, CM and Fickian diffusion coefficients are calculated using the analytical expressions, derived for this lattice, and kMC simulations. The numerical and theoretical data coincide amazingly well in the whole concentration region and in a very wide interaction region. The lattice inhomogeneity governs the diffusion of the molecules. There are two qualitatively different modes of the molecule jumps: ordinary jumps and jump successions. The jump successions performed by single molecules are the main migrating events in the low concentration region $\theta < p/(p+1)$. The dependencies of the tracer and CM diffusion coefficients are described by the simple and universal law $(1-\theta_p)^{-1}$. In the high concentration region $\theta > p/(p+1)$ the molecule diffusion proceeds by ordinary jumps. The concentration dependencies are described by the analytical expressions derived for the 1D homogeneous lattice. The analytical expressions Eqs 17, 18, and 21 can be used for calculation of the diffusion coefficients in such heterogeneous lattices.

The supposed approach is rather simple. It can be applied for the analysis of the molecule diffusion in the lattice-gas systems of different symmetries and dimensions. In general, the description of the molecule migration as series of jump successions of slow and fast jumps is obvious. The lattice of deep and shallow sites imposes strongly different time scales for the fast and slow jumps, which results in the specific correlation between the jumps. The sequences of jumps arise due to this correlation. The sequences may be treated like long jumps transferring molecules over the lattice. Such simplification of the molecule motion may be useful for simulations of the migration in the complex media like zeolites and other microporous materials having disordered structures and very complex potential relief.

Acknowledgments This work was supported by the Operational Program Research and Development for Innovations–European Social Fund (project CZ.1.05/2.1.00/03.0058 of the Ministry of Education, Youth and Sports of the Czech Republic) and by grants TA01010517 of the TACR and P108/12/1941 of the GACR.

A.1 Appendix

We derive here the analytical expression for the diffusion coefficient for molecules performing the long jump successions. In the low concentration region $(\theta < p/(p+1))$ all molecules occupy deep pocket sites. We need the probability of a molecule jump succession from some initial ith to the final fth d site. The molecule performs slow jump from the ith site, n fast jumps and occupies the fth site at the distance ma from the ith site. The probability of this succession is a product of the probabilities of the elementary migration acts: the probability of the first slow jump from the ith to the channel s site $v\exp(-\varepsilon_p)$, the probability of the sequence n fast jumps $W(n,m)$ and the probability to occupy a d site by the final jump w_d. As we supposed that the probability for a fast jump is equal for all NN sites, then the probability to jump in the NN s site is $w = [p(1-\theta_p)+2]^{-1}$. The probability to occupy a d site is $w_p = p(1-\theta_p)[p(1-\theta_p)+2]^{-1}$. The probability of the fast jump succession is simply the product of the fast jump rates w^n. The number of the different jump successions which transfer the molecule to a distance ma after n jumps is equal to

$$\frac{n!}{\left[\frac{1}{2}(n+m)\right]!\left[\frac{1}{2}(n-m)\right]!} \equiv C^n_{(n+m)/2}, \quad |m| \leq n,$$

where C^n_k is the binomial coefficient.

The square of the length of this jump sequence $L^2(n,m)$ is obviously equal to $(ma)^2$. To obtain the effective jump length L one should average over all sequences

$$L^2 = w_p a^2 \sum_{n=0}^{\infty} w^n \sum_{m=-n}^{n} m^2 C^n_{(n+m)/2} \tag{26}$$

The inner sum is equal $n\,2^n$ [13]. The infinite geometric series is easily calculated. The final result is the following simple expression

$$L^2 = w_p a^2 \sum_{n=0}^{\infty} n(2w)^n = a^2/p(1-\theta_p) \tag{27}$$

References

1. Ala-Nissila, T., Ferrando, R., Ying, S.C.: Adv. Phys. **51**, 949–1078 (2002)
2. Alexandrowicz, Z.: J. Stat. Phys. **13**, 231 (1975)
3. Arkhincheev, V.E.: Phys. A **280**, 304 (2000)
4. Baerlocher, C., Meier, W.M., Olson, D.H.: Atlas of zeolite framework types. http://www.iza-structure.org/databases/books/Atlas 5ed.pdf
5. van Bekkum, H., Flanigen, E.M., Jansen, J.C. (eds.): Introduction to Zeolite Science and Practice. Elsevier, Amsterdam (1991)
6. Bhide, S.Y., Yashonath, S.: J. Chem. Phys. **111**, 1658 (1999)

7. Bortz, A.B., Kalos, M.H., Lebowitz, J.L.: J. Comput. Phys. **17**, 10 (1975)
8. Brandani, S., Ruthven, D.M., Kärger, J.: Zeolites **15**, 494 (1995)
9. Evans, M.G., Polanyi, M.: Trans. Faraday Soc. **31**, 875 (1935)
10. Eyring, H.: J. Chem. Phys. **3**, 107 (1935)
11. Combariza, A.F., Sastre, G., Corma, A.: J. Phys. Chem. C **113**, 11246 (2009)
12. Combariza, A.F., Sastre, G., Corma, A.: J. Phys. Chem. C **115**, 875 (2011)
13. Chandrasekhar, S.: Rev. Mod. Phys. **15**(1), 1 (1943)
14. Chumak, A.A., Tarasenko, A.: Surf. Sci. **91**, 694 (1980)
15. Danani, A., Ferrando, R., Scalas, E., Torri, M.: Int. J. Mod. Phys. B **11**, 2217–2279 (1997)
16. Forster, D.: Hydrodynamic Fluctuations, Broken Symmetry and Correlation Functions. Benjamin, Reading (1975)
17. Gupta, V., Nivarthi, S.S., McCormick, A.V., Davis, H.T.: Chem. Phys. Lett. **247**, 596 (1995)
18. Hahn, K., Kärger, J., Kukla, V.: Phys. Rev. Lett. **76**, 2762 (1996)
19. Harris, T.E.: J. Appl. Probab. **2**, 323 (1965)
20. Kärger, J., Ruthven, D.M.: Diffusion in Zeolites and Other Microporous Solids. Wiley, New York (1992)
21. Kärger, J.: In Single-File Diffusion in Zeolites in Molecular Sieves v.7. In: Karge, H.G., Weitkamp, J., Brandani, S. (eds.) Springer, Berlin (2008)
22. Kärger, J., et al.: Pure Appl. Chem. **61**, 1875 (1989)
23. Metropolis, N., Rosenbluth, A.W., Teller, A.H., Teller, E.: J. Chem. Phys. **21**, 1087 (1953)
24. Moore, J.D., et al.: Appl. Surf. Sci. **256**, 5131 (2010)
25. Novotny, M.A.: In: Annual Reviews of Computational Physics IX. Stauffer, D. (ed) (World Scientific, Singapore, 2001), p. 153. Also see at http://www.arxiv.org/cond-mat/ 010918
26. Pikunic, J., Gubbins, K.E.: Eur. Phys. J. E **12**, 35 (2003)
27. N. Pottier, Nuovo Cimento 16 D (1994)
28. N. Pottier, Physica A 216 D (1995) 1
29. Reed, D.A., Ehrlich, G.: Surf. Sci. **102**, 588 (1981)
30. Tarasenko, A., Jastrabik, L., Muller, T.: Phys. Rev. B **75**, 085401 (2007)
31. Tarasenko, A., Jastrabik, L., Muller, T.: Phys. Rev. B **76**, 134201 (2007)
32. Tarasenko, A., Jastrabik, L.: J. Phys. C **20**, 415210 (2008)
33. Tarasenko, A., Jastrabik, L.: Surf. Sci. **602**, 2975 (2008)
34. Tarasenko, A., Jastrabik, L.: Phys. A **388**, 2109 (2009)
35. Tarasenko, A., Jastrabik, L.: Phys. Chem. Chem. Phys. **11**, 7690 (2009)
36. Tarasenko, A., Jastrabik, L.: Phys. A **75**, 085401 (2011)
37. Tarasenko, A., Jastrabik, L., Nieto, F., Uebing, C.: Phys. Chem. Chem. Phys. **1**, 1583 (1999)
38. Tarasenko, A., Nieto, F., Jastrabik, L., Uebing, C.: Eur. Phys. J. D **12**, 311 (2000)
39. Tarasenko, A., Jastrabik, L., Uebing, C.: Phys. Rev. B **57**, 10166 (1998)
40. Tarasenko, A., Jastrabik, L., Nieto, F., Uebing, C.: Phys. Rev. B **65**, 075413 (2001)
41. Tarasenko, A., Nieto, F., Jastrabik, L., Uebing, C.: Surf. Sci. **536**, 1 (2003)
42. Tunca, C., Ford, D.M.: Chem. Eng. Sci. **58**, 3373 (2003)
43. White, S.R., Barma, M.: J. Phys. A **17**, 2995 (1984)
44. Weiss, G., Havlin, S.: Phys. A **134**, 474 (1986)

Transient Diffusion in Arbitrary Shape Porous Bodies: Numerical Analysis Using Boundary-Fitted Coordinates

Vera S. O. Farias, Wilton P. Silva, Cleide M. D. P. S. Silva, J. M. P. Q. Delgado, Severino R. Farias Neto and A. G. Barbosa de Lima

Abstract This chapter provides information about the diffusion phenomenon (heat and mass transfer) in porous materials such as definition, classification, modeling and experiments, with particular reference to capillary-porous body with arbitrary shape. A transient three-dimensional mathematical formulation written in boundary-fitted coordinates and all numerical formalism to discretize the diffusion equation by using the finite-volume method, including grid generation and numerical analysis of the computational solution are presented. Applications to food and ceramic industries have been done with success. An optimization

V. S. O. Farias (✉)
Department of Education, Federal University of Campina Grande, Cuité-PB, Brazil
e-mail: vera-solange@uol.com.br

W. P. Silva · C. M. D. P. S. Silva
Department of Physics, Federal University of Campina Grande, Campina Grande-PB, 58429–900, Brazil
e-mail: wiltonps@uol.com.br

C. M. D. P. S. Silva
e-mail: cleidedps@uol.com.br

J. M. P. Q. Delgado
LFC—Building Physics Laboratory, Civil Engineering Department, Faculty of Engineering, University of Porto, Porto, Portugal
e-mail: jdelgado@fe.up.pt

S. R. Farias Neto
Department of Chemical Engineering, Federal University of Campina Grande, Campina Grande-PB, 58429-900, Brazil
e-mail: fariasn@deq.ufcg.edu.br

A. G. Barbosa de Lima
Department of Mechanical Engineering, Federal University of Campina Grande, Campina Grande-PB, Brazil, 58429-900, Brazil
e-mail: gilson@dem.ufcg.edu.br

J. M. P. Q. Delgado et al. (eds.), *Numerical Analysis of Heat and Mass Transfer in Porous Media*, Advanced Structured Materials 27,
DOI: 10.1007/978-3-642-30532-0_4,

technique has been presented to estimation of transport properties by comparison between numerical and experimental data.

Keywords Mass · Heat · 3D simulation · Porous media · Generalized coordinates · Finite-volume

1 Introduction

What is a porous material? In the microscopic scale, more useful definition is: a material consisting of a solid matrix with an interconnected void [1–3]. In a natural porous material the distribution of porous (void) with respect to shape and size is irregular (random, non-uniform) that are filled with one or more fluids. Further some pores can be disconnected. In the usual situation, the solid matrix is either rigid or it undergoes small deformation. The interconnectedness of the pores allows the heat and mass transfer and flow of fluids through the porous material [4, 5]. Figure 1 illustrates the schematic of porous materials.

Classically, because of variety of porous material that exists (organic and inorganic materials, polymeric foams, biological materials, etc.) we can classify it like macro, micro and nanoporous materials. The question is: When a material is macro, micro or nanoporous? In general, when the pore sizes are 10 μm or more (large pore size), one is likely to use the term macroporous. For pore sizes less than 1 nm (appreciably less than the small pore size) we can use the term nanoporous. Further when the equivalent pore diameter is less than 10^{-7} m (pore size comparable to the molecular dimensions), the material is defined like capillary-porous material. Examples of these materials include wood, clay, sand, grains, fruits and vegetables [4, 6]. In the capillary-porous bodies, capillary and gravitational forces act in the fluid phase into the pores being that the capillary forces are more than the gravitational forces. If the meniscus of the liquid inside the pores is formed mainly due to action of the surface tension and less than by gravitational force, then, the voids are termed capillaries [7, 8]. On the other hand, if the capillary forces are comparable with the gravitational forces then the body is called simply a porous body [9].

Many authors, for example, Strumillo and Kudra [6, 8] and Keey [10], classify the porous bodies still like hygroscopic and non-hygroscopic taking as a basis the states of moisture (liquid and/or vapor) into the material. The following types of moisture exist: surface moisture, unbound, free or capillary moisture and bound, hygroscopic or dissolved moisture. Unbound water in a hygroscopic material is the water in excess which can be removed by drying. The electrical, mechanical, thermal and physical properties of the capillary-porous bodies are affected by the water contained in them (form of bonding and content), nature (for example, grain, fruits, wood, etc.) and states of aggregation of the particles (dispersed system, granular material) (for example, clay, sand, etc.).

2 Diffusion in Capillary-Porous Bodies

2.1 Theoretical Background

Moist materials are capillary-porous bodies filled with moisture and inert gas. Then, transfer of non-condensable gases, vapor and liquids can occur into capillary-porous bodies. In general, the following main mechanisms of moisture migration can be cited: liquid diffusion, vapor diffusion, effusion, thermodiffusion, by capillary forces, by osmotic pressure, due to pressure gradient, by gravity and by shrinkage. Details about these topics are reported, for example, in [6] and [11]. In this context, the term diffusion specifically refers to atomic or molecular transport.

Tortuosity is defined like the ratio of body dimension in a given direction to length of the path traversed by the component in the diffusion process [6]. Then, because of tortuosity of interconnected porous and internal forces into the material, mass transport occurs mainly by activation, for example, heat. Some diffusing matter (atom and molecules) through these pores can be absorbed or immobilized (adsorbed) simultaneously with the heat absorption or sometimes accompanied by heat generation due to phase change. This is a cross-effect due to heat transfer.

From this discussion, we have one question. How can a complex process like diffusion be modeled? This question presents two answers: In pore-scale or in a phenomenological manner. The theory of heat and mass transfer in capillary-porous materials on the basis of molecular and molar transfer mechanisms (microscopic scale) presents great difficulties. Then, many theoretical studies are directed to macroscopic scale where it is possible to obtain overall information of the process and variation of transport coefficients as a function of the moisture content and/or temperature of the body.

Mass diffusion is a phenomenon by which matter is transported from one region in space to another due to molecular motions. Thus, the diffusion theory states that diffusional flux is proportional to the concentration gradient (Fick's first law) [12–14]. The Fick's second law states that accumulated matter rates into the capillary-porous material equal to net mass transport by diffusion (effects of mass convection and mass generation aren't significant).

Lewis [15], Sherwood [16–18], and Newman [19], and many other researchers (after Fick's second law of diffusion) states that the only moisture migration mechanism in capillary-porous bodies is the liquid diffusion [20]. Evaporation can take place at the surface of them. Thus, the Fick's second law has been frequently used to describe mass transfer inside capillary-porous bodies during drying and wetting processes.

Heat diffusion (conduction) is defined as thermal energy transport within a single-phase medium (solid, liquid or gas) or porous material (which a multiphase medium) by molecular interaction (contact among molecules in thermal non-equilibrium) due to temperature gradient. Heat conduction occurs in the direction of decreasing temperature (negative temperature gradient) (Fourier's law) [21, 22].

By analyzing the definition of Fick's (mass transfer) and Fourier's (heat transfer) laws we can say that both two laws presents very similar way, including

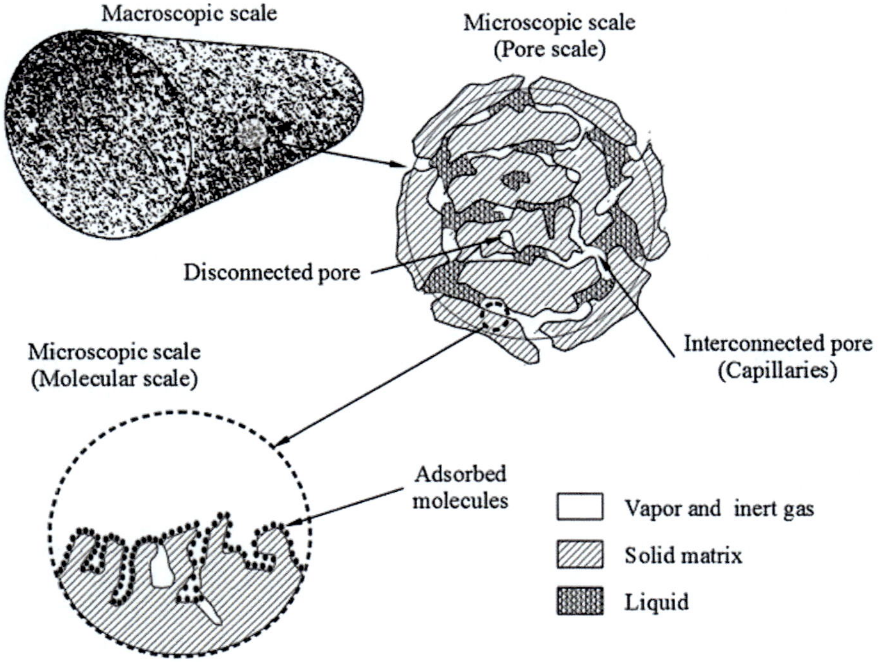

Fig. 1 Schematic of a porous body at different levels of analysis

the negative signal to account the fluxes in the direction of negative gradients (of heat and mass). Thus, it is evident the analogy between heat conduction and mass diffusion processes.

By applying an energy balance in an infinitesimal control-volume within a single-phase medium it is possible to demonstrate that accumulated energy rate equal to net heat transfer by conduction. It is true since that others phenomena like energy generation and heat convection be neglected.

According to Fig. 1, in a capillary-porous material different interfaces between the phases presents (solid matrix and gas; solid matrix and liquid; gas and liquid) exists into the pores. In general, in a microscopic scale, the local temperature in each phase separated by a surface (interface) is different. However, to study heat transfer in a macroscopic scale, it is considered that in the separation surface the temperature of the phases has the same values (continuity of temperature across the boundary). This condition is so-called local thermal equilibrium between the phases. The validate of the assumption of local equilibrium remains in discussion, especially when the timescale of transient diffusion is short and the transport properties ratio between the fluids and solid matrix is very different from unity [23].

The complexity of real processes taking place in a capillary-porous materials not only leads to uncertainties and difficulties with their mathematical formulation, but also cause considerable problem with the numerical and analytical treatment of resulting equation. The assumptions of the local thermal equilibrium is an idealization, simplify the theoretical treatment and be used when the dynamic and thermal behaviors are relatively slow, thus providing ample time for efficient heat transfer between the individual phases [24].

In the macroscopic scale, diffusion modeling approach presents like advantages: less mathematics than the microscopic transport analysis, external effects can be incorporated in boundary conditions, easy to obtain analytical solution for some particular cases. However, like disadvantages we can cite: many questionable theoretical assumptions are required, extrapolation to areas outside region of experimental data are questionable [25]. From this information we have the following question: Is the diffusion model physically realistic? In fact, in despite of the limitations many researchers prefer to use this approach in different physical situations (for examples, in cooling, heating, wetting and drying processes).

The solution of the diffusion problems for various physical situations of interest often requires the need to establish certain assumptions in describing the physical process. One of them is related to the geometry of the body in which occurs the transport of matter or energy. Several studies have been reported in literature using the diffusive model to describe the physical process, and consider the geometric shape of the bodies as cylinder, sphere or infinite slab [26–30]. On the basis of many researchers in early works, for these simpler geometries, the heat and mass diffusion within a capillary-porous body can be predicted by mean of different techniques such as: numerical (finite-difference, finite-element, boundary-element and finite-volume), analytical (separation of variables, Laplace transform and Galerkin-based integral method) and semi-analytical (Generalized integral transform), for several cases (appropriate considerations and boundary conditions), and often constant thermophysical properties for the medium are assumed. These geometric simplifications facilitate the solution of the diffusion problem. However, although these procedures usually presents good results, sometimes it does not adequately describe the processes involved, if the geometric shape of the solid under study is too different from the one considered.

Analytical and numerical solutions to predict heat and mass diffusion in capillary-porous bodies are also reported for parallelepipeds and prolate and oblate spheroids [31–40] and others solutions using generalized coordinates by means of 2D models [41, 42]. However, there is a lack of studies that take into account the heat and mass transfer in bodies with complex geometric shape through three-dimensional models, which is necessary in order to describe more precisely the process. Hence, rigorous solutions to the heat and mass diffusion in other geometrical shapes are essential, for the predictions of both the performance and control of the processes like heating, cooling, wetting and drying, for examples.

The practical interest in diffusive heat and mass transport in porous media is expanding quickly, due to a wide range of applications in several areas of science and technology for several geometries and different processes.

Heat and mass diffusion creates temperature and moisture gradients inside the solid, which in general, depends on the internal and external conditions of the porous body. In the context of diffusion, the calculation of the potential profiles requires the transport coefficients which are available in the engineering literature or determined by comparison with experimental data of a particular process of heat and mass transfer. For examples, to obtain temperature distribution inside the porous material it is necessary to know thermal conductivity, density, specific heat and/or thermal diffusivity of them. The calculation of the moisture profiles requires knowledge of the diffusion coefficient or water diffusivity of these materials. Besides, in some situations, convective heat and mass transfer coefficients are required too. With the knowledge of this information it is possible to determine themo-hydrical-mechanical stresses and to predict different problems such as crack, fissures, and deformations into the solid during the diffusion process. Depending of nature of the material, for example, fruits and vegetables, during heat and mass transfer process it is possible to have altering of color, texture and flavor, reducing in this way, the quality of them.

Now that we have discussed about porous bodies and diffusion, the next step is to build a mathematical model to predict diffusion process (heat and/or mass) inside capillary-porous bodies with arbitrary shape on an overall basis.

2.2 *Macroscopic Mathematical Modeling in Non-Orthogonal Curvilinear Coordinates*

The knowledge of the heat and mass transfer processes in bodies with arbitrary geometry is important in several engineering applications such as cooling, heating, freezing, soaking and drying of porous solids.

In many areas of production involving heat and mass transfer, there are many situations in which the porous solids under study do not present regular geometry. Some examples can be cited: sanitary ware, some types of coating, tile, rice, wheat, lentil, pumpkin, cashew, among many others. These solids have a geometry which should be considered as arbitrary. In this sense, the rigorous description of the heat and mass transfer in these bodies must consider an arbitrary domain.

2.2.1 Three-Dimensional Diffusion Equation: Physical and Transformed Domain

In a general way, the diffusion equation can be written as [27, 43–45]:

$$\frac{\partial}{\partial t}(\lambda\Phi) = \nabla \cdot (\Gamma^{\Phi}\, \nabla\Phi) + S, \tag{1}$$

where Φ is the dependent variable, t is the time, λ and Γ^{Φ} are the process parameters and S is a source term. In Cartesian coordinates, Eq. 1 is written as:

$$\frac{\partial(\lambda\,\Phi)}{\partial t} = \frac{\partial}{\partial x}\left(\Gamma^{\Phi}\frac{\partial\Phi}{\partial x}\right) + \frac{\partial}{\partial y}\left(\Gamma^{\Phi}\frac{\partial\Phi}{\partial y}\right) + \frac{\partial}{\partial z}\left(\Gamma^{\Phi}\frac{\partial\Phi}{\partial z}\right) + S, \tag{2}$$

Fig. 2 Cartesian and generalized coordinate systems

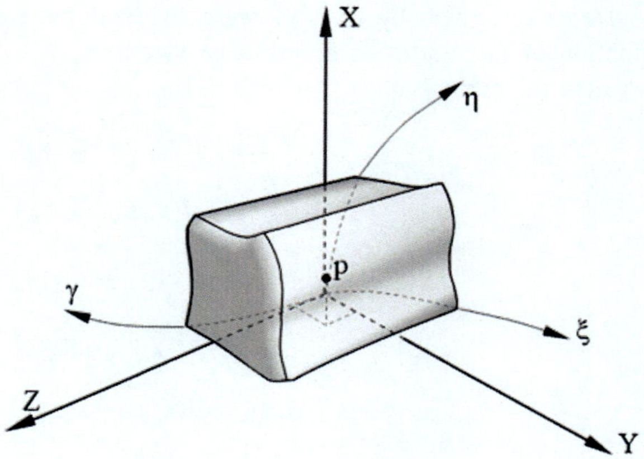

where x, y and z are the Cartesian coordinates of position. Equation 2 is frequently named diffusion equation in the physical domain.

However, in general, Cartesian coordinates are not appropriate to solve diffusion problems for solids with arbitrary shape. Thus, it is necessary to create a coordinate system whose axes are coincident with the borders of the studied solid (transformed domain).

2.2.2 Generalized Coordinate Systems

The study of the diffusion processes in a body of arbitrary shape requires the numerical solution of diffusion equation in a coordinate system that should be adaptable to its geometry. This means that the new axes, denoted by ξ, η and γ, defining a curvilinear, non-orthogonal coordinates system must be created, as shown in Fig. 2, [46–49].

Through Fig. 2 it is verified that the point P can be identified by two systems of axes. One of them is a system of orthogonal axes in which the position of point P is defined by Cartesian coordinates x, y and z; while the other is a system in generalized coordinates ξ, η and γ. As can be seen in Fig. 2, the generalized coordinate system has curvilinear axes that can be arranged in any direction.

The curvilinear coordinate system ξ, η and γ can be expressed as a function of x, y and z through transformations of the type:

$$\xi = \xi(x,y,z) \tag{3}$$

$$\eta = \eta(x,y,z) \tag{4}$$

$$\gamma = \gamma(x,y,z) \tag{5}$$

Equations (3)–(5) contain a simplification which is to assume that the axis systems do not exhibit change over time. Moreover, the relationships between the two axis systems are given by the following expressions, called metrics of the transformation of the Cartesian coordinate system (x, y, z) to a generalized coordinate system (ξ, η, γ):

$$\xi_x = J(y_\eta z_\gamma - y_\gamma z_\eta) \tag{6}$$

$$\xi_y = -J(x_\eta z_\gamma - x_\gamma z_\eta) \tag{7}$$

$$\xi_z = J(x_\eta y_\gamma - x_\gamma y_\eta) \tag{8}$$

$$\eta_x = -J(y_\xi z_\gamma - y_\gamma z_\xi) \tag{9}$$

$$\eta_y = J(x_\xi z_\gamma - x_\gamma z_\xi) \tag{10}$$

$$\eta_z = -J(x_\xi y_\gamma - x_\gamma y_\xi) \tag{11}$$

$$\gamma_x = J(y_\xi z_\eta - y_\eta z_\xi) \tag{12}$$

$$\gamma_y = -J(x_\xi z_\eta - x_\eta z_\xi) \tag{13}$$

$$\gamma_z = J(x_\xi y_\eta - x_\eta y_\xi) \tag{14}$$

In the equations above, the symbol g_m means partial derivative of g with respect to m; and J is the Jacobian of the transformation, defined by:

$$\frac{1}{J} = \det\begin{bmatrix} x_\xi & x_\eta & x_\gamma \\ y_\xi & y_\eta & y_\gamma \\ z_\xi & z_\eta & z_\gamma \end{bmatrix} \tag{15}$$

The inverse metrics are given by [46–50]:

$$x_\xi = \frac{\eta_\gamma \gamma_z - \eta_z \gamma_y}{J} \tag{16}$$

$$x_\eta = -\frac{\xi_\gamma \gamma_z - \xi_z \gamma_y}{J} \tag{17}$$

$$x_\gamma = \frac{\xi_\gamma \eta_z - \xi_z \eta_y}{J} \tag{18}$$

$$y_\xi = -\frac{\eta_x \gamma_z - \eta_z \gamma_x}{J} \tag{19}$$

$$y_\eta = -\frac{\xi_x \gamma_z - \xi_z \gamma_x}{J} \tag{20}$$

$$y_\gamma = -\frac{\xi_x \eta_z - \xi_z \eta_x}{J} \tag{21}$$

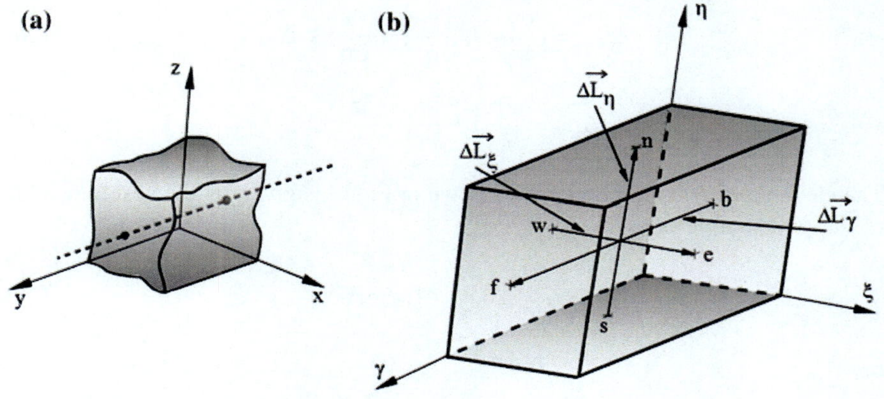

Fig. 3 Elementary volume ΔV: **a** Physical domain; **b** transformed domain highlighting the vectors: west-east, south-north and back-front

$$z_\xi = \frac{\eta_x \gamma_y - \eta_y \gamma_x}{J} \tag{22}$$

$$z_\eta = -\frac{\xi_x \gamma_y - \xi_y \gamma_x}{J} \tag{23}$$

$$z_\gamma = \frac{\xi_x \eta_y - \xi_y \eta_x}{J} \tag{24}$$

2.2.3 Calculation of Volume, Area and Length Using Generalized Coordinates

The numerical solution of the diffusion equation requires the calculation of volumes, areas and lengths in elements called control volumes. In the case of irregular geometries, such calculations are made based on Fig. 3, which shows an elementary volume in the physical and transformed domains. In order to calculate the volume of an elementary solid, the west-east vector ($\overrightarrow{\Delta L}_\xi$) that starts from the centre of the west face and finishes in the centre of the east face must be defined. In a similar way, the vectors $\overrightarrow{\Delta L}_\eta$ (south-north) and $\overrightarrow{\Delta L}_\gamma$ (back-front) must be obtained.

The expressions for $\overrightarrow{\Delta L}_\xi$, $\overrightarrow{\Delta L}_\eta$ and $\overrightarrow{\Delta L}_\gamma$ are given by:

$$\vec{\Delta L}_\xi = \left(\frac{\partial x}{\partial \xi}\vec{i} + \frac{\partial y}{\partial \xi}\vec{j} + \frac{\partial z}{\partial \xi}\vec{k}\right)\Delta\xi \tag{25}$$

$$\vec{\Delta L}_\eta = (\frac{\partial x}{\partial \eta}\ \vec{i} + \frac{\partial y}{\partial \eta}\ \vec{j} + \frac{\partial z}{\partial \eta}\ \vec{k})\Delta \eta \tag{26}$$

$$\vec{\Delta L}_\gamma = (\frac{\partial x}{\partial \gamma}\vec{i} + \frac{\partial y}{\partial \gamma}\vec{j} + \frac{\partial z}{\partial \gamma}\vec{k})\Delta \xi \tag{27}$$

The elementary volume of the solid shown in Fig. 3 is given by the absolute value of the scalar triple product [51]:

$$\Delta V = \left| \vec{\Delta L}_\xi \cdot (\vec{\Delta L}_\eta \times \vec{\Delta L}_\gamma) \right| \tag{28}$$

or

$$\Delta V\ = \det \begin{bmatrix} x_\xi & x_\eta & x_\gamma \\ y_\xi & y_\eta & y_\gamma \\ z_\xi & z_\eta & z_\gamma \end{bmatrix} \Delta\xi\Delta\eta\Delta\gamma \tag{29}$$

Comparing Eqs. 29 with 15, it is observed that the determinant in Eq. 29 is the inverse of Jacobian defined by Eq. 15. Thus, Eq. 29 can be rewritten as:

$$\Delta V = \frac{1}{J}\Delta\xi\,\Delta\eta\,\Delta\gamma \tag{30}$$

in which $\Delta\xi\,\Delta\eta\,\Delta\gamma$ can be defined as the volume of a elementary volume in the transformed domain.

A geometric interpretation can be obtained from Eq. 30. In general, the generalized coordinates ξ, η and γ are identified by consecutive integers and therefore $\Delta\xi$, $\Delta\eta$ and $\Delta\gamma$ are equal to unity. Thus, the inverse of the Jacobian represents the value of the volume of the elementary volume in the physical domain.

For a quadrilateral located in a plane defined by the lines η and γ (Fig. 4), for instance, the area is determined in a manner similar to that used for the volume. The vectors south-north $\left(\vec{\Delta L}_\eta\right)$ and back-front $\left(\vec{\Delta L}_\gamma\right)$, are given by Eqs. 26 and 27, respectively. According to Beer and Johnston [52], the elementary area is given by the modulus of the cross product between these vectors:

$$\Delta S = \left| \vec{\Delta L}_\eta \times \vec{\Delta L}_\gamma \right| \tag{31}$$

Thus, the expression to calculate a elementary area is given in the following way:

$$\Delta S = \sqrt{(y_\eta z_\gamma - z_\eta y_\gamma)^2 + (z_\eta x_\gamma - x_\eta z_\gamma)^2 + (x_\eta y_\gamma - y_\eta x_\gamma)^2}\ \Delta\eta\Delta\gamma \tag{32}$$

in which $\Delta\eta\,\Delta\gamma$ can be defined as the area in the transformed domain.

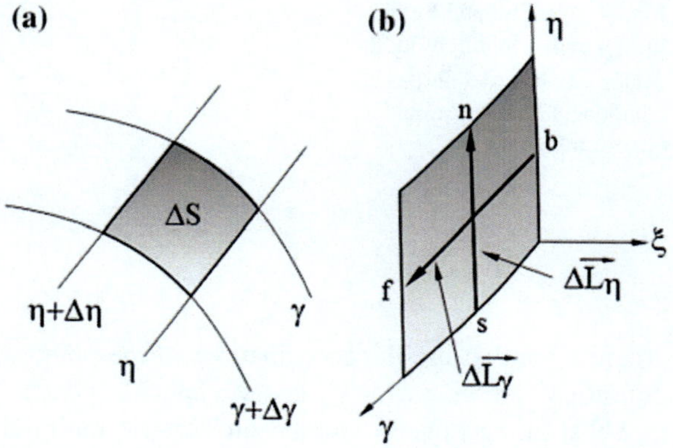

Fig. 4 Elementary area ΔS: **a** Physical domain; **b** Transformed domain highlighting the vectors: south-north $\overrightarrow{\Delta L_\eta}$ and back-front $\overrightarrow{\Delta L_\gamma}$.

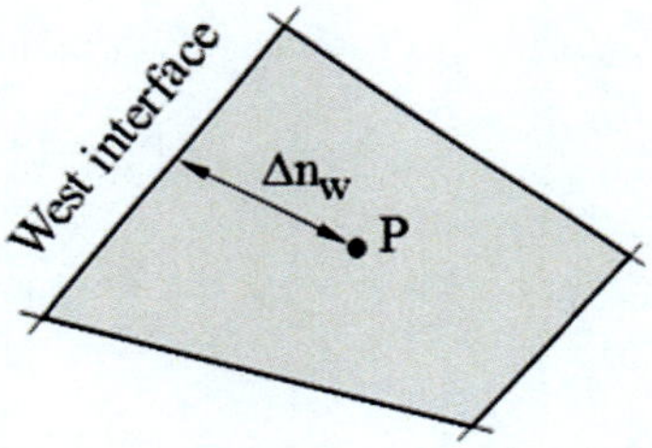

Fig. 5 Profile in the plane $\xi\eta$ of an elementary volume with arbitrary geometry in the physical domain

In order to calculate the distance between a point and a face, we will use the Fig. 5, as an example, where the distance Δn_w between the point P and the west face is shown.

The distance Δn_w can be obtained by the absolute value of the inner product between a unitary vector orthogonal to the west face ($\overrightarrow{VU_w}$) and the vector ($\overrightarrow{r_{Pw}}$) defined by the distance between P and w, in which w is a point of the west face. The unitary vector perpendicular to the west face (located at the plane $\eta\gamma$ and perpendicular to this plane) can be calculated as:

$$\overrightarrow{VU_w} = \frac{\overrightarrow{\Delta L_\eta^w} \times \overrightarrow{\Delta L_\gamma^w}}{\left|\overrightarrow{\Delta L_\eta^w} \times \overrightarrow{\Delta L_\gamma^w}\right|} \tag{33}$$

Thus, the value Δn_w is calculated through the inner product given by:

$$\Delta n_w = \left|\overrightarrow{r_{Pw}} \,.\, \overrightarrow{VU_w}\right| \tag{34}$$

in which $\overrightarrow{r_{Pw}}$ is given in the following way:

$$\overrightarrow{r_{Pw}} = (x^w - x^P)\,\vec{i} + (y^w - y^P)\,\vec{j} + (z^w - z^P)\,\vec{k} \tag{35}$$

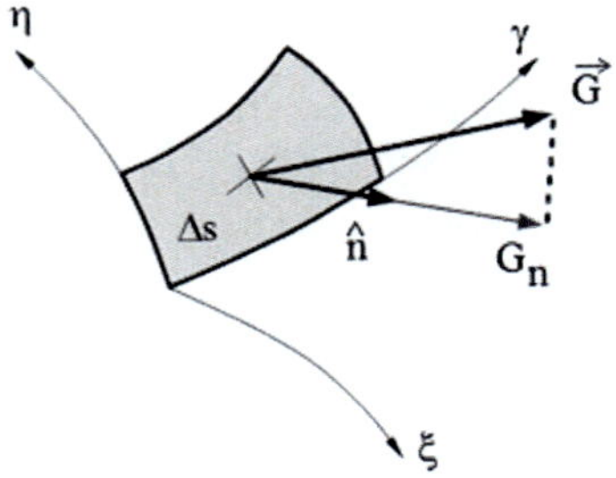

Fig. 6 Area defined by the lines η and γ, highlighting the vector quantity $\vec{G}$ and its component in the normal direction G_n

where x^w and x^P are the coordinates x of the points w and P, respectively. Similar definitions are given for y^w and y^P; and z^w and z^P.

Substituting Eqs. 35 and 33 into Eq. 34, the final expression for Δn_w is:

$$\Delta n_w = \Big[(x^w \text{ minus; } x^P)(y^w_\eta z^w_\gamma \textit{minus;} z^w_\eta y^w_\gamma) + (y^w \text{ minus; } y^P)(z^w_\eta x^w_\gamma - x^w_\eta z^w_\gamma) + (z^w \text{ minus; } z^P)(x^w_\eta y^w_\gamma - y^w_\eta x^w_\gamma)\Big] \frac{\Delta\eta\Delta\gamma}{\Delta S_w} \tag{36}$$

In a similar way, the distances of P to the other faces can be calculated.

2.2.4 Definition of Flux

In transport phenomena, flux is defined as "flux per unit area", where flow is the movement of some quantity per unit time. In other fields of the physics, flux is the integral of a vector quantity (flux density) over a finite surface. In this sense, the general expression for flux φ of a quantity $\vec{G}$ over a surface ΔS is given in the following way:

$$\varphi = \int_{\Delta S} \vec{G}.\hat{n} dS \tag{37}$$

where $\hat{n}$ is the unitary vector orthogonal to the area ΔS.

In Fig. 6, a vector quantity $\vec{G}$ and its component G_n, perpendicular to the area ΔS defined by the lines η and γ (placed in a line of ξ constant) is shown.

From Fig. 6 it is observed that:

$$G_n = \vec{G}.\hat{n} \tag{38}$$

From Eq. 37, it can be concluded that Eq. 38 gives the flux per unit area φ'', i.e.:

$$\varphi'' = G_n = \vec{G}.\hat{n} \tag{39}$$

Thus, for an elementary area ΔS, the flux of the vector quantity $\vec{G}$ can be calculated by:

$$\varphi = \varphi'' \Delta S = \vec{G}.\hat{n}\, \Delta S \tag{40}$$

For diffusion problems, the component G_n is the flux per unit area of the quantity defined by the potential Φ, and is given by:

$$G_n = \varphi'' = -\Gamma^{\Phi}\frac{\partial\Phi}{\partial n} \tag{41}$$

where n defines a direction normal to the area. In this sense, substituting Eq. 41 into Eq. 40, the flux of a quantity Φ in the elementary area ΔS can be rewritten as:

$$\varphi = -\Gamma^{\Phi}\frac{\partial\Phi}{\partial n}\Delta S \tag{42}$$

2.2.5 Diffusion Equation in the Transformed Domain

The Eq. 2 is the diffusion equation in Cartesian coordinates and, therefore, the independent variables are x, y, z and t. In order to write this equation in a generalized domain, defined by the independent variables ξ, η, γ and τ, where τ is time in the transformed domain, it is necessary several transformations involving the metrics defined by Eqs. 6–24. Details of these transformations can be found in Maliska [45] and Silva [48, 49]. These procedures result in the diffusion equation in the transformed domain, given by:

$$\begin{aligned}\frac{\partial}{\partial\tau}\left(\frac{\lambda\Phi}{J}\right) = &\frac{\partial}{\partial\xi}\left[\left(\alpha_{11}\frac{\partial\Phi}{\partial\xi} + \alpha_{12}\frac{\partial\Phi}{\partial\eta} + \alpha_{13}\frac{\partial\Phi}{\partial\gamma}\right)J\Gamma^{\Phi}\right]\\ &+\frac{\partial}{\partial\eta}\left[\left(\alpha_{21}\frac{\partial\Phi}{\partial\xi} + \alpha_{22}\frac{\partial\Phi}{\partial\eta} + \alpha_{23}\frac{\partial\Phi}{\partial\gamma}\right)J\Gamma^{\Phi}\right]\\ &+\frac{\partial}{\partial\gamma}\left[\left(\alpha_{31}\frac{\partial\Phi}{\partial\xi} + \alpha_{32}\frac{\partial\Phi}{\partial\eta} + \alpha_{33}\frac{\partial\Phi}{\partial\gamma}\right)J\Gamma^{\Phi}\right] + \frac{S}{J}\end{aligned} \tag{43}$$

where the coefficients α_{ij} are given by the expressions:

$$\alpha_{11} = \frac{1}{J^2}(\xi_x^2 + \xi_y^2 + \xi_z^2); \tag{44}$$

$$\alpha_{22} = \frac{1}{J^2}(\eta_x^2 + \eta_y^2 + \eta_z^2); \tag{45}$$

$$\alpha_{33} = \frac{1}{J^2}(\gamma_x^2 + \gamma_y^2 + \gamma_z^2); \tag{46}$$

$$\alpha_{12} = \alpha_{21} = \frac{1}{J^2}(\xi_x\eta_x + \xi_y\eta_y + \xi_z\eta_z); \tag{47}$$

$$\alpha_{23} = \alpha_{32} = \frac{1}{J^2}(\eta_x\gamma_x + \eta_y\gamma_y + \eta_z\gamma_z); \tag{48}$$

$$\alpha_{13} = \alpha_{31} = \frac{1}{J^2}(\xi_x\gamma_x + \xi_y\gamma_y + \xi_z\gamma_z). \tag{49}$$

The coefficients α_{ij} in which the index "i" is different of the index "j" are equal to zero for orthogonal grids; and are different of zero for non-orthogonal grids.

2.2.6 Initial and Boundary Conditions

The diffusion equation relates time and space variations of the potential Φ; it governs the transfer process in a body. To determine the distribution of Φ inside a body at any instant, i.e., to solve the diffusion equation, we necessitate to know the distribution of Φ at the initial instant (initial conditions), the geometry of the body, and the law of interaction between the surrounding medium and the body surface (boundary condition).

The initial condition describes the distribution of Φ inside a body at the initial instant:

$$\Phi(x,y,z,0) = f_1(x,y,z) \tag{50}$$

In several problems, where $f_1(x,y,z)$ is a unknown function, a uniform distribution is assumed at the initial instant; then, $\Phi(x,y,z,0) = \Phi_0 = \text{const.}$

The usual boundary conditions are:

Boundary Condition of the First Kind (Dirichlet Boundary Condition)

For the boundary condition of the first kind, the distribution of Φ over the body surface at any instant is prescribed, i.e.:

$$\Phi(x,y,z,t) = f_2(x,y,z,t)|_s. \tag{51}$$

A particular example is $\Phi(x,y,z,t) = \Phi_{eq} = \text{const.}$, i.e., the value of Φ on surface is constant during the whole diffusion process.

Boundary Condition of the Second Kind (Neumann Boundary Condition)

A boundary condition of the second kind occurs when the flow per unit area on surface is given as a function of the time:

$$\Phi''(x,y,z,t) = f_3(x,y,z,t)|_s. \tag{52}$$

The simplest case of the boundary of the second kind is that in which $\Phi''(x,y,z,t)$ is constant.

Boundary Condition of the Third Kind (Robin Boundary Condition)

Usually a boundary condition of the third kind is characteristic of the law of convective transfer between the solid surface and the surrounding medium. In this case, the inner diffusion flux per unit area on the surface equal the convective flux per unit area that arrives on the same surface from the outside. In this case:

$$h(\Phi_{eq} - \Phi|_s) = -\Gamma^{\Phi} \left.\frac{\partial \Phi}{\partial n}\right|_s \tag{53}$$

where h is the convective transfer coefficient and Φ_{eq} represents the value of Φ at the equilibrium condition with the surrounding.

2.3 Numerical Treatment

2.3.1 Discretization of the Diffusion Equation

In this work, the method used in the discretization of the diffusion equation is the finite volume method with a fully-implicit formulation. The reason for this choice is that the solution is unconditionally stable for any time interval [[46, 48].

In order to solve numerically Eq. 43, it is necessary to discretize the transformed domain. Then, the three-dimensional transformed domain was divided into elementary control volumes, termed control volumes that are differentiated by location in the generated mesh. The distinction between the control volumes is given by the number of faces in contact with the external environment. The transformed domain shown in Fig. 8 was divided into 27 types of control volumes. Thus, it is possible to distinguish 8 types of control volumes with three faces in contact with the medium, 12 types of control volumes with 2 faces in contact with the ambient, and 6 types of control volumes having only one of faces in contact with the external medium. The last type of control volume (internal volume) does not have any face in contact with the medium. Obviously, each control volume shown in Fig. 7, generate one algebraic equation distinct of the other.

Integrating Eq. 43 about space $\Delta\xi\Delta\eta\Delta\gamma$ and time $(\Delta\tau)$ for each control volume, we obtain the following result:

$$\frac{\lambda_P \Phi_P - \lambda_P^0 \Phi_P^0}{J_P} \frac{\Delta\xi\Delta\eta\Delta\gamma}{\Delta\tau} = \alpha_{11e} J_e \Gamma_e^{\Phi} \Delta\eta\Delta\gamma \left.\frac{\partial \Phi}{\partial \xi}\right|_e +$$

$$\alpha_{12e} J_e \Gamma_e^{\Phi} \Delta\eta\Delta\gamma \left.\frac{\partial \Phi}{\partial \eta}\right|_e + \alpha_{13e} J_e \Gamma_e^{\Phi} \Delta\eta\Delta\gamma \left.\frac{\partial \Phi}{\partial \gamma}\right|_e -$$

$$\alpha_{11w} J_w \Gamma_w^{\Phi} \Delta\eta\Delta\gamma \left.\frac{\partial \Phi}{\partial \xi}\right|_w - \alpha_{12w} J_w \Gamma_w^{\Phi} \Delta\eta\Delta\gamma \left.\frac{\partial \Phi}{\partial \eta}\right|_w -$$

Fig. 7 Different types of control volumes in a three-dimensional mesh at transformed domain

$$\alpha_{13w}J_w\,\Gamma^{\Phi}_{w}\Delta\eta\Delta\gamma\left.\frac{\partial\Phi}{\partial\gamma}\right|_{w} + \alpha_{21n}J_n\,\Gamma^{\Phi}_{n}\Delta\xi\Delta\gamma\left.\frac{\partial\Phi}{\partial\xi}\right|_{n} +$$

$$\alpha_{22n}J_n\Gamma^{\Phi}_{n}\Delta\xi\Delta\gamma\left.\frac{\partial\Phi}{\partial\eta}\right|_{n} + \alpha_{23n}J_n\,\Gamma^{\Phi}_{n}\Delta\xi\Delta\gamma\left.\frac{\partial\Phi}{\partial\gamma}\right|_{n} -$$

$$\alpha_{21s}J_s\,\Gamma^{\Phi}_{s}\Delta\xi\Delta\gamma\left.\frac{\partial\Phi}{\partial\xi}\right|_{s} - \alpha_{22s}J_s\Gamma^{\Phi}_{s}\Delta\xi\Delta\gamma\left.\frac{\partial\Phi}{\partial\eta}\right|_{s} -$$

$$\alpha_{23s}J_s\,\Gamma^{\Phi}_{s}\Delta\xi\Delta\eta\left.\frac{\partial\Phi}{\partial\gamma}\right|_{s} + \alpha_{31f}J_f\,\Gamma^{\Phi}_{f}\Delta\xi\Delta\eta\left.\frac{\partial\Phi}{\partial\xi}\right|_{f} +$$

$$\alpha_{32f}J_f\Gamma^{\Phi}_{f}\Delta\xi\Delta\eta\left.\frac{\partial\Phi}{\partial\eta}\right|_{f} + \alpha_{33f}J_f\,\Gamma^{\Phi}_{f}\Delta\xi\Delta\eta\left.\frac{\partial\Phi}{\partial\gamma}\right|_{f} -$$

$$\alpha_{31b}J_b\,\Gamma^{\Phi}_{b}\Delta\xi\Delta\eta\left.\frac{\partial\Phi}{\partial\xi}\right|_{b} - \alpha_{32b}J_b\Gamma^{\Phi}_{b}\Delta\xi\Delta\eta\left.\frac{\partial\Phi}{\partial\eta}\right|_{b} -$$

$$\alpha_{33b}J_b\,\Gamma^{\Phi}_{b}\Delta\xi\Delta\eta\left.\frac{\partial\Phi}{\partial\gamma}\right|_{b} + \frac{S_P\Phi_P}{J_P}\Delta\xi\Delta\eta\Delta\gamma + \frac{S_C}{J_P}\Delta\xi\Delta\eta\Delta\gamma \tag{54}$$

where the superscript 0 means "old time" and its absence means "current time". The subscripts "e", "w", "n", "s" "f" and "b" mean the east, west, north, south, front and back borders, respectively, of an elementary control volume, while P is the nodal point of this volume. All the elements described above are shown in Fig. 8.

An observation of Eq. 54 indicates that the term which represents the Jacobian J, must be calculated at point P for each control volume and also, as well as α_{ij}, at the interfaces east, west, north, south, front and back of the same control volume.

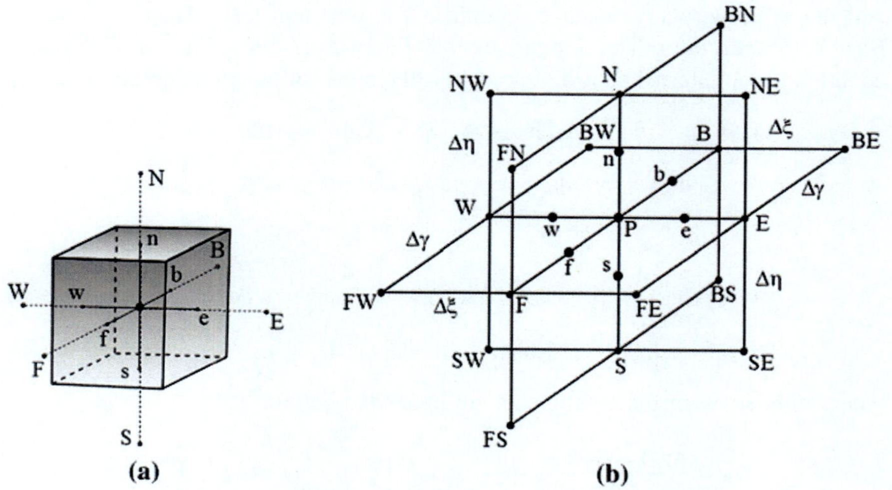

Fig. 8 **a** Elementary control volume with nodal point P and faces "e", "w", "n", "s", "f" e "b"; **b** Nodal point P and its neighbors

In order to complete the discretization of Eq. 54, it is necessary to define each type of control volume for which the derivatives of the variable Φ are calculated.

Internal Control Volumes

For an internal control volume, the derivatives of Φ related to the generalized axes for each interface are calculated. As an example, two direct derivatives are shown below:

$$\left.\frac{\partial \Phi}{\partial \xi}\right|_{e} = \frac{\Phi_E - \Phi_P}{\Delta \xi}; \tag{55}$$

$$\left.\frac{\partial \Phi}{\partial \xi}\right|_{w} = \frac{\Phi_P - \Phi_W}{\Delta \xi}. \tag{56}$$

As an example, two expressions for the cross derivatives are given in the following way:

$$\left.\frac{\partial \Phi}{\partial \eta}\right|_{e} = \frac{\Phi_N + \Phi_{NE} - \Phi_S - \Phi_{SE}}{4\Delta \eta}; \tag{57}$$

$$\left.\frac{\partial \Phi}{\partial \eta}\right|_{w} = \frac{\Phi_N + \Phi_{NW} - \Phi_S - \Phi_{SW}}{4\Delta \eta}. \tag{58}$$

and the other derivatives can be found in [46, 49] and [50]. Thus, substituting Eqs. 55–58 and the others derivatives into Eq. 54, the discretization of diffusion equation for an internal control volume results in the following algebraic equation:

$$A_p\Phi_P = A_e\Phi_E + A_w\Phi_W + A_n\Phi_N + A_s\Phi_S + A_f\Phi_F + A_b\Phi_B +$$
$$A_{ne}\Phi_{NE} + A_{nw}\Phi_{NW} + A_{se}\Phi_{SE} + A_{sw}\Phi_{SW} + A_{fe}\Phi_{FE} +$$
$$A_{fw}\Phi_{FW} + A_{be}\Phi_{BE} + A_{bw}\Phi_{BW} + A_{fn}\Phi_{FN} + A_{fs}\Phi_{FS} +$$
$$A_{bn}\Phi_{BN} + A_{bs}\Phi_{BS} + B\ , \tag{59}$$

where only some of the coefficients are presented below:

$$A_P = \frac{\lambda_P}{J_P}\frac{\Delta\xi\Delta\eta\Delta\gamma}{\Delta\tau} + \frac{\Delta\eta\Delta\gamma}{\Delta\xi}\alpha_{11e}J_e\,\Gamma_e^{\Phi} + \frac{\Delta\eta\Delta\gamma}{\Delta\xi}\alpha_{11w}J_w\,\Gamma_w^{\Phi} +$$
$$\frac{\Delta\xi\Delta\gamma}{\Delta\eta}\alpha_{22n}J_n\,\Gamma_n^{\Phi} + \frac{\Delta\xi\Delta\gamma}{\Delta\eta}\alpha_{22s}J_s\,\Gamma_s^{\Phi} +$$
$$\frac{\Delta\xi\Delta\eta}{\Delta\gamma}\alpha_{33f}J_f\,\Gamma_f^{\Phi} + \frac{\Delta\xi\Delta\eta}{\Delta\gamma}\alpha_{33b}J_b\,\Gamma_b^{\Phi} - \frac{S_P}{J_P}\Delta\xi\Delta\eta\Delta\gamma; \tag{60}$$

$$A_e = \frac{\Delta\eta\Delta\gamma}{\Delta\xi}\alpha_{11e}J_e\,\Gamma_e^{\Phi} + \frac{\Delta\gamma}{4}\alpha_{21n}J_n\,\Gamma_n^{\Phi} - \frac{\Delta\gamma}{4}\alpha_{21s}J_s\,\Gamma_s^{\Phi} + \frac{\Delta\eta}{4}\alpha_{31f}J_f\,\Gamma_f^{\Phi} - \frac{\Delta\eta}{4}\alpha_{31b}J_b\Gamma_b^{\Phi}; \tag{61}$$

$$A_{ne} = \frac{\Delta\gamma}{4}\alpha_{12e}J_e\,\Gamma_e^{\Phi} + \frac{\Delta\gamma}{4}\alpha_{21n}J_n\,\Gamma_n^{\Phi}; \tag{62}$$

$$A_{nw} = -\frac{\Delta\gamma}{4}\alpha_{12w}J_w\,\Gamma_w^{\Phi} - \frac{\Delta\gamma}{4}\alpha_{21n}J_n\,\Gamma_n^{\Phi}; \tag{63}$$

$$A_{fe} = \frac{\Delta\eta}{4}\alpha_{13e}J_e\,\Gamma_e^{\Phi} + \frac{\Delta\eta}{4}\alpha_{31f}J_f\,\Gamma_f^{\Phi}; \tag{64}$$

$$A_{fs} = -\frac{\Delta\xi}{4}\alpha_{23s}J_s\,\Gamma_s^{\Phi} - \frac{\Delta\xi}{4}\alpha_{32f}J_f\,\Gamma_f^{\Phi}; \tag{65}$$

$$B = \frac{\lambda_P^0\Phi_P^0}{J_P}\frac{\Delta\xi\Delta\eta\Delta\gamma}{\Delta\tau} + \frac{S_C}{J_P}\Delta\xi\Delta\eta\Delta\gamma. \tag{66}$$

Commonly, the ξ, η and γ lines of a grid in the transformed domain are identified by consecutive integers, and therefore $\Delta\xi$, $\Delta\eta$ and $\Delta\gamma$ are equal to 1.

Metrics in the Nodal Points and at the Interfaces of the Internal Control Volumes

In order to determine the terms α_{ij} and the Jacobian J, the partial derivatives x_ξ, x_η, x_γ, y_ξ, y_η, y_γ, z_ξ, z_η and z_γ must be known. Therefore, expressions for these derivatives must be established for both the nodal point P, and the faces of all control volumes. Thus, for a nodal point P of an internal control volume, the derivatives of the coordinates x with respect to ξ, for example is given by:

$$x_\xi^P = \frac{x_e - x_w}{\Delta\xi}; \tag{67}$$

where x_e and x_w are the coordinates of the midpoints of the interfaces "e" and "w", respectively, and are calculated as the arithmetic mean of the coordinates of the four vertices that form the face. Thus, for the east interface, for example, we have:

$$u_e = \frac{u_{i+1,j,k+1} + u_{i+1,j,k} + u_{i+1,j+1,k} + u_{i+1,j+1,k+1}}{4}, \tag{68}$$

where u can be x, y or z and the indexes i, j and k indicate, respectively, the position of the planes ξ, η and γ that define the control volume. Similarly, we calculate the coordinates of the midpoints of the others faces. The derivatives y_ξ^P, y_η^P, y_γ^P, z_ξ^P, z_η^P and z_γ^P are obtained through the same steps of Eqs. 67 and 68. On the other hand, the derivatives of the coordinate x with respect to ξ, η and γ for the east interface of an internal control volume are given by the equation:

$$x_\xi^e = \frac{x_E - x_P}{\Delta\xi} = \frac{1}{2\,\Delta\xi}\left[\frac{x_{i+2,j,k} + x_{i+2,j+1,k} + x_{i+2,j+1,k+1} + x_{i+2,j,k+1}}{4} - \frac{x_{i,j,k+1} + x_{i,j+1,k+1} + x_{i,j+1,k} + x_{i,j,k}}{4}\right], \tag{69}$$

$$x_\eta^e = \frac{1}{\Delta\eta}\left[\frac{x_{i+1,j+1,k+1} + x_{i+1,j+1,k}}{2} - \frac{x_{i+1,j,k+1} + x_{i+1,j,k}}{2}\right], \tag{70}$$

$$x_\gamma^e = \frac{1}{\Delta\gamma}\left[\frac{x_{i+1,j,k+1} + x_{i+1,j+1,k+1}}{2} - \frac{x_{i+1,j,k} + x_{i+1,j+1,k}}{2}\right], \tag{71}$$

where x_P and x_E are, respectively, the coordinates x of the nodal point P and its neighbor to east (E), which can be calculated by the arithmetic mean of the coordinates of the eight vertices of the parallelepiped representing the control volume in the transformed domain. Similar expressions can be found for the coordinates of the others nodal points located inside the solid. More details are given in [49] and [50].

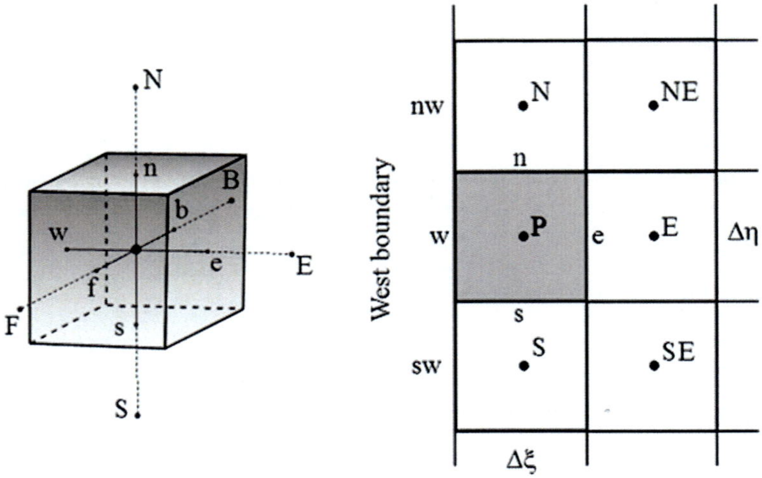

Fig. 9 Control volume with nodal point P at the west boundary

Boundary Control Volumes

The discretization for the volumes located, for instance, on the inner west boundary, is performed through a balance of the potential Φ at the surface of body. The nodal point P of this volume, as well as its neighbors, can be seen in Fig. 9, which shows the control volume located near the west face. Note that when boundary condition of the first kind is used the values Φ_{nw}, Φ_w, Φ_{sw}, Φ_{fw}, and Φ_{bw} must be known.

Some derivatives at the west boundary that are different from those obtained for the internal control volumes and valid only for the boundary condition of the first kind are written as:

$$\left.\frac{\partial\Phi}{\partial\xi}\right|_w = \frac{\Phi_P - \Phi_w}{\frac{\Delta\xi}{2}}; \tag{72}$$

$$\left.\frac{\partial\Phi}{\partial\eta}\right|_w = \frac{\Phi_{nw} - \Phi_{sw}}{2\Delta\eta}; \tag{73}$$

$$\left.\frac{\partial\Phi}{\partial\xi}\right|_n = \frac{1}{\Delta\xi}\left[\frac{\Phi_P + \Phi_N + \Phi_E + \Phi_{NE}}{4} - \frac{(\Phi_w + \Phi_{nw})}{2}\right] \tag{74}$$

Replacing the derivatives mentioned in Eq. 54, we obtain the following algebraic equation for these control volumes:

$$\begin{gathered}A_p\Phi_P = A_e\Phi_E + A_n\Phi_N + A_s\Phi_S + A_f\Phi_F + A_b\Phi_B + A_{ne}\Phi_{NE} + \\ A_{se}\Phi_{SE} + A_{fe}\Phi_{FE} + A_{be}\Phi_{BE} + A_{fn}\Phi_{FN} + A_{fs}\Phi_{FS} + \\ A_{bn}\Phi_{BN} + A_{bs}\Phi_{BS} + B.\end{gathered} \tag{75}$$

In Eq. 76, some coefficients differ from those determined for the internal control volumes. The Ap coefficient is given by:

$$A_P = \frac{\lambda_P}{J_P}\frac{\Delta\xi\Delta\eta\Delta\gamma}{\Delta\tau} + \frac{\Delta\eta\Delta\gamma}{\Delta\xi}\alpha_{11e}J_e\,\Gamma_e^{\Phi} + \frac{\Delta\eta\Delta\gamma}{\Delta\xi}2\alpha_{11w}J_w\,\Gamma_w^{\Phi} +$$
$$\frac{\Delta\xi\Delta\gamma}{\Delta\eta}\alpha_{22n}J_n\,\Gamma_n^{\Phi} + \frac{\Delta\xi\Delta\gamma}{\Delta\eta}\alpha_{22s}J_s\,\Gamma_s^{\Phi} + \frac{\Delta\xi\Delta\eta}{\Delta\gamma}\alpha_{33f}J_f\,\Gamma_f^{\Phi} +$$
$$\frac{\Delta\xi\Delta\eta}{\Delta\gamma}\alpha_{33b}J_b\,\Gamma_b^{\Phi} - \frac{\Delta\gamma}{4}\alpha_{21n}J_n\,\Gamma_n^{\Phi} + \frac{\Delta\gamma}{4}\alpha_{21s}J_s\,\Gamma_s^{\Phi} -$$
$$\frac{\Delta\eta}{4}\alpha_{31f}J_f\,\Gamma_f^{\Phi} + \frac{\Delta\eta}{4}\alpha_{31b}J_b\,\Gamma_b^{\Phi} - \frac{S_P}{J_P}\Delta\xi\Delta\eta\Delta\gamma; \tag{76}$$

and the other coefficients A and B are given in Farias [50].

With a similar procedure to that shown for the internal control volumes we can also determine the metrics for the control volume located on the inner boundary to the west. Algebraic equations like the equation above can be found for each control volume located on the boundaries of the solid. Thus, for each time step, we obtain a linear system of algebraic equations in Φ that can be solved, for example, using Gauss–Seidel method, with a tolerance of 1×10^{-8}.

2.3.2 Average of Φ

If the values of Φ are known for each control volume at time t, its mean value at the same time can be determined by the equation [46, 50]:

$$\overline{\Phi} = \frac{1}{\sum \frac{1}{J_P}} \sum \Phi_P \frac{1}{J_P} \tag{77}$$

where the summation must be applied over all control volumes. The term $1/J_P$ corresponds to the volume of the control volume with nodal point P. Hence, the mean value of Φ is determined by a weighted mean in which the volume of each control volume is used for the weighting.

2.4 Grid Generation

According to Tannehill et al. [53] and Ferziger and Perić [54], for complex geometries, is recommended the use of a set of regular grids fitted to the irregular regions, combining cylindrical, spherical and non-orthogonal grids next the boundaries with Cartesian grids in the rest of the domain. In this technique, the biggest problem is the coupling among the grids, which complicates the computational implementation. To overcome these difficulties, non-orthogonal structured grids that are fitted to the boundaries of the domain can be used [54–56]. The advantage of these grids is that they can be adapted to any geometry, and are determined through a coordinate system coinciding with the boundaries. In this case, it is simpler to implement the boundary conditions.

There are several ways available in the literature to generate a non-orthogonal structured grid both two and three-dimensional. Among these ways, it is common to use elliptical equations in the process of generation, as the Laplace's equation (and Poisson). Besides the use of partial differential equations, there are also several algebraic methods for the generation of grids as, for example, Lagrange's interpolation (and Hermite). Herein, the use of the Poisson equation will be highlighted to generate two-dimensional grids. These grids are particularly important in diffusion studies involving solids which can be obtained by extrusion or revolution of plane areas [42, 43, 50, 57]. Besides, their generation is relatively simple.

For the two-dimensional case, the Poisson equation is given in the following way [46, 55]:

$$\frac{\partial^2 \xi}{\partial x^2} + \frac{\partial^2 \xi}{\partial y^2} = P(\xi,\eta), \tag{78}$$

and

$$\frac{\partial^2 \eta}{\partial x^2} + \frac{\partial^2 \eta}{\partial y^2} = Q(\xi,\eta), \tag{79}$$

in which ξ and η are axes of a non-orthogonal curvilinear coordinate system, named generalized coordinates, and represent lines of the grid in a domain frequently called transformed domain. Lines ξ connect the north and south boundaries while a line η connects the east and west boundaries. Among the possibilities of source terms (P and Q) for the attraction of lines by other lines or points, Maliska [46] suggests:

$$\begin{aligned} P(\xi,\eta) = &- \sum_{j=1}^{N_\xi} a_{\xi j} \frac{(\xi - \xi_j)}{|\xi - \xi_j|} \exp(-c_{\xi j}|\xi - \xi_j|) \\ &- \sum_{i=1}^{M_\xi} b_{\xi i} \frac{(\xi - \xi_i)}{|\xi - \xi_i|} \exp\left\{ -d_{\xi i} \left[(\xi - \xi_i)^2 + (\eta - \eta_i)^2 \right]^{1/2} \right\} \end{aligned} \tag{80}$$

and

$$\begin{aligned} Q(\xi,\eta) = &- \sum_{j=1}^{N_\eta} a_{\eta j} \frac{(\eta - \eta_j)}{|\eta - \eta_j|} \exp(-c_{\eta j}|\eta - \eta_j|) \\ &- \sum_{i=1}^{M_\eta} b_{\eta i} \frac{(\eta - \eta_i)}{|\eta - \eta_i|} \exp\left\{ -d_{\eta i} \left[(\xi - \xi_i)^2 + (\eta - \eta_i)^2 \right]^{1/2} \right\} \end{aligned} \tag{81}$$

The first term on the right side of Eq. 80 represents the attraction factors of all lines ξ of the grid by the N_ξ lines of index j (ξ_j), previously stipulated. Moreover, the second term on the right side of this equation represents the factors that attract all lines of the grid for the M_ξ points of index i, (ξ_i, η_i), previously stipulated. The terms of Eq. 81 are interpreted in a manner similar to the interpretation of the terms of Eq. 80.

The solutions of the Eqs. 78 and 79 give, respectively, ξ and η for values stipulated of x and y. However, to create a grid, it is necessary to know the values of x and y as a function of the values stipulated for ξ and η. Thus, it is necessary to transform Eqs. 78 and 79 from the physical domain (where x and y are the independent variables) to the transformed domain (where ξ and η are the independent variables). This procedure results in equations given in the following [46, 48, 49]:

$$g_{22}x_{\xi\xi} + g_{11}x_{\eta\eta} - 2g_{12}x_{\xi\eta} + \frac{1}{J^2}[x_\xi P(\xi,\eta) + x_\eta Q(\xi,\eta)] = 0, \tag{82}$$

and

$$g_{22}y_{\xi\xi} + g_{11}y_{\eta\eta} - 2g_{12}y_{\xi\eta} + \frac{1}{J^2}[y_\xi P(\xi,\eta) + y_\eta Q(\xi,\eta)] = 0, \tag{83}$$

where factors of the type h_m mean partial derivative of h with respect to m, h_{mn} represents the second derivative of h with respect to m and n. The factors g_{11}, g_{22} and g_{12} are the components of the metric tensor for the two-dimensional case, and are given by

$$g_{11} = x_\xi^2 + y_\xi^2; \tag{84}$$

$$g_{22} = x_\eta^2 + y_\eta^2; \tag{85}$$

$$g_{12} = x_\xi x_\eta + y_\xi y_\eta. \tag{86}$$

In Eqs. 82 and 83, J is the Jacobian of the transformation given by:

$$J = \frac{1}{x_\xi y_\eta - x_\eta y_\xi} \tag{87}$$

Equations 82 and 83 are named generation equations, and can be numerically solved, for instance, by the finite difference method. These equations and this method were implemented in the software called "2D Grid Generation" available on Internet at http://zeus.df.ufcg.edu.br/labfit/gg.htm [58].

2.5 Determination of the Transport Parameter Γ^Φ by Optimization

To determine the parameter Γ^Φ by optimization, the objective function was defined by the Chi-square referring to the fit of the simulated curve to the experimental data of the drying kinetics. The expression for the Chi-square involving the fit of a simulated curve to the experimental data is given by [59]:

$$\chi^2 = \sum_{i=1}^{N_P} (\overline{\Phi}_i - \overline{\Phi}^{sim})^2 \frac{1}{\sigma_i^2} \tag{88}$$

where $\overline{\Phi}_i$ is the mean value of Φ referring to the ith experimental point, $\overline{\Phi}^{sim}$ is the correspondent simulated mean value of Φ, N_P is the number of experimental points, $1/\sigma_i^2$ is the statistical weight referring to the ith experimental point. In general, in the absence of information, the statistical weights are made equal to 1. In Eq. 88, supposing the use of boundary condition of the first kind, the Chi-square depends on $\overline{\Phi}^{sim}$, which only depends on Γ^{Φ}. On the other hand, performing a numerical simulation for which experimental data are available, the Chi-square inherent to the simulation can be calculated.

With respect to minimizing the Chi-square, the sequence of calculations to determine the parameter Γ^{Φ}, basically have followed, the steps reported by Silva et al. [57], which considered Γ^{Φ} variable. When we consider a value constant for Γ^{Φ}, the steps are as follows:

a) Step 1 Inform the initial values of the parameter Γ^{Φ}. Solve the diffusion equation and determine the Chi-square;
b) Step 2 Inform the value for the correction of Γ^{Φ};
c) Step 3 Correct the parameter Γ^{Φ}. Solve the diffusion equation and calculate the Chi-square;
d) Step 4 Compare the latest calculated value of the Chi-square with the previous one. If the latest value is smaller, return to the step 2; otherwise, decrease the last correction of the value of Γ^{Φ} and proceed to step 5;
e) Step 5 Begin a new cycle coming back to the step 2 until the stipulated convergence for the parameter Γ^{Φ} is reached.

According to [57], in each cycle, the value of the correction of each parameter can be initially modest, compatible with the tolerance of convergence imposed to the problem. Then, for a given cycle, in each return to the step 2, the value of the new correction can be multiplied by the factor 2. If the modest correction initially informed does not minimize the objective function, in the next cycle its value can be multiplied by the factor -1.

3 Applications

3.1 Pasteurization of Tomato Puree (Heat Transfer)

The proposed numerical solution can be used to study the heat conduction in porous solids if we impose: $\Phi = T$ (temperature), $\Gamma^{\Phi} = k$ (effective thermal conductivity), $\lambda = \rho c_p$ and $S = q'''$. Herein, such solution will applied to describe a heating process of tomato puree ($S = 0$).

An example of the use of heat transfer in the industry is the pasteurization of foodstuffs. In order to describe the process, thermo-physical parameters referring to convection, radiation, phase change, heat sources and conduction should be known. For solid products with no heat source and no phase change, in general only the conduction and radiation are involved in the transfer process. Many times, solely the conduction mechanism is used to describe heat transfer and, consequently, the parameters involved in the process are considered as apparent. The knowledge of these thermo-physical parameters for the product to be heated and/or cooled is necessary in order to describe the process and define the time and cost of these operations. Due to this fact, the determination of these parameters is widely found in the literature [57, 60–63]. Besides, other aspect that deserves attention is the food quality, which is greatly influenced by the thermo-physical properties. The safety of pasteurized products directly depends on the values of the thermo-physical parameters used in the simulation of the heat transfer [64]. It is necessary guarantee that the correct temperature be reached in all points of the product in order to inactivate pathogenic agents. Thus, with the knowledge of the thermo-physical parameters, simulations of the heat penetration can be performed in strategic points, enabling to determine the needed time in which the temperature of inactivation of pathogenic agents is reached.

3.1.1 Experimental Data

In order to evaluate the potential of the proposed numerical solution applied to describe an experimental dataset, one of the experiments performed by Betta et al. [62] will be used. The experiment refers to heat penetration kinetics in tomato puree placed within a cylindrical tin-plated can with inner radius R = 40.0 and height L = 125.0 mm. Two thermocouples were inserted in the product. One of them was placed in the geometric centre of the cylinder, and the other was adhered to the internal wall of the can, in the interface with the product, in the median plane. The system, with an initial temperature T = 5.2 °C, was dipped into a preheated bath filled with water at T = 80.0 °C. Then, in regular time intervals, the temperatures were measured during the entire process. An inspection in the measurements obtained by Betta et al. [62] makes it possible to consider that the inner wall of the cylinder almost instantaneously assumes the temperature of 80.0 °C. Thus, the boundary condition of the first kind was used to describe the heating process. Additionally, in the experiments, the dimensions of the can were measured only in the beginning of the process and, therefore, their variations were disregarded.

3.1.2 Results and Analysis

Silva et al. [57] analyzed the dataset from [62], investigating several expressions for the thermal diffusivity $\hat{\alpha}$ ($m^2\ s^{-1}$) as a function of the local temperature T (°C), in order to describe the heating process. To this end, Silva et al. [57] used

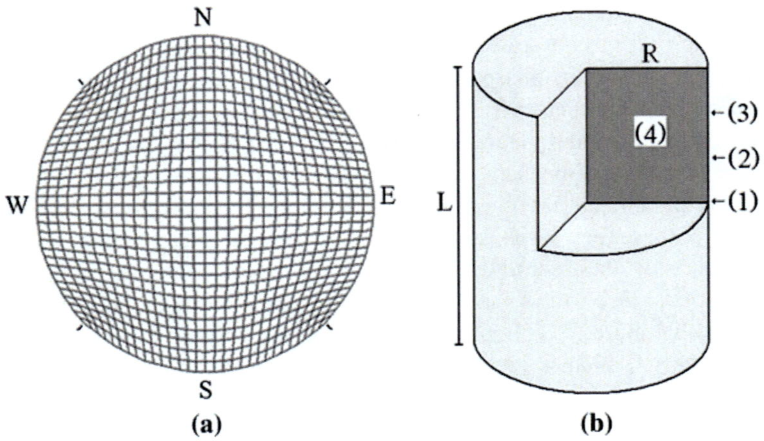

Fig. 10 **a** Circular mesh with 32 × 32 elements showing the boundaries: North (N); South (S); East (E) and West (W); **b** Cylinder (with no scale) obtained by extrusion of a circular mesh highlighting: (1) circular section in the centre; (2) circular section to 1/3 of the centre; (3) circular section to 2/3 of the centre; (4) vertical plane

cylindrical coordinates and an optimization algorithm. The authors found that, among the investigated expressions, the best of them to describe the heating kinetics is given by:

$$\hat{\alpha} = \mathrm{b}\,\mathrm{e}^{\mathrm{a}\mathrm{T}^{1/2}} \tag{89}$$

with $\mathrm{a} = 2.96 \times 10^{-2}{}^{\circ}\mathrm{C}^{-1/2}$ and $\mathrm{b} = 1.21 \times 10^{-7}\ \mathrm{m}^2\ \mathrm{s}^{-1}$.

In the present work, the heat penetration in tomato puree is described through the proposed 3D numerical solution, using the results from Silva et al. [57]. To this end, using 2D Grid Generation software, a two-dimensional circular grid with radius R = 40 mm and 32 × 32 elements was generated as shown in Fig. 10a. An inspection in this figure enables to observe that the mesh is structured and non-orthogonal. Therefore, in the numerical solution, the cross-terms $\alpha_{ij}(i \neq j)$ that appear in Eq. 43 must be considered.

The 3D-grid representing the can was obtained by extrusion of the 2D-grid in the direction of the axis perpendicular to its plane, and the length L referring to the extrusion was divided into 62 elements, generating a three-dimensional grid with 32 × 32 × 62 control volumes. The solid obtained with this procedure is shown in Fig. 10b. In this figure, the positions of the circular planes where the temperature distribution will be analyzed are indicated. Additionally, a vertical plane is also shown with the same objective.

The simulation of the heating kinetics in the centre of the cylinder, using the expression for the thermal diffusivity obtained by Silva et al. [57] and the numerical solution presented in this work, is showed in Fig. 11a. On the other hand, defining the error of an experimental temperature "i" as

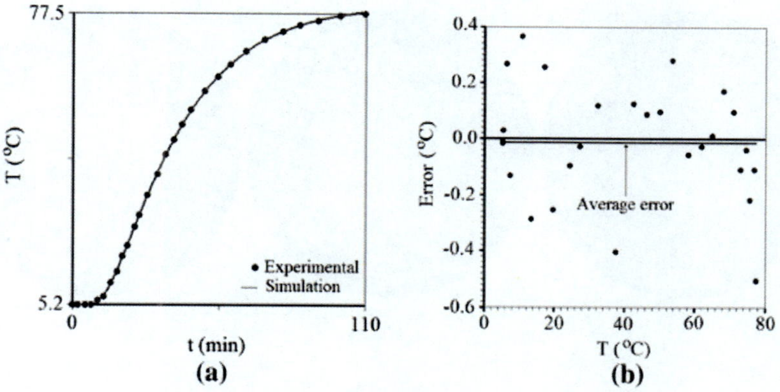

Fig. 11 **a** Simulation of the heating kinetics in the centre of the cylinder; **b** Error as a function of the experimental temperature

$$\mathrm{Error}_i = T_i^{exp} - T_i^{sim}, \tag{90}$$

the plot of the errors as function of the experimental temperatures, as well as the average error, are given in Fig. 11b.

Note that the time of heating, equal to 110 min, was divided into 2,000 time steps. In each time step, the system of equations obtained by the discretizations was solved through the Gauss–Seidel method, with a tolerance of 10^{-8}. The statistical indicators of the simulation are $R^2 = 0.999947$ (determination coefficient) and $\chi^2 = 1.1082$ (Chi-square). As the uncertainty of each experimental point was not provided by Betta et al. [62], the statistical weights to determine the Chi-square were considered with the arbitrary value equal to 1. This means that the same weight was considered for all the points.

An inspection in Fig. 11b indicates that the errors can be considered randomly distributed and the average error is very close to zero (−0.015 °C), as is expected.

For an analyse of the temperature inside the product, temperature distributions at the circular sections indicated in the parts 1–3 of Fig. 10b are presented in Fig. 12a for the instant t = 11 min. In order to understand why the temperature distribution in the planes 1 and 2 are similar and significantly different of the case 3, the temperature distribution at the vertical plane highlighted in the part 4 of Fig. 10b is shown in Fig. 12b at instant t = 11 min.

A careful observation of the Fig. 12b helps to understand the three temperature distributions shown in the circular sections of Fig. 12a: in the central region of the cylinder, heating is slowest than at the extremities. Due to this fact, frequently the central point of the cylinder is chosen so that its transient state is accompanied during heating. When this point reaches the temperature in which pathogenic agents are inactivated, all other points have already reached this temperature. On the other hand, essentially, the obtained results herein are similar to those found by Silva et al. [57], and this fact makes it possible to validate the proposed mathematical formulation.

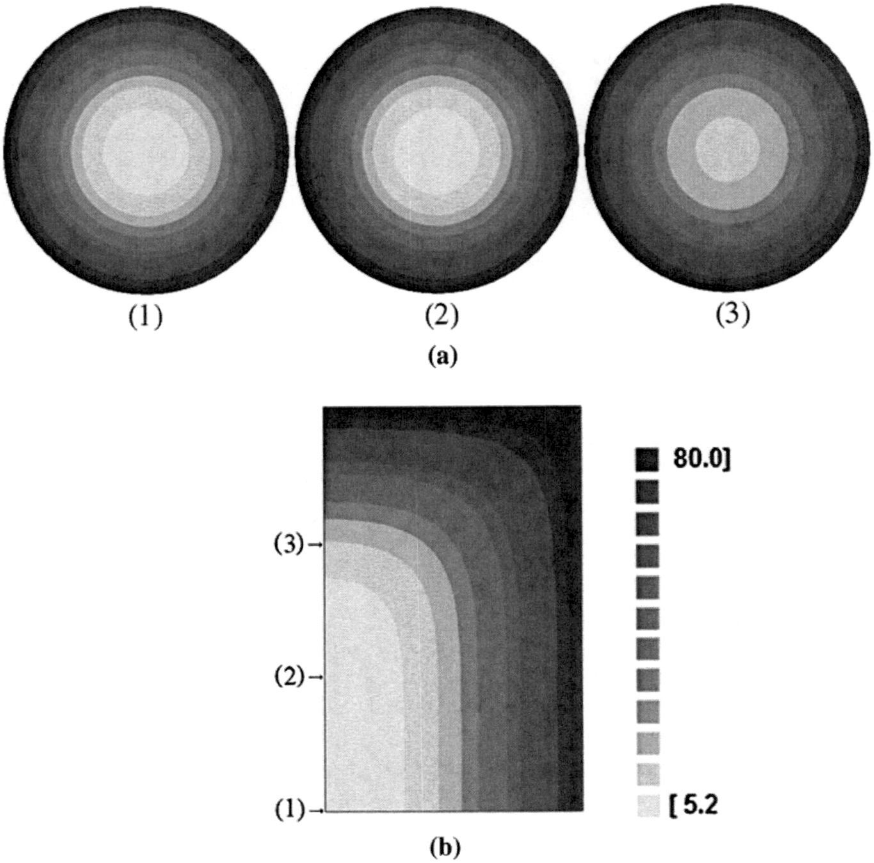

Fig. 12 **a** Temperature distributions for t = 11 min at the circular sections located in the position: (1) centre; (2) 1/3 of the centre; (3) 2/3 of the centre; **b** Temperature distributions for t = 11 min, highlighting the positions in the vertical plane: (1) centre; (2) 1/3 of the centre; (3) 2/3 of the centre

3.2 Drying of Ceramic Roof Tile (Mass Transfer)

The proposed numerical solution can be used to study the mass diffusion in porous solids if we impose: $\Phi = M$ (moisture content), $\Gamma^{\Phi} = D_{ef}$ (effective diffusivity), $\lambda = 1$ and $S = \dot{M}'''$. As an example, such solution will applied to describe a drying process of ceramic roof tiles ($S = 0$).

3.2.1 Experimental Data

The ceramic roof tiles were shaped from red clay that is used for brick and tiles. The test bodies were processed by extrusion. Samples of the product were placed inside an oven with mechanical circulation of air and digital temperature

Table 1 Temperature, initial and equilibrium moisture content, and dimensions of the tiles

T (°C)	M_0 (db)	M_{eq} (db)	E (m)	L (m)	C (m)
72.8	0.2345	0.0049	6.49×10^{-3}	45.49×10^{-3}	83.36×10^{-3}

controller. The drying was accomplished at T = 72.8 °C. The tiles were weighed at time intervals ranging from 5 min at the beginning of the drying, up to about 1 h, at the end of the process. Measurement of the mass was accomplished with a electronic device. The process took place until the mass reached its equilibrium value. At the end of drying, the oven temperature was set at 105 °C and the test bodies remained there for 24 h, enabling the measurement of dry matter. Experimental data about the accomplished dryings are summarized in Table 1.

Such dataset was used to determine value of the effective mass diffusivity, by optimization, using the optimization algorithm described in Sect. 2.5 [57]. Once the effective mass diffusivity was determined, the drying kinetic can be simulated.

3.2.2 Results and Analysis

The description of the moisture migration in ceramic roof tiles using the diffusion equation obeys the following hypotheses: (a) the dimensions of the solid do not vary during the water diffusion; (b) liquid diffusion is the only transport mechanism of water inside the solid; (c) the initial moisture distribution must be uniform; (d) the solid is considered homogeneous and isotropic; (e) the effective water diffusivity does not vary during the diffusion; (f) the prescribed boundary condition is the same in all external faces of the tiles, and (g) the field moisture content is axi-symmetric related to axes x and z. Thus, the generated mesh represents one symmetrical quarter of the tile, resulting in lower computational effort.

Figure 13a shows a sample of the ceramic roof tile, while the Fig. 13b shows a two-dimensional grid with 20 × 20 elements that was generated by software 2D Grid Generation, mentioned in Sect. 2.4.

The symmetrical quarter of the tile was obtained by extrusion of the 2D mesh in the direction of the axis perpendicular to its plane, until the half of the length C. This procedure has generated a three-dimensional grid with 20 × 20 × 20 control volumes.

For the numerical simulation, the total time of the drying process was divided into 2,000 time step. In each time step, the system of equations obtained by the discretization was solved through the Gauss–Seidel method, with a tolerance of 10^{-8}. The result obtained by optimization for the effective mass diffusivity was $D_{ef} = 1.074 \times 10^{-9}\ m^2\ s^{-1}$. With this result, the drying kinetics can be simulated.

To get an idea about how well the liquid-diffusion model with constant diffusivity matches experimental data for the drying kinetic of ceramic roof tiles, the result of simulation is plotted together with experimental data in Fig. 14. The statistical indicators of the simulation are $R^2 = 0.9896$ (determination coefficient) and $\chi^2 = 2.0667 \times 10^{-2}$ (Chi-square), which can be considered reasonable.

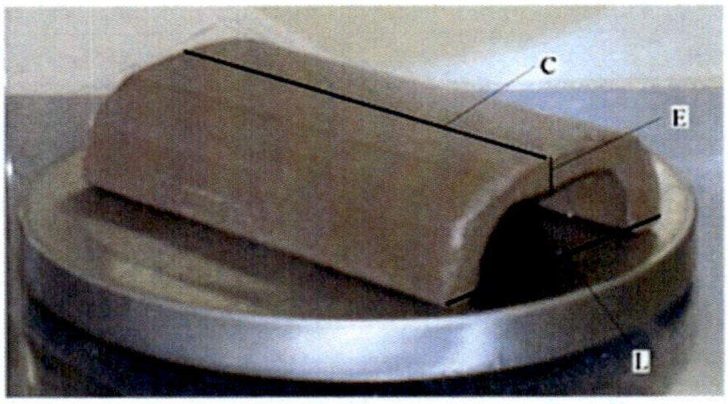

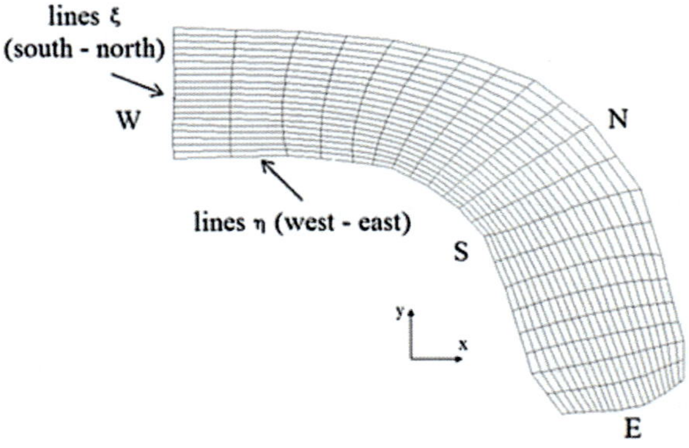

Fig. 13 **a** A sample of the ceramic roof tile; **b** non-orthogonal structured 2D grid (with no scale) to obtain 3D grid, showing the boundaries: North (N); South (S); West (W) and East (E)

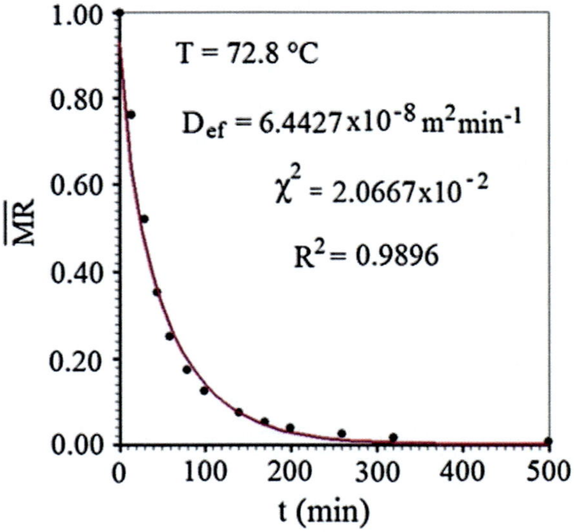

Fig. 14 Simulation of the drying kinetics at T = 72.8 °C and experimental dataset. $\bar{M}\bar{R} = \left(\frac{\bar{M}-M_{eq}}{M_0-M_{eq}}\right)$

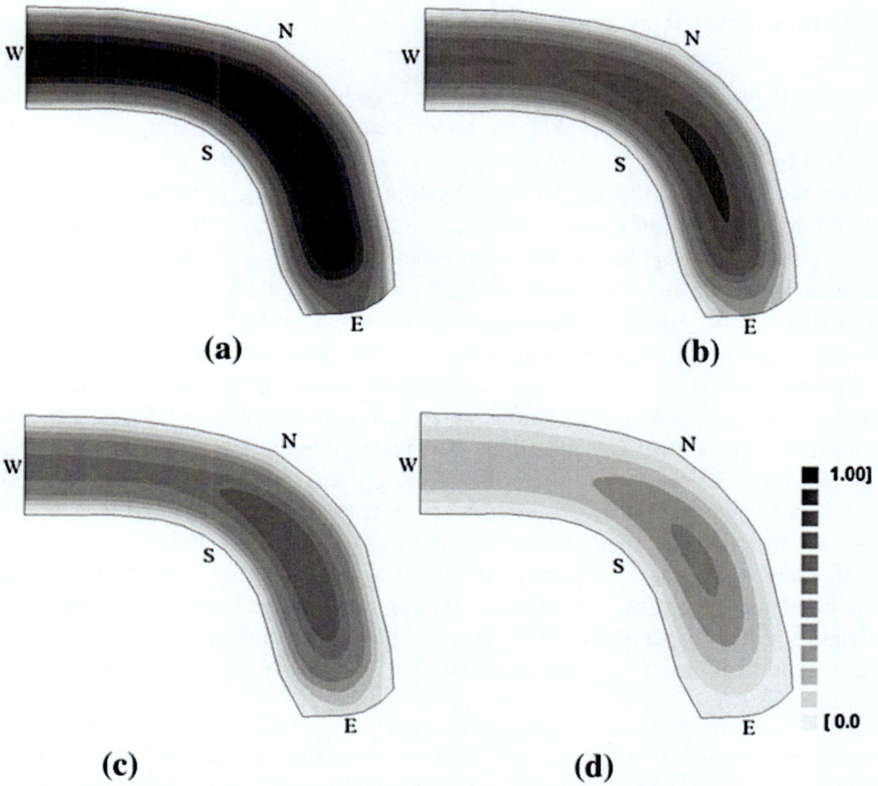

Fig. 15 Contour plots showing moisture distribution within the ceramic roof tiles (central plane) at T = 72.8 °C in different elapsed times **a** 10.0; **b** 30.0; **c** 50 and **d** 100 min

The knowledge of the difference between the moisture content in the center and at the boundary is important, because this difference causes stresses that may damage the product. Therefore, it was analyzed the water distribution inside the ceramic roof tile and contour plots showing this distribution for the central plane are presented in Fig. 15.

Through an inspection of Fig. 15, one can see the difference between the drying kinetics of the control volumes located inside the solid those situated on the borders. Note that the control volumes that are located in the east side dry more quickly than those are located inside the solid, during all process. At initial instants there is also a slight asymmetry in the distribution of water in relation to the south and north boundaries. The inner face (S) is closed and thus dry more slowly than the north face, since in this face (N), the surface area in contact with the environment is more open than the south side (see Fig. 13).

Note further that the internal control volumes located near the curvature in the interior of the tile have the lowest drying rate than other control volumes, throughout the whole process.

4 Concluding Remarks

In this chapter, diffusion phenomenon (heat conduction and mass diffusion) in solids has been explored, whit particular reference to capillary-porous body. Interest in this type of problems is motivated by its importance in many practical situations related to drying, wetting, heating and cooling processes. Here, our attention is focused on unsteady-state and local thermal equilibrium conditions.

A consistent three-dimensional mathematical model to predict diffusion phenomena inside porous solids with arbitrary shape is proposed. A general numerical formalism for the governing equation, which is based on the finite-volume method, has been applied to the solution of the diffusion equation in a boundary-fitted coordinate system. The analyses of the obtained results using this model following predictions are made by including agreement with experimental data of two industrial applications: pasteurization of tomato puree and drying of ceramic roof tile. An optimization technique to determine process parameters (transport properties) has been presented and tested with success. Finally, we would like to cite the fact that models, such as this outlined in this work, can be used with great confidence to elucidate unknown features of several complex systems in porous solids related to diffusion processes including new physical situation considering variable properties, dimensions variations and other boundary conditions under small modifications, so present in actual heat and mass transfer problems.

Acknowledgments The authors would like to express their thanks to Brazilian agencies CNPq (Conselho Nacional de Desenvolvimento Científico e Tecnológico) and CAPES (Coordenação de Aperfeiçoamento de Pessoal de Nível Superior) for supporting this work, and are also grateful to the authors of the references in this chapter that helped in the improvement of quality. Sincere thanks to Editor Prof. João M.P.Q. Delgado by the opportunity given to present our research in this book.

References

1. Nield, D.A., Bejan, A.: Convection in Porous Media. Springer-Verlag, New York (1992)
2. Bejan, A.: Convection Heat Transfer. Wiley, New York (1995)
3. Nield, D.A.: Modeling fluid flow in saturated porous media and at interfaces. In: Ingham, D.B., Pop, I. (eds.) Transport Phenomena in Porous Media II, Elsevier, Oxford (2002)
4. Lima, A.G.B., Farias Neto, S.R., Silva, W.P.: Heat and mass transfer in porous materials with complex geometry: Fundamentals and applications. In: Delgado, J.M.P.Q. (ed.) Heat and Mass Transfer in Porous Media. Springer-Verlag, Berlin (2012)
5. Lima, A.G.B., Oliveira, L.G., Lima, W.C.P.B.: Heat transfer in a packed-bed cylindrical reactor of elliptic cross section: mathematical modeling and simulation. In: Öchsner, A., Silva, L.F.M., Altenbach, H. (eds.) Materials with Complex Behavior II: Properties, Non-classical Materials and New Technologies. Spriger-Verlag, Berlin (2012)
6. Strumillo, C., Kudra, T.: Drying: Principles, Science and Design. Gordon and Breach Science Publishers, New York (1986)
7. Kowalski, S.J.: Thermomechanics of drying processes. Springer-Verlag, Berlin (2003)
8. Datta, A.K.: Porous media approaches to studying simultaneous heat and mass transfer in food processes. I: Problem formulations. J. Food Eng. **80**, 80–95 (2007)

9. Luikov, A.V.: Heat and Mass Transfer in Capillary Porous Bodies. Pergamon Press, New York (1966)
10. Keey, R.B.: Drying of Loose and Particulate Materials. Hemisphere Publishing Corporation, New York (1992)
11. Brooker, D.B., Bakker-Arkema, F.W., Hall, C.W.: Drying and storage of grains and oilseeds. AVI Book, New York (1992)
12. Fick, A.: Ann. Phys. **170**, 59 (1855)
13. Hwang, S-.T, Kammermeyer, K.: Membranes in separations. Wiley, New York (1975)
14. Shukla, K.N.: Diffusion Processes During Drying of Solids. World Scientific, Singapore (1990)
15. Lewis, W.K.: The rate of drying of solid materials. J. Ind. Eng. Chem. **13**, 427–432 (1921)
16. Sherwood, T.K: The drying of solid, Pt. I. Ind. Eng. Chem. **21**, 12–16 (1929a)
17. Sherwood, T.K.: The drying of solids, Pt. II. Ind. Eng. Chem. **21**, 976–980 (1929)
18. Sherwood, T.K.: The drying of solids, Pt. III, Mechanism of the drying of pulp and paper. Ind. Eng. Chem. **22**, 132–136 (1930)
19. Newman, A.B.: Diffusion and surface emission equation. Trans. AIChe. **27**, 203–220 (1931)
20. Fortes, M., Okos, M.R.: Drying theories: Their bases and limitations as applied to foods and grains. In: Mujumdar, A.S. (ed.) Advances in Drying. Hemisphere Publishing Corporation, Washington (1980)
21. Fourier, J.B.: Théorie analytique de la chaleur, English Translation by A. Freeman. Dover Publishing, New York (1822)
22. Kaviany, M.: Principles of Heat Transfer. Wiley, New York (2002)
23. Chin-Tsau, Hsu: Dynamic modeling of convective heat transfer in porous media. In: Vafai, K. (ed.) Handbook of Porous Media, 2nd edn. CRC Press/Taylor & Francis, Boca Raton (2005)
24. Vafai, K., Amiri, A.: Non-Darcian effects in confined forced convection flows. In: Ingham, D.B., Pop, I. (eds.) Transport Phenomena in Porous Media. Elsevier, Oxford (1998)
25. Rosen, H.N.: Recent advances in the drying of solid wood. In: Mujumdar, A.S. (ed.) Advances in Drying, Vol 4, Hemisphere Publishing Corporation, Berlin (1987)
26. Carslaw, H.S., Jaeger, J.C.: Conduction of Heat in Solids. University Press, New York (1959)
27. Luikov, A.V.: Analytical heat diffusion theory. Academic Press, Inc. Ltd., London (1968)
28. Gebhart, B.: Heat Conduction and Mass Diffusion. McGraw-Hill, Inc., New York (1993)
29. Silva, W.P., Precker, J.W., Silva, C.M.D.P.S., Silva, D.D.P.S.: Determination of effective diffusivity via minimization of the objective function by scanning: application to drying of cowpea. J. Food Eng. **95**, 298–304 (2009)
30. Silva, W.P., Precker, J.W., Silva, C.M.D.P.S., Gomes, J.P.: Determination of effective diffusivity and convective mass transfer coefficient for cylindrical solids via analytical solution and inverse method: application to the drying of rough rice. J. Food Eng. **98**, 302–308 (2010)
31. Silva, W.P., Silva, C.M.D.P.S., Farias, V.S.O., Gomes, J.P.: Diffusion models to describe the drying process of peeled bananas: optimization and simulation. Drying Tech. **30**(2), 164–174 (2012a)
32. Haji-Sheikh, A., Sparrow, E.M.: The solution of heat conduction problems by probability methods. Trans. ASAE. J. Heat Transf. **89**, 121–131 (1967)
33. Beck, J.V., Cole, K.D., Haji-Sheikh, A., Litkouhi, B.: Heat Conduction Using Green's Functions. Hemispheric Publishing Corporation, New York (1992)
34. Alassar, R.S: Heat conduction from spheroids. J. Heat Transf. **121**, 497–499 (1999)
35. Wu, B., Yang, W., Jia, C.: A three-dimensional numerical simulation of transient heat and mass transfer inside a single rice kernel during the drying process. Biosystems Eng. **87**(2), 191–299 (2004)
36. Lima, D.R., Farias, S.N., Lima, A.G.B.: Mass transport in spheroids using the Galerkin method. Braz. J. Chem. Eng. **21**, 667–680 (2004)
37. Nascimento, J.J.S., Lima, A.G.B., Teruel, B.J., Belo, F.A.: Heat and mass transfer including shrinkage during ceramics bricks drying. Información Tecnológica **17**(6), 145 (2006). (In Spanish)

38. Hacihafizoglu, O., Cihan, A, Kahveci, K., Lima, A.G.B.: A liquid diffusion model for thin-layer drying of rough rice. European Food Res. Tech. **226**, 787–793 (2008)
39. Cihan, A., Kahveci, K., Hacihafizo?lu, O., Lima, A.G.B.: A diffusion based model for intermittent drying of rough rice. Heat Mass Transf. **44**, 905–911 (2008)
40. Carmo, J.E.F., Lima, A.G.B.: Mass transfer inside oblate spheroidal solids: modelling and simulation. Braz. J. Chem. Eng. **25**(1), 19–26 (2008)
41. Salinas, C., Ananias, A., Alvear, M.: Wood drying simulation. Maderas, Ciência e Tecnologia. **6**(1), 3 (2004). (In Spanish)
42. Silva, W.P., Precker, J.W., Silva, D.D.P.S., Silva, C.D.P.S., Lima, A.G.B.: Numerical simulation of diffusive processes in solids of revolution via finite volume method and generalized coordinates. Int. J. Heat Mass Transfer **52**(21–22), 4976–4985 (2009)
43. Silva, W.P., Silva, C.M.D.P.S., Silva, D.D.P.S., Neves, G.A., Lima, A.G.B.: Mass and heat transfer study in solids of revolution via numerical simulations using finite volume method and generalized coordinates for the Cauchy boundary condition. Int. J. Heat Mass Transfer **53**(5–6), 1183–1194 (2010)
44. Crank, J.: The Mathematics of Diffusion. Clarendon Press, Oxford (1992)
45. Bird, R.B., Stewart, W.E., Lightfoot, E.N.: Transport phenomena. John Wiley & Sons, Inc., New York (2001)
46. Maliska, C.R.: Computational heat transfer and fluid mechanics. LTC Editora S. A. Rio de Janeiro (2004). (In Portuguese)
47. Patankar, S.V.: Numerical Heat Transfer and Fluid Flow. Hemisphere Publish Corporation, New York (1980)
48. Silva, W.P.: Diffusion transport in solids with arbitrary shape using generalized coordinates. Doctoral Thesis. Process Engineering, Center of Science and Technology, Federal University of Campina Grande, Campina Grande-PB, Brazil (2007). (In Portuguese)
49. Silva, W.P.: Discretization in computational transport phenomena, Curse 1 and 2 PowerPoint slides. http://zeus.df.ufcg.edu.br/labfit/TCMFC.htm. Accessed may/2010. (2009)
50. Farias, V.S.O.: 3D diffusion in solids with arbitrary shape using generalized coordinates. Doctoral Thesis. Process Engineering, Center of Science and Technology, Federal University of Campina Grande, Campina Grande-PB, Brazil (2011). (In Portuguese)
51. Farias, V.S.O., Silva, W.P., Silva, C.M.D.P.S., Lima, A.G.B.: Three-dimensional diffusion in arbitrary domain using generalized coordinates for the boundary condition of the first kind: application in drying. Defect Diff. Forum **326–328**, 120–125 (2012)
52. Beer, F.P., Johnston, E.R.J.: Vector mechanics for engineers: Static and dynamics. McGraw-Hill São Paulo, Makron (1990)
53. Tannehill, J.C., Anderson, D.A., Pletcher, R.H.: Computational Fluid Mechanics and Heat Transfer. Philadelphia: Taylor & Francis, p. 792 (1997)
54. Ferziger, J.H, Perić M.: Computational method for fluid dynamics. Springer, Berlin (2002)
55. Thompson, J.F., Warsi, Z.U.A., Mastin, C.W.: Numerical Grid Generation. Elsevier Science Publishing Co, New York (1985)
56. Fletcher, C.A.J.: Computational techniques for fluid dynamics, vol. 2. Springer, Berlin (2003)
57. Silva, W.P., Silva, C.M.D.P.S., Lins, M.A.A.: Determination of expressions for the thermal diffusivity of canned foodstuffs by the inverse method and numerical simulations of heat penetration. Int. J. Food Sci. Technol. **46**(4), 811–818 (2011)
58. Silva, W.P.: Software 2D Grid Generation, V 5.1, online: http://zeus.df.ufcg.edu.br/labfit/gg.htm, Accessed december/2011. (2008)
59. Taylor, J.R.: An Introduction to Error Analysis. University Science Books, Sausalito, California (1997)
60. Bairi, A., Laraqi, N.: García de María, J.M.: Determination of thermal diffusivity of foods using 1D Fourier cylindrical solution. J. Food Eng. **78**(2), 669–675 (2007)
61. Kumar, R., Kumar, A., Murthy, U.N.: Heat transfer during forced air precooling of perishable food products. Biosystems Eng. **99**, 228–233 (2008)
62. Betta, G., Rinaldi, M., Barbanti, D., Massini, R.: A quick method for thermal diffusivity estimation: Application to several foods. J. Food Eng. **91**, 34–41 (2009)

63. Silva, W.P., Silva, C.D.P.S., Gama, F.J.A.: An improved technique for determining transport parameters in cooling processes. J. Food Eng. doi:10.1016/j.jfoodeng.2012.02.003 (2012b)
64. Mohamed, I.O.: Numerical investigation on the effect of uncertainty in thermal diffusivity on thermal process lethality of canned foods. Int. J. Food Eng. **6**(6), Article 2. doi:10.2202/1556-3758.1862 (2010)

Resin Transfer Molding Process: Fundamentals, Numerical Computation and Experiments

Felipe Ferreira Luz, Sandro Campos Amico, Jeferson Ávila Souza, Enivaldo Santos Barbosa and Antonio Gilson Barbosa de Lima

Abstract Resin Transfer Molding (RTM) is one of the most widely known composite manufacturing technique of the liquid molding family, being extensively studied and used to obtain advanced composite materials comprised of fibers embedded in a thermoset polymer matrix. The fibrous reinforcement is considered a porous medium regarding its infiltration by the polymer resin. In this sense, this chapter aims to briefly discuss multiphase flow and heat transfer theory in RTM process, focusing on a multifluid model and the Control Volume/Finite Element (CV/FE) method. Finally, computational analysis was developed on the basis of ANSYS CFX® and PAM-RTM commercial software's for the investigation of the fluid flow in RTM composite molding. In order to show the versatility and performance of the commercial codes, RTM experiments were carried under distinct injection pressure and fiber volume fraction conditions using plain-weave

F. F. Luz · S. C. Amico
Federal University of Rio Grande do Sul, Porto Alegre, RS 91501-970, Brazil
e-mail: felipe.luz@ufrgs.br

S. C. Amico
e-mail: amico@ufrgs.br

J. Á. Souza
Federal University of Rio Grande, Rio Grande, RS 96201-900, Brazil
e-mail: jasouza@furg.br

E. S. Barbosa
Federal University of Campina Grande, Campina Grande, PB 58429-900, Brazil
e-mail: enivaldo.sb@gmail.com

A. G. B. de Lima (✉)
Department of Mechanical Engineering, Federal University of Campina Grande, Campina Grande, PB 58429-900, Brazil
e-mail: gilson@dem.ufcg.edu.br

J. M. P. Q. Delgado et al. (eds.), *Numerical Analysis of Heat and Mass Transfer in Porous Media*, Advanced Structured Materials 27,
DOI: 10.1007/978-3-642-30532-0_5,

glass fiber cloth as the porous media. The transient numerical simulations provided information about volume fraction, pressure and velocity distribution of the phases (resin and air) inside the porous media.

Keywords LCM · RTM · Modelling · Simulation · Experimentation · Composites

1 Introduction

1.1 Composite Materials

Composite materials are considered to be combinations of materials differing in composition or form on a macroscale [1]. The constituents, the dispersed phase and the matrix (continuous phase), retain their identities in the composite and they also exhibit an interface between one another [2]. In general, the advantages of this class of materials include weight reduction, flexibility in design, reduction of manufacturing cost and corrosion resistance.

According to the principle of the combined action, better property combinations are fashioned by the judicious combination of two or more distinct materials. However, for a hybrid material to be regarded as a composite, it must satisfy at least four criteria, including: (i) Being made up of two or more materials, inherently immiscible; (ii) The properties of the composite should be markedly different from those of the constituents; (iii) The components must be present in reasonable proportions, and (iv) The different materials must be separated by an interface.

Technologically, the most important composites are those wherein the dispersed phase is in the form of a fiber, whose strength and/or rigidity characteristics are often desired. The fibrous reinforcement is usually comprised of fabrics produced with continuous strands or tows of a few hundred fibers. The most common synthetic fibers currently used are glass, aramid and carbon, which can be used in a variety of architectures such as random mat, plain weave and stitched fabric. Natural fibers, such as hemp, flax, bamboo, sisal, coir, jute, pineapple, among others, can also be employed due to their renewability, biodegradability, light weight, reasonably high specific mechanical properties and low thermal conductivity properties [3]. For the selection of the resin (matrix), reasonable cost, durability under anticipated exposure conditions, adhesion to the aggregate, handling characteristics and ease of curing are very important considerations [4] and the most commonly used thermoset resins nowadays are epoxy, polyester, phenolic and vinyl ester.

Most composites are developed in order to exploit mechanical properties such as stiffness, strength and toughness, which are sometimes combined with other properties such as resistance to environmental conditions and high temperature. Furthermore, the interface is known to have large influence on the mechanical performance of the composites, especially with regard to strength and fracture toughness [5]. The final properties of the composite materials are dependent on several factors including the properties of the phases, size and orientation of the fibers and fiber content. In addition, their characteristics are highly susceptible to the manufacturing process whereby they were produced. Therefore, investigation of the process is of uttermost importance.

1.2 Composite Manufacturing

The Liquid Composite Molding (LCM) family, or simply liquid molding, comprises a number of manufacturing techniques for composite materials where a liquid thermosetting resin is injected into a closed mold containing a dry fibrous perform (the only existing absorbed moisture into the preform is due to water vapor in the air), impregnating the reinforcement. Some of the most popular LCM molding techniques are detailed below.

1.2.1 Resin Transfer Molding

Among the several advanced composite materials manufacturing processes, Resin Transfer Molding (RTM) is considered state-of-the-art technology for producing textile reinforced composite parts [6]. It was mainly developed for general aero-space applications in the 1980s, following some limited earlier applications such as aircraft radomes [7]. Nowadays, RTM is used by many industrial sectors such as automotive, aerospace, civil and sporting equipment.

RTM processing consists in positioning the preform (i.e., fibrous reinforcement preform in the shape of the piece) within the mold. This preform is a dry porous/ fibrous medium through which fluid flow occurs. The mold is closed and resin is injected into the mold cavity until the fibrous reinforcement is fully impregnated. After resin curing, the mold is opened and the part is removed from the mold. This part may still require finishing operations and/or to go through a post-curing process. The RTM mold must have at least one inlet port for resin injection and one outlet to enable, during resin injection, the output of air from inside the mold. However, in larger pieces, the presence of multiple inputs and outputs is common [8].

The main advantages of RTM include: non-expensive process equipment, excellent control on mechanical properties, closed mold process, low filling pressures, ability to incorporate metal inserts and attachments, possibility of producing large and complex parts with low labor costs [9], use of different weave and mat fabrics architectures, high fiber volume content, use of various types of resins, low injection pressure (10 bar max.), good dimensional tolerances, relatively low equipment cost, short production cycle times, process automation and low solvent emission. RTM molding is successfully used in medium-to-high-volume applications, but the need for tooling of relative high cost and difficulty of controlling fiber orientation limits their application in non-structural situations. RTM was identified with the potential to minimize/eliminate manufacturing difficulties involved in processing parts with medium and high fiber content, being a processing route with acceptable cost [10].

The fiber volume content and the fibrous preform architecture influence the mechanical properties of the composites manufactured by liquid molding. In many cases, the mechanical properties can be improved by increasing the fiber volume fraction. Higher fiber volume fraction, however, requires higher injection pressures, which may adversely affect the properties of the final molded component [11]. The injection pressure is between 1 to 10 bar, since higher values can cause fiber movement inside the mold (i.e., fiber wash), deflection of the mold and the appearance of voids in the composite. For a good impregnation and to reduce injection time, the resin must have low viscosity (typically $\leq$1 cP).

To avoid defects (i.e., voids due to partial impregnation) in the molded components, the resin should fill all spaces not occupied by the fibers and the air within the reinforcement must be totally eliminated [12]. Some works in the literature discuss a few techniques for void reduction in the composites through the study of different reinforcement architectures [12, 13] and injection parameters, especially fluid pressure and front flow orientation [14].

RTM is still considered underutilized and reproducibility of the finished parts [13] and the need to understand the resin flow through the preform are among the major barriers. For example, if the input and output channels are not properly positioned in the mould, some perform areas may not be reached by the resin being injected.

Indeed, understanding of the flow process is of uttermost importance for RTM and all liquid molding processes. However, it is very difficult to intuitively determine the behavior of the resin flow, so there is the need to perform a numerical analysis of the mold filling process, contemplating the development of a mathematical model and the simulation of this model [15]. The behavior of a fluid in a porous media has been studied for many years and numerical analysis is recognized as an effective method to analyze it and to propose optimum processing conditions [16]. However, to allow a proper prediction of the mold filling, these simulation codes need correct input values such as mold geometry, fluid viscosity, reinforcement porosity and reinforcement permeability [17].

1.2.2 Vacuum-Assisted Resin Transfer Molding

In the Vacuum-Assisted Resin Transfer Molding (VARTM) process, dry fabrics are impregnated with a thermoset resin using vacuum bag and a vacuum pump. VARTM is widely used for molding of complicated composite structures because of its high drapability and low cost, being applied to marine vessels and next generation airplanes [18], refrigerated cargo boxes, etc. [19]. The VARTM process offers several advantages over the competing composite manufacturing processes such as: (i) low tooling cost; (ii) low emission of volatile organic chemicals; (iii) process flexibility; (iv) low void-content in the molded parts, and

(v) potential for the production of relatively large (surface area ~150–200 m^2) and thick (0.1–0.15 m) composite parts, containing high reinforcement content (75–80 wt%) [19].

1.2.3 Infusion

In this process, the reinforcement is laid on the rigid mold with a layer of removable flow-enhancement medium, used to reduce filling time, along with a peel ply to facilitate disposal. Inlets and outlets are placed and the vacuum bag is sealed to the mold using a sealant tape. Air is extracted from the cavity to compact the preform between the vacuum bag and the mold and to subject the resin to a pressure differential that drives resin flow through the compacted preform. Infusion is a closed mold resin infiltration production process that presents many similarities with RTM. The most striking difference, however, is that the fully enclosing, two-part rigid mold used in RTM is replaced with a one-part rigid mold sealed with a vacuum bag [20].

1.2.4 RTM Light

In RTM Light, a variant of traditional RTM, a more flexible mold material (e.g., a composite) and an upper translucent component are used, allowing observation of the progress of the resin during impregnation. For LCM processes utilizing flexible molds such as RTM light, local laminate thickness evolution is governed by the compaction characteristics of the reinforcement, and improved compaction models are vital for enhanced filling simulations [21]. RTM light uses a peripheral vacuum to seal the mold, whereas a low positive pressure along with partial vacuum in the mold cavity drives resin flow inside the cavity industries [22]. RTM Light has been widely used in automotive, aerospace, civil and sports [9].

1.2.5 Seemann Composites Resin Infusion Molding

In this process, a network of grooves or channels is used to distribute the resin, and reduce flow resistance and filling time. The resin fills those grooves/channels first by vacuum pressure and then infuses into the fiber preform. The Seemann Composites Resin Infusion Molding (SCRIMP) is often used to produce large composites of complex geometry such as boat hulls and decks [23]. Considering factors like environmental protection, relatively inexpensive tooling and the ability to fabricate large parts with complex geometry, the SCRIMP process may be used to replace traditional open molding processes [24].

Now that we have discussed about composite materials and manufacturing of these materials, the next step is to address discussions about modeling and simulation applied to LCM process. Following a mathematical modeling to predict fluid flow and heat transfer inside porous media with particular reference to RTM process on an overall basis is presented.

2 Multiphase Flow and Heat Transfer Theory in RTM Process

2.1 General Mathematical Modeling in Porous Media

2.1.1 Macroscopic Conservation Equation

Non-isothermal fluid flow in porous media can be modeled on the basis of fluid velocity, pressure and temperature into the media. For very small velocity, many researchers have utilized the so called Darcy's model, which will be presented and discussed later.

In macroscopic scale, the porous model is a generalization of both, the Navier–Stokes equations and Darcy's law. The model retains both advection and diffusion terms and can therefore be used for flows in which such effects are important. In deriving the continuum equations, it is assumed that control volumes are infinitesimal and that surfaces are large in relation to the interstitial spacing of the porous medium but small relative to the scales in study. Thus, given control cells and control surfaces are assumed to contain both solid and fluid regions. This is a concept of continuum mechanics. This way, the particular details of the porous structure such as dispersion, tortuosity and interfacial transfer between phases at the pore level are neglected.

Dispersion is the transport due to hydrodynamic mixing of the interstitial fluid at the pore scale (spreading phenomena). At this point, it is important to distinguish between dispersion and diffusion. Diffusion occurs due to random molecular motion whereas dispersion is caused by fluctuations of the bulk flow. Then, dispersion occurs when convection occurs [25]. Discussion about this topic can be found in related works [25–29].

Tortuosity is a very important property of a porous medium. It is defined as the ratio between the straight line distance and the curved path length between two points [25]. Allen III et al. [27] defined this parameter as the average ratio between the distance traveled by the fluid in the microscopic pores of the medium and the net macroscopic distance traveled. Other definitions of tortuosity are given by Vallabh et al. [30] and Epstein [31]. According to the former author, higher tortuosity value indicates longer, more complicated and sinuous path, resulting in greater resistance to fluid flow.

In this context, by assuming that the continuum approach provides the necessary information, the general scalar advection–diffusion equation in a porous medium is given by [32]:

$$\frac{\partial(\rho\phi\Phi)}{\partial t} + \nabla \bullet (\rho\mathbf{K} \cdot U\Phi) - \nabla \bullet (\Gamma^{\Phi}\mathbf{K} \cdot \nabla\Phi) = \phi S \tag{1}$$

where: t is the time, ρ is the density, U is the velocity vector, Φ is the potential of transport, Γ^{Φ} represents effective transport property and S the source term.

In Eq. (1), the volume porosity, ϕ, at a point is the ratio between the volume V' available to flow in an infinitesimal control cell surrounding the point, and the physical volume, V, of the cell. Hence:

$$V' = \phi V \tag{2}$$

Nield and Bejan [33] and Bear [26] defined porosity as the fraction of the total volume of the porous media that is occupied by pores. However, disconnected pores can exist in the material, therefore the term effective porosity has been defined as the ratio between the volume occupied by connected pores and overall volume of the porous material.

In this formulation, it is assumed that the vector area available to flow, A', through an infinitesimal planar control surface of vector area A is given by:

$$A' = K.A \tag{3}$$

where: **K** is a symmetric second rank tensor called the area porosity tensor. The inner product of a symmetric rank two tensor and a vector produces a vector.

In addition to the usual production and dissipation terms, the source term S may contain transfer terms from the fluid to the solid parts of the porous medium.

2.1.2 General Initial and Boundary Conditions

For a problem that involves partial differential equations to be well-posed, the solution to the problem must exist, must be unique and must depend on the initial and/or boundary data [34, 35]. In general, the following initial and boundary conditions can be specified for the unknown Φ.

Initial conditions

$$\Phi = f_1(D, t = 0) \tag{4}$$

where D is the studied domain.

Boundary conditions

- Prescribed value (Dirichlet or first kind condition)

$$\Phi = f_2(S) \tag{5}$$

where S is the boundary of the domain.

- Prescribed flux (Newmann or second kind condition)

$$\frac{\partial \Phi}{\partial n} = f_3(S) \tag{6}$$

$$\frac{\partial \Phi}{\partial s} = f_4(S) \tag{7}$$

where n and s represent the normal and tangential directional vectors to the boundary S, respectively.

- Mixed (Robin or third kind condition)

$$f_5(S)\frac{\partial \Phi}{\partial n} + f_6(S)\Phi = f_7(S) \tag{8}$$

where $f_5(S) \neq 0$ and $f_6(S) \neq 0$ on the boundary S.

The choice of one or more conditions is dependent on the problem to be solved.

Solution of a partial differential equation applied to physical problems requires physically realistic initial and boundary conditions. Fluid flow equations in porous media comprise a complicated formulation that involves elliptic, parabolic and hyperbolic terms. These different behaviors affect the way like the initial and boundary conditions in the domain must be specified.

For a physical problem where fluid-fluid (liquid–gas) interface exists, the momentum flux and hence the velocity gradient in the liquid is negligible compared to that in the gas phase (because liquid viscosity is much greater than that of the gas) [36]. In addition, in transient problems the initial condition, i.e., the values of the variables, must be defined at the start of the process (t = 0).

2.2 Heat Transfer and Fluid Flow in RTM

The reinforcement (porous) media inside the mold is initially filled with air. Resin is injected into the mold through one or more points (inlets) and air is forced out of the mold through one or more outlets. In standard RTM, the use of a single resin inlet and a single air/resin outlet is common. Resin injection through the reinforcement media can be mathematically described as a viscous fluid moving through a porous medium. The problem consists of two inviscid fluids (resin + air) flowing through a porous media contained inside the mold cavity. Different approaches can be used to model the problem. In the present work, three formulations will be discussed and compared:

1. Multifluid model;
2. Control Volume/Finite Element (CV/FE) based model;
3. Volume of Fluid (VOF) model.

2.2.1 Multifluid Model

Fluid Flow Model

From the general equation, the following macroscopic equations can be derived:

1. Mass conservation equation:

$$\frac{\partial}{\partial t}(\phi\rho) + \nabla \bullet (\rho \mathbf{K} \bullet U) = 0 \tag{9}$$

2. Momentum conservation equation:

$$\frac{\partial(\rho\phi U)}{\partial t} + \nabla \bullet [\rho(\mathbf{K} \bullet U) \otimes U] = \nabla \bullet \{\mu_e \mathbf{K} \bullet [\nabla U + (\nabla U)^T]\} - \phi R \bullet U - \phi \nabla p + B \tag{10}$$

In these equations, **U** is the true velocity, **B** is the volumetric body force, p is the pressure, μ_e is the effective viscosity—either the laminar viscosity or a turbulent quantity, and **R** represents the resistance to flow in the porous medium. This is, in general, a symmetric positive definite second rank tensor, in order to account for possible anisotropies in the resistance.

Modeling of the fluid flow strives to obtain the velocity, pressure and volume fraction (saturation) distribution of the fluid phases within the porous medium. If there is a large resistance to flow, a large pressure gradient must be set up to balance that resistance. In an extreme situation, both terms on the right-hand side of Eq. (10) are large and of opposite sign, whereas the convective and diffusive terms on the left-hand side are negligible. Neglecting body force effects, Eq. (10) reduces to Eq. (11). Hence, in the limit of large resistance, an anisotropic version of Darcy's law is obtained, with permeability proportional to the inverse of the resistance tensor.

$$U = -R^{-1} \bullet \nabla p \tag{11}$$

Heat Transfer Model

Heat transfer into the porous media can be modeled with an equation similar to:

$$\frac{\partial}{\partial t}(\phi\rho H) + \nabla \bullet (\rho \mathbf{K} \bullet UH) = \nabla \bullet (\lambda_e \mathbf{K} \bullet \nabla T) + \phi S^H \tag{12}$$

where: H is the specific enthalpy, T is the temperature, λ_e is an effective thermal conductivity and S^H contains a heat source or sink to or from the porous medium.

A heat transfer model attempts to obtain the temperature distribution within the porous medium. It is important to understand the temperature profile during RTM process for different reasons: (a) temperature for resin cure is elevated, (b) the curing process is usually very exothermic, (c) resin viscosity is dependent on the temperature thus, influencing mold filling time, and (d) depending of the nature of the preform (e.g., vegetal fibers are very sensitive to heat), high temperature may affect the final composite.

Momentum sources can be used to model isotropic losses in porous regions, directional losses in porous regions, or other processes. For a better understanding, consider the general momentum equation (Eq. 8) for a fluid domain in the form:

$$\frac{\partial(\rho\tilde{U})}{\partial t} + \nabla \bullet (\rho\tilde{U}\tilde{U}) = -\nabla p + \rho g + \nabla \bullet \tau + S^M \qquad (13)$$

where $\boldsymbol{\tau}$ represents the stress tensor, and $\mathbf{g}$ is the gravity acceleration vector.

The momentum source, S^M, can be represented by:

$$S^M = -C_{LR}\tilde{U} - C_{QR}|\tilde{U}|\tilde{U} + S^{spec} \qquad (14)$$

where: C_{LR} is a linear resistance coefficient due to viscous losses, C_{QR} is a quadratic resistance coefficient referring to inertial losses; S^{spec} contains other momentum sources (which may be directional) and $\tilde{U}$ is superficial velocity (Darcy velocity). The relationship between true (interstitial) velocity and superficial velocity, at any point in the porous medium, is given as follows:

$$\tilde{U} = \phi U \qquad (15)$$

True velocity corresponds to the mean fluid velocity through the pores whereas superficial velocity is the velocity that a fluid would exhibit if no reduction in cross-sectional area of flow occurred due to the presence of the solid material. Usually, the superficial velocity is expressed by the ratio between volumetric flow rate and cross-sectional area of the porous medium. The particulate material within the porous medium reduces the area available for fluid flow, therefore true velocity is greater than superficial velocity.

On the basis of Eq. (11), Darcy's law is only valid when the pore Reynolds number is of the order of 1 [37]. When the microscopic Reynolds number is greater than that (flow occurs at high velocity), Darcy's equation is supplemented on the right-hand side by a quadratic non-linear term corresponding to form drag within the medium to correct the advection term (Forchheimer term) [25, 33, 38–40]. Other versions of the Darcy's equation exist; for example in Liu and Masliyah [25], Nield and Bejan [33], Magyari et al. [38] and Hsu [40].

The following generalized form of Darcy's law has been reported in the literature:

$$-\nabla p = \frac{\mu}{K_P}\tilde{U} + K_{EL}\frac{\rho}{2}|\tilde{U}|\tilde{U} \qquad (16)$$

where: μ is the dynamic viscosity; K_P is the permeability and K_{EL} is the empirical loss coefficient.

Permeability is a measure of how easily a single-phase fluid moves through a porous medium under the influence of a dynamic pressure gradient [41]. Permeability, defined by Ye et al. [42] as a measure of the resistance to fluid flow into a porous material, can be obtained through experiments or analytical predictions and different equations to calculate this parameter have been report. According to Hsu [40], Nield [43], Lage [44], Chen and Hsiao [45], Vafai and Amiri [46], and Nield and Bejan [33], permeability may be obtained as follows (Kozeny equation):

$$K_P = \frac{\phi^3 d_p^2}{\hat{a}(1-\phi)^2} \tag{17}$$

where: d_p is particle diameter and $\hat{a}$ is a constant to characterize the microscopic geometry of the porous material (morphology of the pores). When the porous medium is made of particles or fibers, $\hat{a}$ is usually set to 180 [47].

When two or more distinct fluid phases (in contact) simultaneously flow through a porous medium, the medium exhibits distinct permeability for the different fluids. This dependence reduces the effective permeability of each phase depending on the degree to which each fluid phase affects the flow of the other. Thus, fluid permeability is different of the porous medium permeability, introducing the concept of relative permeability. This parameter, which a value between 0–1, appears in Darcy equation multiplying the porous medium permeability and therefore alters this equation. More details about this topic can be found in Allen III et al. [27].

According to the cited literature, the empirical loss coefficient in Eq. (16) can be obtained as follows:

$$K_{EL} = \frac{2C_{FD}}{\sqrt{K_P}} \tag{18}$$

where: C_{FD} is a dimensionless form-drag constant.

Comparing Eq. (14) with Eq. (16), the following coefficients can be obtained:

$$C_{LR} = \frac{\mu}{K_P} \tag{19}$$

$$C_{QR} = \frac{K_{EL}\rho}{2} \tag{20}$$

Equations (19) and (20) use superficial velocity. However, data is often expressed in terms of true velocity and, in this case; the coefficients would be better represented by:

$$C_{LR} = \frac{\mu}{\phi K_P} \tag{21}$$

$$C_{QR} = \frac{K_{EL}\rho}{2\phi^2} \tag{22}$$

Considering $K_{EL} = 0$ and comparing Eq. (16) with Eq. (11), one can write:

$$R = \frac{\mu}{K_P} \tag{23}$$

when superficial velocity is used in Eq. (16), and

$$R = \frac{\mu}{\phi K_P} \tag{24}$$

when true velocity is used.

In practical terms, heat and mass transfer phenomena dominate resin impregnation and curing (exothermic chemical reaction) in RTM. Therefore, the process can be treated as non-isothermal reactive flow [48]. In Eq. (12), the source term represents the energy generation by volume unit that occurs into the porous material, e.g., heat release due to the curing process. Sometimes, impregnation of resin into the porous material is comparatively fast and resin curing has not yet started to alter the temperature of the fluid and, in such case, the RTM process may be considered isothermal.

In this chapter, the medium is considered rigid, i.e., relative positions of solid structure are constant and dilation or contraction within the porous material is negligible. Besides, no sink phenomenon (absorption of liquid by the pores of the fiber) occurs within the porous medium. More information about sink models can be found in Parnas and Phelan [49] and Pillai [50].

The transient formulation presented here can be applied to 1D, 2D and 3D situations and also to all phases present into the preform, for example, resin and air. Thus, there is a local thermal non-equilibrium condition between the phases. This mathematical treatment yields more accurate results than the homogeneous treatment under local thermal equilibrium conditions. More discussion on local thermal equilibrium condition can be found in previous works [25, 33, 40, 46, 51–59].

In addition, since the volumetric heating rate release due to curing is a function of the degree to which the chemical reaction occurred, a curing model is required for a more complete formulation and to obtain more realistic results. Discussion on the cure kinetic model can be found in Advani and Hsiao [48].

2.2.2 Control volume/Finite element Based Model

In this formulation, the problem is simplified by only considering the resin. The influence of the air flow is assumed negligible and is not taken into account in the model. Another simplification regards the fact that the transient problem of determining resin position as a function of time inside the mold is obtained by successive solutions of a simpler steady-state problem.

The resin velocity is formulated using Darcy's Law, which can be mathematically described as shown in Eq. (11). Typical polymeric resins used in the RTM process show a non-Newtonian behavior. However, for the purpose of this chapter (description of the basic formulation of the problem), a Newtonian approach is suitable, being, in fact, largely used in the numerical investigation of RTM [60–62]. Thus, assuming a Newtonian incompressible fluid we can write:

$$\nabla \bullet U = 0 \tag{25}$$

Combination of Eqs. (11), (23) and (25) results in

$$\nabla \bullet \left(\frac{K_P}{\mu} \nabla P \right) = 0 \tag{26}$$

The boundary conditions to be used with Eq. (26), (see Fig. 1), are given by

1. $p = P_0$ at the injection point;
2. $\partial p / \partial n = 0$ at the mold walls (n is the direction normal to the wall), and
3. $p = P_f$ at the resin front line, where P_f is the front line preset pressure which is usually zero (manometric).

The numerical solution of Eq. (26) is quite simple, but it must be solved many times during the simulation. Since Eq. (26) does not include a transient term, the transient problem is solved by finding a steady state solution for this equation for each time step. The solution procedure for a rectilinear flow can be summarized in five steps schematically shown in Fig. 1. The methodology used to track the resin front position is presented in Phelan [63], being called Flow Analysis Network (FAN).

The **first step** in solution consists on the discretization of the computational domain. Figure 1a shows a simple computational domain discretized with irregular quadrangular elements. In this method, besides the elements, it is also necessary to create control volumes (detailed in Fig. 1a) around the nodes. The control volumes are used to evaluate resin flow advance.

The **second step** consists in using Eq. (26) to calculate the pressure gradient between the inlet section and the flow front volumes. As shown in Fig. 1b, for a partially filled domain, a prescribed pressure P_0 is assigned to the inlet section (left wall) and $p = P_f$ is set to all elements not yet filled with resin. Neumann boundary condition ($\partial p / \partial n = 0$) is assigned to all others frontiers of the computational domain (top, bottom and right wall). The gray gradient in Fig. 1b is a schematic representation of the pressure field obtained with the solution of Eq. (26). Figure 1b also shows the flow front volumes and the adjacent (to the front) volumes.

Solution of Eq. (26) is usually obtained with a finite element method [10, 64]. However, other methods like finite volume and finite differences are also used. Detailing of such solutions will not be included here, but can be found in references [10, 64–67].

In the **third step**, the pressure gradient determined in step two is used to calculate the velocity field. In the simplified solution presented in Fig. 1c, the

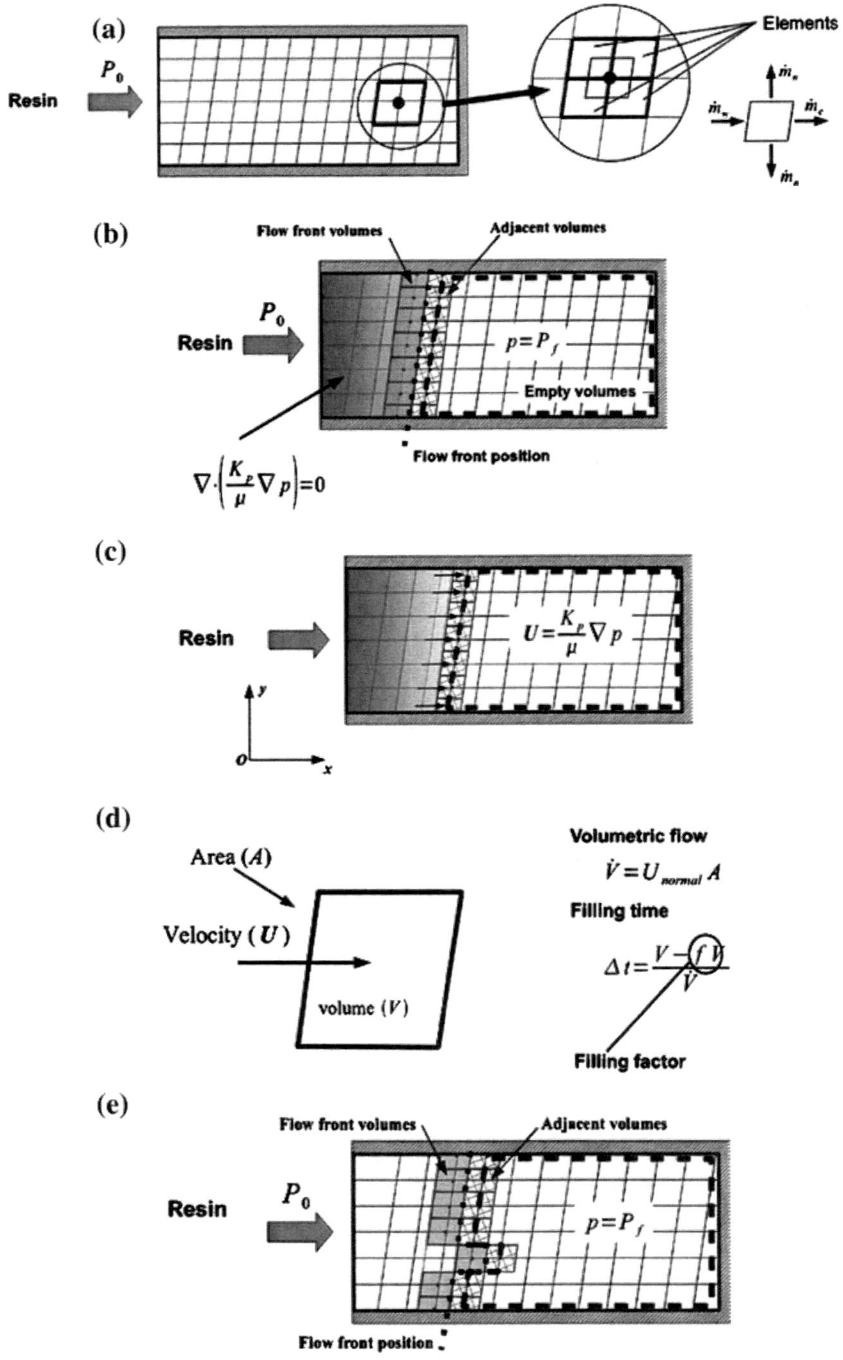

Fig. 1 CV/FE solution. **a** *Step 1* Discretization **b** *Step 2* Pressure field calculation **c** *Step 3* Velocity field calculation **d** *Step 4* Filling time calculation and **e** *Step 5* Determination of the new flow front

velocity field is parallel to the x axis, thus only the u component of the velocity vector is presented, however depending on the problem and geometry, both u and v components of **U** are calculated with Eq. (11). It is also possible for the resin to enter the control volume by two or more faces and leave the control volume through one or more faces when unstructured grids are used. The same analysis is valid for 3D solutions, though all three velocities components (u, v and w) must be considered in the formulation. In Fig. 1c, only the velocity vectors that are entering the volumes adjacent to the flow front are shown. These velocities are needed in the calculations, but it is possible to use Eq. (11) to calculate the flow velocity for all volumes.

The filling time for all volumes adjacent to the flow front is calculated in the **fourth step** (Fig. 1d). With the velocity field **U** and the volume V of each control volume, the filling time of volume i can be calculated by

$$\Delta t_i = \frac{V_i - V_i^f(t)}{\dot{V}(t)} \tag{27}$$

where: V_i is the overall volume of the control volume i, $V_i^f(t) = fV_i(t)$ is the filled volume at time t (current time step), and $\dot{V}_t(t) = U_{normal}A$ is the volumetric flow rate into volume i. Additionally, U_{normal} is the velocity component in the direction normal to the face with area A.

The advance in the flow front position can be calculated by determining the control volume that will be the first to be completely filled with resin. This can be done by using Eq. (27) to identify the control volume with minimum Δt. Calculation of Δt_{min} guarantees that at least one control volume is completely filled and that none of them will overflow (to receive more resin than its volume) in every solution step.

All other volumes adjacent to the flow front will be partially filled. The filling factor of these volumes for the next time step is calculated by

$$f_i(t + \Delta t) = \frac{V_i^f(t) - \Delta t_{\min} \dot{V}_i(t)}{V_i} \tag{28}$$

Finally, in the **fifth step**, the new flow front position is determined. All adjacent volumes (Fig. 1e) with f = 1 will be considered as flow front volumes in the next time step. Figure 1e shows a particular case where only one new volume is added to the flow front and the flow front line (dashed line) moves to include this volume.

The CV/FE formulation is quite easy to understand and of simple computational implementation. It has a major advantage in the fact that only one differential equation (Eq. 26) must be solved. Eq. (26) is linear and, even though it may present problems related to the discretization of complex geometries, its solution is fast and stable even for coarse grids. However, accuracy in flow front position and filling time predictions will increase for a larger number of elements in the discretization.

The fast solution of Eq. (26) will not always represent an overall fast solution of the problem. Depending on the discretization, it is possible that only one new volume is completely filled in every time step. When this happens, the number of time steps needed to solve the problem will be equal to the number of volumes in the grid (N_v), implying that Eq. (26) will be solved N_v times. In an attempt to increase the number of volumes completely filled in every time step, it is recommended to use grids with elements of the same size and, if possible, aligned with the resin flow. The use of rectangular (2D) or hexahedral (3D) elements is also recommended.

2.2.3 Volume of Fluid Model

In general, the VOF method [68] can be used to model multiphase flows with two or more inviscid fluids. In this formulation, all phases are well defined and the volume occupied by one phase cannot be occupied by the others. The volume fraction concept is used to represent the existence of different phases inside each control volume. All fractions are assumed continuous in space and time and their sum, inside each control volume, is equal to one. If the volume fraction of a particular phase inside a control volume is denoted as f_i, then the following three conditions are possible:

1. $f_i = 0$: the cell is empty with fluid of phase i;
2. $f_i = 1$: the cell is full with fluid of phase i;
3. $0 < f_i < 1$: the cell contains the interface between phase *i* and one or more other phases.

For the particular case of modeling resin and air flowing through a porous media (e.g., in RTM), only two phases are considered in the formulation. In the VOF method, only a single set of momentum and continuity equations is applied to both fluids, and the volume fraction of each fluid in every computational cell (control volume) is tracked throughout the domain. The model is composed of continuity, volume fraction and momentum equations as follows:

$$\frac{\partial \rho}{\partial t} + \nabla \bullet (\rho U) = 0 \tag{29}$$

$$\frac{\partial (\rho f)}{\partial t} + \nabla \bullet (\rho f U) = 0 \tag{30}$$

$$\frac{\partial (\rho U)}{\partial t} + \nabla \bullet (\rho U) = -\nabla p + \nabla \bullet \tau + \rho g + B \tag{31}$$

where: f is the volume fraction of resin and τ is the stress tensor.

Since a single set of equations is used for both phases, average properties for ρ and μ need to be defined. These properties can be approximated as [69]

$$\rho = f\rho_{resin} + (1 - f)\rho_{air} \quad (32)$$

$$\mu = f\mu_{resin} + (1 - f)\mu_{air} \quad (33)$$

The porous media (reinforcement) resistance to the flow is modeled by adding a source term to the standard momentum equation such as

$$B = -\frac{\mu}{K_P}U \quad (34)$$

Combining Eqs. (31) and (34), and considering that K_P is very small ($\sim 1 \times 10^{-10}$), it may be inferred that **B** is very large, allowing simplification of Eq. (31) as shown in Eq. (35). This equation may be written as Eq. (11) which is actually the Darcy's Law.

$$\nabla p = -\frac{\mu}{K_P}U \quad (35)$$

In contrast to the CV/FE formulation where only one partial differential equation must be solved, in the VOF model, four partial differential equations (five in a 3D formulation) must be simultaneously solved. Due to that, solution normally suffers from convergence issues, what brings grid refinement and time discretization constraints.

An important characteristic of the VOF method is that it can be used to solve problems whose computational domains are divided into two or more regions, some of them filled with the reinforcement and the others are empty. This is possible by setting the permeability close to infinity (high value) in Eq. (35) for the regions without reinforcement. This will result in a resistance term **B** in Eq. (31) equal to zero. This kind of solution is of particular interest for the solution of RTM-Light problems, due to the presence of empty channels, and cannot be performed with the CV/FE method.

2.3 Preform Geometry and Grid Generation

When numerical methods are applied to solve a particular physical problem, it is necessary to define the grid (mesh), i.e., the numerical representation of the physical domain in the computational domain where the conservation equation can be applied. When the domain presents a complex geometry, grid generation can be a very time-consuming task. However, a well-constructed grid improves the quality of the results.

According to Shafer [70] and Ferziger and Perić [71], the grid can be classified as: boundary-fitted, Cartesian and overlapping grid. Another well known classification is based on the logical arrangement of the control-volume as structured and unstructured grids (Fig. 2) [34, 70, 71]. Structured grids are characterized by a regular arrangement of grid control-volumes. For unstructured grids, generally

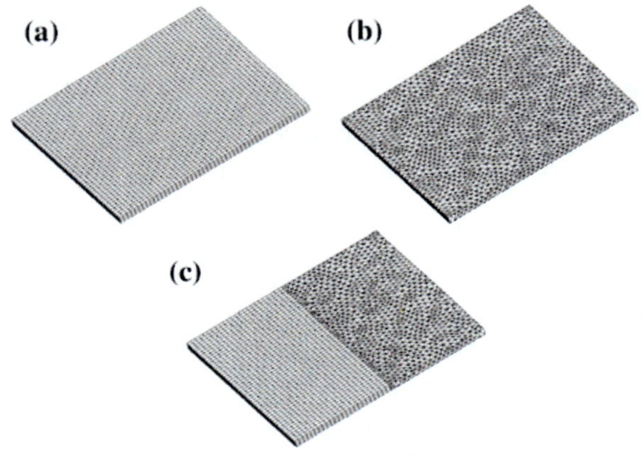

Fig. 2 Types of grids: **a** structured, **b** unstructured and **c** hybrid

used for very complex geometries, there is no regularity in the arrangement of the control-volumes (cells). Methods of grid generation can be found in related books [71, 72]. The grid can present uniform or non-uniform size distribution. For higher accuracy of a particular physical problem in regions where a large gradient is expected, the non-uniform grid size distribution is more effective [36].

For liquid composite molding, flow is transient, and mold and preform may present a very irregular shape. In this case, an adequate grid is essential to the accuracy of the solution, mainly when we apply a three-dimensional treatment. In addition, depending on the specific liquid molding technique and the particular mold design and construction being investigated, the mold may display deformation (change in relative position between the upper and lower parts of the mold) during infiltration. This brings significant challenges related to grid generation along the transient process and, because of that, researches usually consider the mold and grid to be rigid during RTM process. In spite of the importance, just a few works on this the research area are available [73–75].

3 RTM Applications: Numerical Computation and Experiments

3.1 RTM Experimentations

Luz [76] and Luz et al. [77] conducted several experiments of resin injection into a parallelepipedic porous preform (300 × 300 × 2 mm). The Newtonian fluid used in the experiments was a commercial vegetable (soybean) oil with density of 914 kg/m^3. The fluid viscosity at 23 °C (same temperature of the experiments), measured in a Brookfield viscometer HBDV-II + C/P with the S40 spindle, was 37.1 cP. As the fibrous reinforcement, plain-weave (0/90) E-glass fiber cloth from Owens Corning (300 g/m^2) was used.

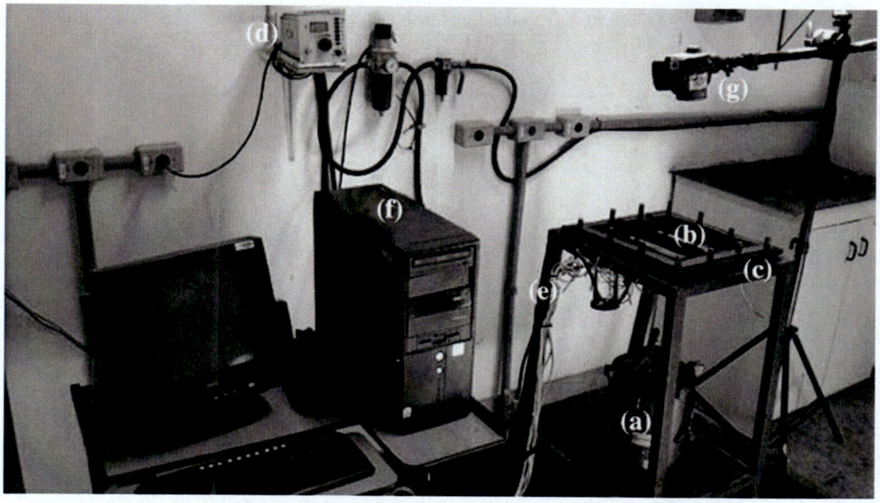

Fig. 3 Radial RTM system available at UFRGS (Brazil): **a** pressure pot, **b** strengthened-glass top mold, **c** steel bottom mold, **d** pressure controller, **e** pressure transducers, **f** data acquisition system and **g** camera

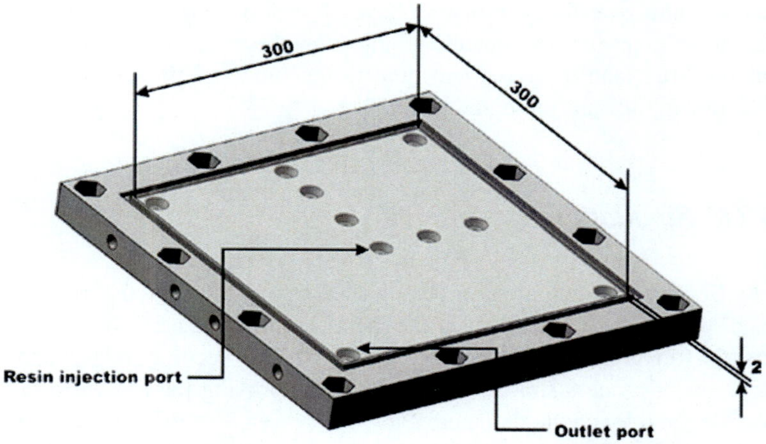

Fig. 4 Geometry of the injection mold with the dimensions of the studied area of the mold (300 × 300 × 2 mm

The radial experiments were conducted in a RTM system (shown in Fig. 3) with one injection point and four outlet points (vents) (Fig. 4). For radial infiltrations, data comprising the time required (t_{ff}) for the fluid which passes through an inlet to fill a region of radius r_{ff} in the mold may be used to calculate permeability of the medium according to Eq. (36). This equation may be applied only until the flow reaches the wall of the mold (t_{wall}).

Table 1 Permeability and mold filling time for several experiments

Case	V_f (%) (pre-set)	V_f (%) (real)	P_{inj} (bar) (pre-set)	P_{inj} (Pa) (real)	K_P ($\times 10^{-11}$ m^2)	t_{wall} (s)
P01V20	20	24.0	0.10	7420	136.8	130
P01V30	30	30.3	0.10	7780	30.9	420
P01V40	40	43.4	0.10	7460	4.4	2280
P03V40	40	41.7	0.30	28570	3.0	860
P05V40	40	41.6	0.50	48260	2.5	700
P07V40	40	42.2	0.70	67240	2.2	640
P09V40	40	41.7	0.90	88450	2.0	560
P10V40	40	41.9	1.00	99800	3.7	380

$$K = \frac{\phi\mu}{2P_{inj}t_{ff}}\left[r_{ff}^2 \ln\left(\frac{r_{ff}}{r_{inj}}\right) - \frac{1}{2}\left(r_{ff}^2 - r_{inj}^2\right)\right] \quad (36)$$

where: r_{inj} is the radius of the injection port, ϕ is the porosity of the porous fibrous media ($\phi = 1-V_f$, V_f is the fiber volume fraction) and P_{inj} is the pressure (gauge) in the resin injection gate. During infiltration, position (radius) of the flow front was recorded with a camera at different times.

The injection pressure of the resin in the mold was monitored with pressure transducers during each experiment. Several experiments were performed by varying the volumetric fiber content and injection pressure. The conditions of each experimental run, and the observed mold filling time and the calculated permeability data (Eq. 36) are presented in Table 1.

3.2 RTM Simulations

In the LCM process, simulations are being used mainly to verify the trial-and-error approach, which is still prevalent in the manufacturing industry. However, as the composite structures being manufactured by LCM become larger and more complex, use of process simulation may aid in improving the process design and increase the yield by counterbalancing any unforeseen disturbances that may arise during the impregnation phase [48].

For the application of the models and equations presented, it is necessary the use of a software to solve them. In modeling by RTM, simulation software are used to (i) predict the resin injection profile and then be able to evaluate the filling time, that should remain below the resin curing time, (ii) predict impregnation deficient (dry) points, and (iii) determine the more suitable injection and outlet points of air/resin.

Flow simulation software may be used, increasing the confidence in the mold design and fabrication process parts. There are different software dedicated exclusively to study RTM process such as PAM-RTM from ESI Group, the RTM-WORX from Polywork, the LIMS from the University of Delaware that are

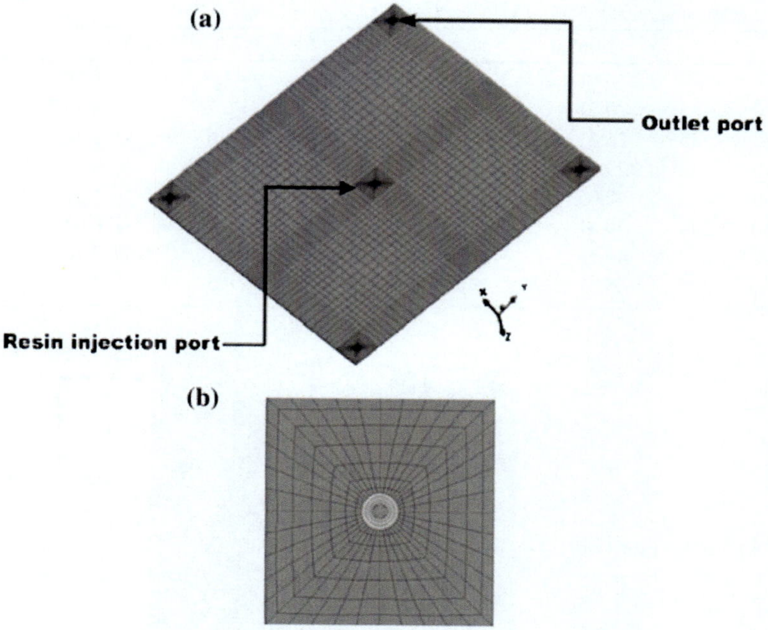

Fig. 5 **a** The grid used in the ANSYS CFX® simulations, **b** Details of the outlet port

very commonly used by industry for having a simple usage and focused only on this process. However, dedicated software may end up limiting some more profound studies on the process behavior in adverse situations. In fact, it is possible to use commercial software, not only dedicated to the study of composite processing, but computational tools that allow customization of the problem being studied. However, adapting such software to solve specific problems of RTM can be quite laborious. The non-dedicated commercial software commonly used to study CFD are Ansys CFX® and FLUENT, both from ANSYS and Abaqus CFD from Simula Abaqus.

Herein, RTM modeling was applied to the manufacturing of a composite with paralellepipedic shape by considering the isothermal system, i.e., no heat transfer. The geometry and dimensions of the mold are shown in Fig. 4. For the 3D numerical analysis, it was used two commercial softwares: ANSYS CFX® (non RTM-dedicated) and PAM-RTM (RTM-dedicated).

3.2.1 ANSYS CFX® Model

In the ANSYS CFX® model, the system is composed of air, resin and solid material (glass fiber). Simulations were performed in a structured 3D mesh with 17,532 elements and 23,876 nodes. It was built using ANSYS ICEM CFD software version 12.0.1 and time step 0.05 s. Figure 5 illustrates the mesh used in the

Table 2 Constants used in Eq. (37)

Case	t_e (s)	a_1 (mbar $s^{-0.25}$)	a_2 (mbar)	a_3 (s^{-1})	a_4 (-)	a_5 (s^{-1})
P01V20	40	13.61	6.01×10^{-4}	7.09	1.29	0.62
P01V30	85	20.21	1.62×10^{-4}	6.76	1.55	0.56
P01V40	35	25.21	2.74×10^{-5}	8691.63	1232.42	637.45
P03V40	132	29.19	1.30×10^{-1}	7.02	9.93	0.87
P05V40	168	42.05	3.68×10^{-2}	1.42	1.39	0.14
P07V40	240	30.41	5.29×10^{-1}	−72.28	-178.62	-9.67
P09V40	294	45.20	0.18	−3099.14	−6960.16	−352.70
P10V40	325	45.87	3.99	4.54	32.35	0.75

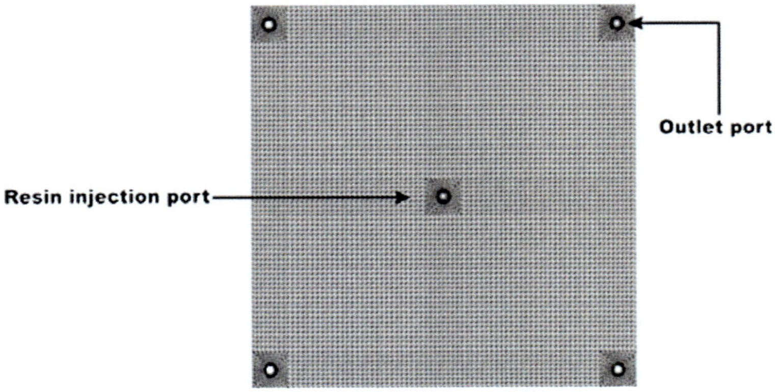

Fig. 6 The grid used in the PAM-RTM simulations

simulations. From the collected pressure data, a non-linear regression was made using the Mathematica® software, yielding a pressure equation as a function of the process time (Eq. 37):

$$P_{inj}(t) = \begin{cases} P_0 + a_1 t^{0.25} + a_2 \exp\left(\frac{a_3 t}{a_4 + a_5 t}\right), & for\ 0 \leq t \leq t_e \\ P_e, & for\ t > t_e \end{cases} \quad (37)$$

where: P_0 is the atmospheric pressure, P_e is the final pressure (stable pressure) achieved in the experiment and t_e is the instant of time where P is now considered P_e. The parameters of Eq. (37) are presented in Table 2.

3.2.2 PAM-RTM Model

In the PAM-RTM model, the system is composed of resin and solid material (glass fiber). It was used a 2D mesh with 14,819 elements and 16,775 nodes. Figure 6 illustrates the mesh used in the simulations. PAM-RTM, unlike the ANSYS CFX®, is a commercial software dedicated to the study of RTM process, so it was not

Table 3 Boundary conditions used

Model	Boundary	f_a	f_r	P (mbar)	U_a (m/s)	U_r (m/s)
Multifluid (ANSYS CFX®)	Inlet	0	1	P(t), Eq. (37)	–	–
	Outlet	–	–	1013.25	–	–
	Walls	–	–	–	0	0
CV/FE (PAM-RTM)	Inlet	–	1	P(t), Collected data table (manometric)	–	–
	Flow front	–	–	0 (manometric)	–	–
	Walls	–	–	$\frac{\partial P}{\partial n} = 0$	–	–

f_a and f_r are the volumetric fractions of air and resin, respectively; $\boldsymbol{U_a}$ and $\boldsymbol{U_r}$ velocity vector of air and resin, respectively; P is the static pressure at the inlet and outlet boundaries of the preform; n is the normal direction to the boundaries

necessary make adjustments and create regression equations for the pressure data. The collected pressure data was directly implemented in the software.

3.2.3 Initial and Boundary Conditions

In all simulations, the preform was initially considered as having pressure (P_i), temperature (T_i), and air saturation (S_{ai}) homogeneously distributed throughout the mold with the following values: P_i = 1013.23 mbar, T_i = 300 K and S_{ai} = 1.0 (consequently, resin saturation is S_{ri} = 0.0). The adopted boundary conditions are presented in Table 3.

3.3 Results Analysis

The variation of the volume fraction of fluid in the experiments and simulations in ANSYS CFX® and PAM-RTM was analyzed. Figure 7 shows a comparison of the evolution of the volume fraction of fluid in the case P09V40 obtained by experiment and numerical simulation generated by the ANSYS CFX® and the PAM-RTM at three different times 60, 150 and 360 s.

It is observed that the flow front moves in radial direction in the experiments and in the simulations the same pattern is observed, but the impregnated region is further defined and there is a regular circle. This is because in the simulations is considered a homogeneous distribution of fibers in the cloth, and constant properties (permeability, porosity and viscosity), which are not always the cases, experimentally. Where there is a local variation of the concentration of fibers generates a forward heterogeneous flow. This problem can be corrected by including in the simulation permeability as a function of saturation resin, for example, or yet porosity as a function of position into the preform.

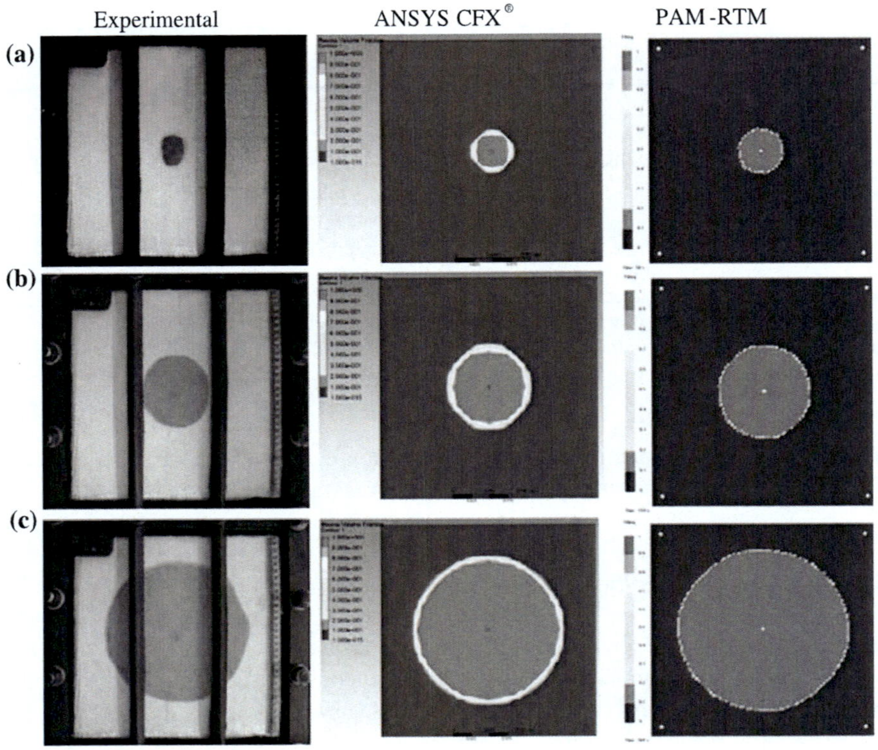

Fig. 7 Evolution of the fluid flow front (resin) in the case P09V40 in **a** 60 s, **b** 150 s and **c** 360 s

It is shown in Fig. 8 a comparison between the r_{ff} experimental, the r_{ff} predicted by ANSYS CFX® and PAM-RTM at the moment t_{wall} for the experimental cases P07V40, P09V40 and P10V40. Table 4 presents a comparison between the t_{wall} experimental and t_{wall} predicted in ANSYS CFX® and PAM-RTM for all the simulated cases, showing their percentage deviations compared to t_{wall} experimental.

Comparing the t_{wall}, it is observed that there is no errors exceeding 10 %, indicating a strong relationship between the numerical solution of the equations of porous medium and the experimental procedure. Considering the cases P01V20, P01V30 and P01V40 observes a relation between the error and the change in V_f, because as V_f is increased while keeping P_{inj} constant, the error is increased. This behavior is not observed in other cases, when increasing the P_{inj} and V_f remains constant.

As seen in Table 1, the K_P highest values are found in these same three cases, thus the deviations of comparing the numerical simulations may be related to an experimental errors, as in the K_P determining, which is a crucial property to the success of the numerical simulation. Therefore, the deviations indicated in Table 4 can be linked to the experimental procedure and, for example, the following

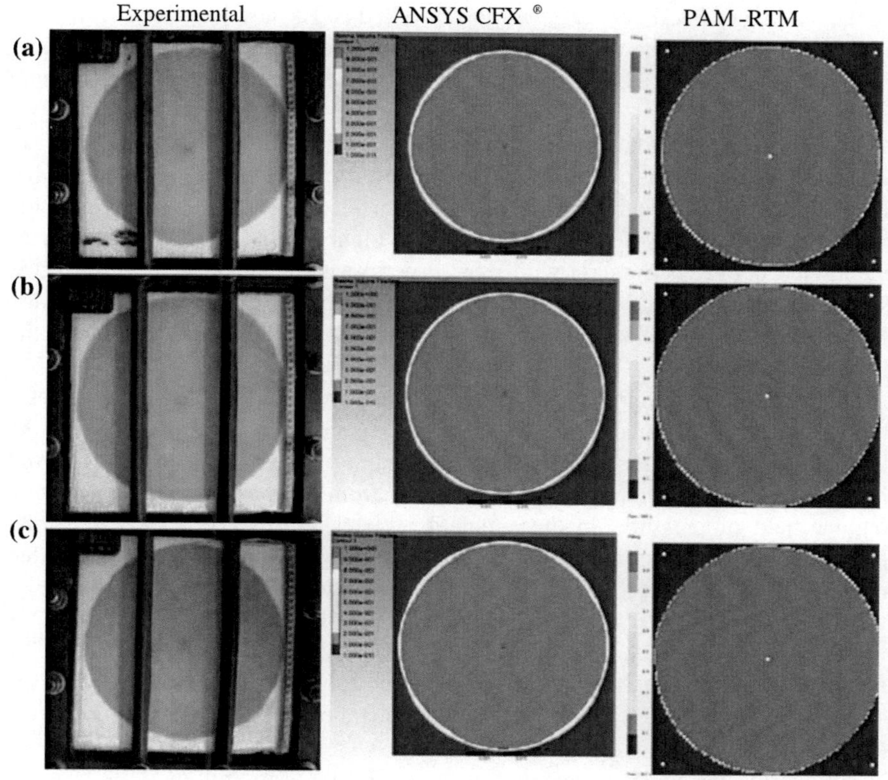

Fig. 8 Comparison between r_{ff} experimental and r_{ff} simulated by ANSYS CFX® and PAM-RTM at the moment t_{wall} experimental for the cases **a** P07V40, **b** P09V40 and **c** P10V40

Table 4 Comparison between the t_{wall} experimental and t_{wall} simulated in ANSYS CFX® and PAM-RTM for all the simulated cases

Case	Experimental	ANSYS CFX®		PAM-RTM	
	t_{wall} (s)	t_{wall} (s)	Deviation (%)	t_{wall} (s)	Deviation (%)
P01V20	130	138	6.15	131	0.77
P01V30	420	450	7.14	430	2.38
P01V40	2280	2506	9.91	2389	4.78
P03V40	1000	1098	9.80	1031	3.10
P05V40	700	768	9.71	722	3.14
P07V40	640	700	9.38	648	1.25
P09V40	560	588	5.00	568	1.43
P10V40	380	405	6.58	378	−0.53

sources of potential experimental errors, may be mentioned (i) the lack of homogeneity in the cloth properties, (ii) the operator sensitivity to start the time measurements and (iii) the readings of the radius of the flow front. The deviation of the reading of PAM-RTM t_{wall} for the case P10V40 accused a value below zero, indicating that this time was less than the experimental time, as this was the only case where this occurred, it is evident that these errors are related to experimental errors.

By observing the deviations shown in Table 4, it is noted that they are higher in ANSYS CFX® simulations in all cases, ranging from 5.0 to 9.91 %, while in the PAM-RTM these deviations were in the range −0.53 and 4.78 %. In both, the higher deviations was found in the case of higher t_{wall}, the P01V40 case, as mentioned before, is the case with most experimental error. Moreover, since the slightest deviation found for ANSYS CFX® was in the case P09V40 and for PAM-RTM in the P10V40 case, there is no apparent relationship of deviation and the time of molding.

Figure 9 shows the pressure distribution along the mold in the instant t_{wall} for the case P09V40 simulated in ANSYS CFX® (Fig. 9a) and simulated with the PAM-RTM (Fig. 9b). As expected, the maximum pressure is located in the fluid injection port, and decreases as it moves away from this point, establishing into a regular field of pressure. In the simulation by using ANSYS CFX®, the total pressure corresponding to the sum of P_{inj} plus the P_0, so the high pressure value indicated in Fig. 9a; in PAM-RTM, the pressure gauge is displayed. The pressure field seen in a transverse plane is shown in Fig. 10 generated by the ANSYS CFX®. It is noteworthy that this level of detail can not be observed in the 2D model of the PAM-RTM. The pressure in most part of the mold is 1.01 bar, 42.19 % less than the P_{inj} (1.89 bar) showing quantitatively the pressure drop over the mold.

Comparing the images generated by ANSYS CFX® and by PAM-RTM in all images presented above, is observed a better quality in the ANSYS CFX® images, indicating a concern in smooth the variation of values in the range of colors on this software. Therefore, analyse the results is easier because it's possible to observe more details in these regions, especially in the analysis of fluid volumetric fraction.

4 Concluding Remarks

In this chapter advanced topics related to composite manufacturing by RTM process, including multiphase mathematical modeling (fluid flow and heat transfer in porous media) and computer simulation is presented and discussed. Appropriate numerical analysis on the basis of the ANSYS CFX® and PAM-RTM commercial softwares, and experiments of RTM process with radial infiltration of resin into a mould is performed. Different examples dealing with 2D and 3D treatment illustrated the effectiveness, versatility and performance of the commercial codes to the solution of RTM practical problems. Since the formation and migration of gas bubbles near the flow front occurs in practical situation of RTM process, the mathematical modeling presented herein predict the distribution of saturation, pressure and velocity of the fluid phases (resin and air) inside the mold in any

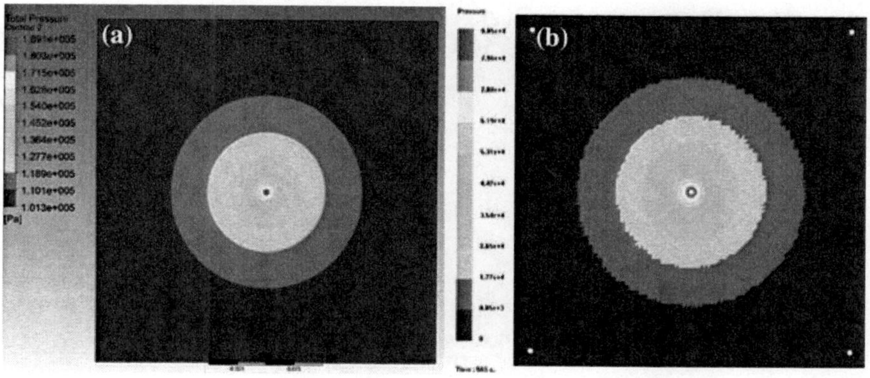

Fig. 9 Total pressure field from the case P09V40 in t_{wall} simulated in **a** ANSYS CFX® and **b** PAM-RTM

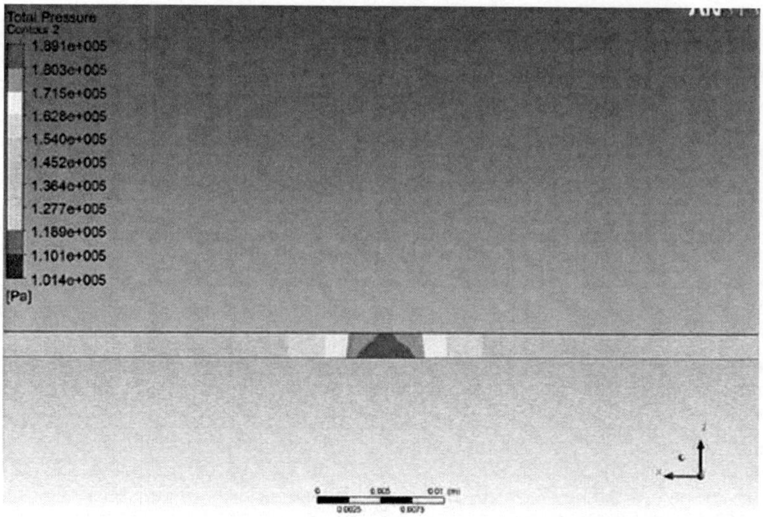

Fig. 10 Total pressure field simulated in ANSYS CFX® in a transversal plane shown the injection port for the case P09V40

moments. According to observations, the ANSYS CFX® software requires greater training in their applications more than the PAM RTM software.

The fluid flow governing equations applied to porous media proved again to be suitable to the study of the RTM process. The developed numerical analysis showed to be very reliable, with errors below 10 % in the measurement of the flow front radius for the various studied cases. The numerical simulation enabled greater understanding of the behavior of the fluid flow during the RTM process, allowing the observation of events that are not possible only by performing laboratory experiments.

Acknowledgments The authors thank to the Brazilian agencies CNPq (Conselho Nacional de Desenvolvimento Científico e Tecnológico) and CAPES (Coordenação de Aperfeiçoamento de Pessoal de Nível Superior) for supporting this work, and are also grateful to the authors of the references in this chapter that helped in the improvement of quality. Sincere thanks to Prof. João M.P.Q. Delgado (Editor) by the opportunity given to present our research in this book.

References

1. Military Handbook—MIL-HDBK-17-1F: Composite Materials Handbook, vol. 1—Polymer Matrix Composites Guidelines for Characterization of Structural Materials. U.S. Department of Defense (2002)
2. Gay, D., Suong, V.H., Tsai, S.W.: Composite Materials: Design and Applications. CRC Press, Boca Raton (2003)
3. Liu, K., Takagi, H., Osugi, R., Yang, Z.: Effect of physicochemical structure of natural fiber on transverse thermal conductivity of unidirectional abaca/bamboo fiber composites. Compos. Part A (2012, in press)
4. Asaad, J.N., Tawfik, S.Y.: Polymeric composites based on polystyrene and cement dust wastes. Mater. Des. **32**, 5113–5119 (2011)
5. Gamstedt, E.K., Sandell, R., Berthold, F., Pettersson, T., Nordgren, N.: Characterization of interfacial stress transfer ability of particulate cellulose composite materials. Mech. Mater. **43**, 693–704 (2011)
6. Morren, G., Bottiglieri, M., Bossuyt, S., Sol, H., Lecompte, D., Verleye, B., Lomov, S.V.: A reference specimen for permeability measurements of fibrous reinforcements for RTM. Compos. Part A **40**, 244–250 (2009)
7. Potter, K.D.: The early history of the resin transfer moulding process for aerospace applications. Compos. Part A **30**, 619–621 (1999)
8. Lawrence, J., Simacek, P., Advani, S.: New advances in modeling and simulation of liquid composite molding processes. In: Proceedings of the International SAMPE Symposium and Exhibition, pp. 1–12, Long Beach (2006)
9. Haider, M., Hubert, P., Lessard, L.: An experimental investigation of class a surface finish of composites made by the resin transfer molding process. Compos. Sci. Techol. **67**, 3176–3186 (2007)
10. Rudd, C.D., Lac, K.N., Mangin, C.G.E.: Liquid moulding, structural reaction injection moulding and related processing techniques. Woodhead Publishing Limited, UK (1997)
11. He, F., Wang, Y., Huang, Y., Wan, Y.: Preparation and mechanical properties of 3-D braided glass fiber reinforced light-cured resin composites. Mater. Lett. **60**, 3339–3341 (2006)
12. Leclerc, J.S., Ruiz, E.: Porosity reduction using optimized flow velocity in resin transfer molding. Compos. Part A **39**, 1859–1868 (2008)
13. Li, J., Zhang, C., Liang, R., Wang, B.: Statistical characterization and robust design of RTM processes. Compos. Part A **36**, 564–580 (2005)
14. Gourichon, B., Deléglise, M., Binetruy, C., Krawczak, P.: Dynamic void content prediction during radial injection in liquid composite molding. Compos. Part A **39**, 46–55 (2008)
15. Gokce, A., Chohra, M., Advani, S.G., Walsh, S.M.: Permeability estimation algorithm to simultaneously characterize the distribution media and the fabric preform in vacuum assisted resin transfer molding process. Compos. Sci. Techol. **65**, 2129–2139 (2005)
16. Han, K., Jiang, S., Zhang, C., Wang, B.: Flow modeling and simulation of SCRIMP for composites manufacturing. Compos. Part A **31**, 79–86 (2000)
17. Hoes, K., Dinescu, D., Sol, H., Vanheule, M., Parnas, R.S., Luo, Y., Verpoest, I.: Compos. Part A **33**, 959–969 (2002)
18. Matsuzaki, R., Kobayashi, S., Todoroki, A., Mizutani, A.: Full-field monitoring of resin flow using an area-sensor array in a VaRTM process. Compos. Part A **42**, 550–559 (2011)

19. Grujicic, M., Chittajallu, K.M., Walsh, S.: Non-isothermal preform infiltration during the vacuum-assisted resin transfer molding (VARTM) process. Appl. Surf. Sci. **245**, 51–64 (2005)
20. Correia, N.C., Robitaille, F., Long, A.C., Rudd, C.D., Simacek, P., Advani, S.G.: Analysis of the vacuum infusion moulding process: I. Analytical formulation. Compos. Part A **36**, 1645–1656 (2005)
21. Buntain, M.J., Bickerton, S.: Modeling forces generated within rigid liquid composite molding tools. Part A: experimental study. Compos. Part A **38**, 1729–1741 (2007)
22. Garay, A.C., Heck, V., Zatera, A., Amico, S.C.: Influence of the calcium carbonate (CaCO3) in the infiltration process and properties of composites molded by RTM and RTM light. In: 19° Brazilian Congress of Materials Sciences and Engineering, Campos do Jordão, Brazil (2010) (In Portuguese)
23. Han, K., Jiang, S., Zhang, C., Wang, B.: Flow modeling and simulation of SCRIMP for composites manufacturing. Compos. Part A **31**, 79–86 (2000)
24. Hudson, A.: Vacuum-assisted RTM increasingly popular. Compos. Techol. March/April, 24–30 (1996)
25. Liu, S., Masliyah, H.: Dispersion in porous media. In: Vafai, K. (ed.) Handbook of Porous Media, 2nd edn. CRC Press, Boca Raton (2005)
26. Bear, J.: Dynamics of Fluid in Porous Media. Dover Publications Inc, New York (1972)
27. Allen III, M.B., Behie, G.A., Trangenstein, J.A.: Multiphase Flow in Porous Media. Springer-Verlag, Berlin (1988)
28. Delgado, J.M.P.D.: Longitudinal and transverse dispersion in porous media. Trans. IChemE Part A, Chem. Eng. Res. Des. **85**(A9), 1245–1252 (2007)
29. Hunt, A.G., Skinner, T.E., Ewing, R.P., Ghanbarian-Alavijeh, B.: Dispersion of solute in porous media. Eur. Phys. J. B **80**(4), 411–432 (2011)
30. Vallabh, R., Banks-Lee, P., Seyam, A.: New approach for determining tortuosity in fibrous porous media. J. Eng. Fiber Fabr. **5**(3), 7–15 (2010)
31. Epstein, N.: Tortuosity and the tortuosity factor in flow and diffusion through porous media. Chem. Eng. Sci. **44**(3), 777–779 (1989)
32. ANSYS Inc.: CFX®-Theory manual (2010)
33. Nield, D.A., Bejan, A.: Convection in Porous Media. Springer-Verlag, New York (2006)
34. Tannehill, J.C., Anderson, D.A., Pletcher, R.H.: Computational Fluid Mechanics and Heat Transfer. Taylor & Francis, Philadelphia (1997)
35. Fletcher, C.A.J.: Computational Techniques for Fluid Dynamics: Fundamental and General Techniques, vol. 1. Springer-Verlag, Berlin (1991)
36. Majumdar, P.: Computational Methods for Heat and Mass Transfer. Taylor & Francis, New York (2005)
37. Mojtabi, A., Charrier-Mojtabi, M.: Double-diffusive convection in porous media. In: Vafai, K. (ed.) Handbook of Porous Media, 2nd edn. CRC Press, Boca Raton (2005)
38. Magyari, E., Rees, D.A.S., Keller, B.: Effect of viscous dissipation on the flow in fluid saturated porous media. In: Vafai, K. (ed.) Handbook of Porous Media, 2nd edn. CRC Press, Boca Raton (2005)
39. Hugo, J., Topin, F.: Metal foams design for heat exchangers: structure and effectives transport properties. In: Delgado, J.M.P.Q. (ed.) Heat and Mass Transfer in Porous Media. Springer-Velag, Berlin (2012)
40. Hsu, C.: Dynamic modeling of convective heat transfer in porous media. In: Vafai, K. (ed.) Handbook of Porous Media, 2nd edn. CRC Press, Boca Raton (2005)
41. McKibbin, R.: Mathematical modeling for heat and mass transport in geothermal systems. In: Ingham, D.B., Pop, I. (eds.) Transport Phenomena in Porous Media. Elsevier, Oxford (1998)
42. Ye, X., Qiu, J., Zhang, C., Liang, R., Wang, B.: A finite element-based heuristic estimation of local perform permeability for resin transfer molding. Transp. Porous Media **76**, 247–263 (2009)
43. Nield, D.A.: Modeling fluid flow in saturated porous media and at interfaces. In: Ingham, D.B., Pop, I. (eds.) Transport phenomena in Porous Media II. Elsevier, Oxford (2002)

44. Lage, J.L.: The fundamental theory of flow through permeable media from Darcy to turbulence. In: Ingham, D.B., Pop, I. (eds.) Transport Phenomena in Porous Media. Elsevier, Oxford (1998)
45. Chen, C.K., Hsiao, S.W.: Transport phenomena in enclosed porous cavities. In: Ingham, D.B., Pop, I. (eds.) Transport Phenomena in Porous Media. Elsevier, Oxford (1998)
46. Vafai, K., Amiri, A.: Non-Darcian effects in confined forced convection flows. In: Ingham, D.B., Pop, I. (eds.) Transport Phenomena in Porous Media. Elsevier, Oxford (1998)
47. Merrikh, A.A., Lage, J.L.: From continuum to porous-continuum: the visual resolution impact on modeling natural convection in heterogeneous media. In: Ingham, D.B., Pop, I. (eds.) Transport Phenomena in Porous Media III. Elsevier, Oxford (2005)
48. Advani, S.G., Hsiao, K.: Transport phenomena in liquid composites molding processes and their roles in process control and optimization. In: Vafai, K. (ed.) Handbook of Porous Media, 2nd edn. CRC Press, Boca Raton (2005)
49. Parnas, R.S., Phelan Jr, F.R.: The effect of heterogeneous porous media on mold filling in resin transfer molding. SAMPE Q. **22**(2), 53–60 (1991)
50. Pillai, K.M.: Modeling the unsaturated flow liquid composite molding processes: a review and some thoughts. J. Compos. Mater. **38**(23), 2097–2118 (2004)
51. Cheng, P., Hsu, C.: Heat conduction. In: Ingham, D.B., Pop, I. (eds.) Transport Phenomena in Porous Media. Elsevier, Oxford (1998)
52. Wang, C.Y.: Modeling multiphase flow and transport in porous media. In: Ingham, D.B., Pop, I. (eds.) Transport Phenomena in Porous Media. Elsevier, Oxford (1998)
53. Bradean, R., Heggs, P.J., Inghan, D.B., Pop, I.: Convective heat flow from suddenly heated surfaces embedded in porous media. I. In: Ingham, D.B., Pop, I. (eds.) Transport Phenomena in Porous Media. Elsevier, Oxford (1998)
54. Kaviany, M.: Principles of Convective Heat Transfer, 2nd edn. Springer, New York (2001)
55. Rees, D.A.S., Pop, I.: Local thermal non-equilibrium in porous medium convective. In: Ingham, D.B., Pop, I. (eds.) Transport Phenomena in Porous Media III. Elsevier, Oxford (2005)
56. de Lemos, M.J.S.: The double-decomposition concept for turbulent transport in porous media. In: Ingham, D.B., Pop, I. (eds.) Transport Phenomena in Porous Media III. Elsevier, Oxford (2005)
57. de Lemos, M.J.S.: Mathematical modeling and applications of turbulent heat and transfer in porous media. In: Vafai, K. (ed.) Handbook of Porous Media, 2nd edn. CRC Press, Boca Raton (2005)
58. Nield, D.A., Kuznetsov, A.V.: Heat transfer in bidisperse porous media. In: Ingham, D.B., Pop, I. (eds.) Transport Phenomena in Porous Media III. Elsevier, Oxford (2005)
59. de Lemos, M.J.S.: Turbulence in Porous Media: Modeling and Applications. Elsevier, Amsterdan (2006)
60. Hu, J.L., Liu, Y., Shao, X.M.: Study on void formation in multi-layer woven fabrics. Compos. Part A Appl. Sci. Manuf. **35**, 595–603 (2004)
61. Shojaei, A.: A numerical study of filling process through multilayer preforms in resin injection/compression molding. Compos. Sci. Techol. **66**, 1546–1557 (2006)
62. Yang, J., Jia, Y.X., Sun, S., Ma, D.J., Shi, T.F., An, L.J.: Enhancements of the simulation method on the edge effect in resin transfer molding processes. Mater. Sci. Eng. A **478**, 384–389 (2008)
63. Phelan, F.R.: Simulation of the injection process in resin transfer molding. Polym. Compos. **18**, 460–476 (1997)
64. Reddy, J.: Introduction to the Finite Element Method. McGraw-Hill Science/Engineering/Math, New York (2005)
65. Patankar, S.V.: Numerical Heat Transfer and Fluid Flow. Hemisphere Publishing Corporation, New York (1980)
66. Maliska, C.R.: Computational Heat Transfer and Fluid Mechanics. LTC, Rio de Janeiro (2004)
67. Versteeg, H., Malalasekra, M.: An Introduction to Computational Fluid Dynamics: The Finite Volume Method. Prentice Hall, London (2007)

68. Hirt, C.W., Nichols, B.D.: Volume of fluid (VOF) method for the dynamics of free boundaries. J. Comput. Phys. **39**, 201–225 (1981)
69. Srinivasan, V., Salazar, A.J., Saito, K.: Modeling the disintegration of modulated liquid jets using volume-of-fluid (VOF) methodology. Appl. Math. Model. **35**, 3710–3730 (2001)
70. Shäfer, M.: Computational Engineering. Springer-Verlag, Berlin (2006)
71. Ferziger, J.H., Perić, M.: Computational Methods for Fluid Dynamics. Springer-Verlag, Berlin (2002)
72. Fletcher, C.A.J.: Computational Techniques for Fluid Dynamics: Specific Techniques for Different Flow Categories, vol. 2. Springer-Verlag, Berlin (1991)
73. Gutowski, T.G., Morigaki, T., Cai, Z.: The consolidation of laminate composites. J. Compos. Mater. **21**(2), 172–188 (1987)
74. Kokx, V.S. Jr.: Modeling the filling phase of liquid composite moulding processes, including dynamic mould deformations. Ph.D. thesis, University of Auckland, New Zealand (2002)
75. Kokx, V.S. Jr., Bickerton, S.: Modeling liquid composite molding processes involving dynamic cavity thickness changes. In: 48th International SAMPE Symposium and Exhibition, Long Beach, 1690–1701 (2003)
76. Luz, F.F.: Comparative analysis of fluid flow in RTM experiments utilizing commercial softwares. Master dissertation, Materials Engineering Department, Federal University of Rio Grande do Sul, Brazil (2011)
77. Luz, F.F., Amico, S.C., Cunha, A.L., Barbosa, E.S., de Lima, A.G.B.: Applying computational analysis in studies of resin transfer moulding. Defect Diffus. Forum **326–328**, 158–163 (2012)

Thermal Dispersion in High-Conductivity Porous Media

Christopher T. DeGroot and Anthony G. Straatman

Abstract Transport in high-conductivity porous media, such as metal foams, has many practical applications in the field of heat transfer. In order to numerically simulate the performance of devices incorporating such materials in a volume-averaged framework, it is necessary to have accurate estimates of all relevant effective properties, including the thermal dispersion conductivity. This chapter focuses on methods for determining the thermal dispersion conductivity, as well as other effective properties, in high-conductivity porous materials. Results are first presented for cylinder arrays with different particle shapes and arrangements. Following this, results for thermal dispersion are presented for an idealized graphite foam pore geometry and are used in volume-averaged simulations to evaluate the impact of the dispersion model on the overall heat transfer predictions. The overall finding of this review is that dispersion behaviour is complicated for all but the simplest pore geometries. Thus, any modelling efforts should consider the Reynolds and Prandtl numbers as separate influences, rather than lumping their effects into a relation based simply on the Péclet number.

Keywords Porous media · Thermal dispersion · Convective heat transfer · Cylinder array · Graphite foam

C. T. DeGroot (✉) · A. G. Straatman
Department of Mechanical and Materials Engineering,
The University of Western Ontario,
London, N6A 5B9, Canada
e-mail: cdegroo@uwo.ca

A. G. Straatman
e-mail: astraatman@eng.uwo.ca

J. M. P. Q. Delgado et al. (eds.), *Numerical Analysis of Heat and Mass Transfer in Porous Media*, Advanced Structured Materials 27,
DOI: 10.1007/978-3-642-30532-0_6,

1 Introduction

Transport in high-conductivity porous media, such as metal foams, is of practical importance in heat transfer applications including compact heat sinks and heat exchangers [1–8]. While the effectiveness of high-conductivity porous materials as enhanced heat transfer materials has been shown, there are significant challenges involved in simulating the flow and thermal performance of devices incorporating these materials. Since there are generally thousands of pores contained within a given domain of interest, it is impractical to compute the pore-level fields directly. Thus, the method of volume-averaging [9, 10] is typically employed to derive transport equations governing the averaged flow and thermal fields. The challenge with averaging methods, however, is that pore-level terms remain in the governing equations and must be properly characterized by effective properties to produce a closed set of equations.

One particular effective property that has been the topic of considerable research is the thermal dispersion conductivity, which arises from volume-averaging the convection term in the energy equation for the fluid phase. From a physical perspective, thermal dispersion describes the apparent enhancement in conduction within the fluid phase, resulting from pore-level mixing, when the flow is viewed in an averaged sense. When considering fluid and solid constituents with similar thermal properties, the two constituents can be assumed to be in local thermal equilibrium with one another and a single energy equation can be used for the two phases combined. In this case, the effective thermal conductivity, which takes into account dispersion and tortuosity effects, is the only effective property required for closure of the energy equation. Since porous media literature has historically been focused on low-conductivity media where the local thermal equilibrium assumption is valid, most work on the closure of volume-averaged equations has been focused on determining the effective thermal conductivity. Furthermore, since dispersion effects typically dominate tortuosity effects at moderate Péclet numbers ($Pe = RePr$ where Re is the Reynolds number and Pr is the Prandtl number), most efforts have been directed towards the dispersion term.

Dispersion was first studied by Taylor [11] and Aris [12] for a chemical species flowing in a tube. These studies showed that the flowing species is dispersed relative to a plane moving at the mean flow speed with an apparent axial diffusivity equal to the molecular diffusivity plus the dispersion diffusivity, which is proportional to the square of the Péclet number. While this is strictly only valid for flow in a tube, similar Péclet number dependencies have been observed and/or assumed in many other cases of significantly different geometry. Since dispersion depends on local mixing, any results obtained are particular to the flow configuration for which they are derived. To this end, many experimental and analytical studies have been undertaken to explore dispersion in other porous media such as packed beds of particles. For example, Yagi et al. [13] found that the axial dispersion conductivity depends linearly on Pe for packed beds of particles. A similar relationship was observed by Hunt and Tien [14] for fibrous porous materials.

More recently, numerical studies have been used to investigate thermal dispersion in porous media. One type of closure method that has been used involves direct calculations of the pore-level flow and thermal fields and integration of the results to obtain effective properties such as the thermal dispersion conductivity. Kuwahara et al. [15, 16] used this type of closure method to investigate both axial and transverse dispersion in arrays of square cylinders, under thermal equilibrium conditions. Pedras and de Lemos [17] used a similar methodology to study the impact of the solid-to-fluid conductivity ratio on both axial and transverse dispersion in both laminar and turbulent flows, also under local thermal equilibrium conditions.

When the thermal properties of the fluid and solid constituents are substantially different, as in the case of high-conductivity metal foams with air or water as working fluids, it is necessary to consider local thermal non-equilibrium between phases. This can be accomplished empirically using an additional exchange term [18] or more rigorously using constitutive equations to express pore-level quantities in terms of volume-averaged quantities [19]. When constitutive equations are used it can be shown that in addition to the effective thermal conductivity term (which includes the effects of dispersion and tortuosity) and an interfacial exchange term, a modification to the convecting velocity is required [19, 20].

To date, there have been only a small number of studies investigating thermal dispersion under general thermal non-equilibrium conditions. Quintard et al. [19] were among the first to investigate this problem and formulated an elegant method for determining all required effective properties for general non-equilibrium energy equations. Their derivation involved the formulation of constitutive equations for the fluid and solid temperature deviations, which resulted in a set of closure problems, the solutions of which could be integrated to obtain the relevant effective properties. Note that the solution of the closure problems given by Quintard et al. [19] are quite complex. Thus, a simplified model formulated specifically for high-conductivity porous media has recently been proposed by DeGroot and Straatman [20]. This model eliminates the need to solve any closure problems in the solid region of the porous medium and thus eliminates coupling of equations at interfaces. Another approach to investigating dispersion under non-equilibrium conditions has been explored by Moghari [21] which is an extended version of the closure models employed by Kuwahara et al. [15, 16].

In this chapter, thermal dispersion in high-conductivity porous media, where thermal non-equilibrium effects are important, will be discussed. First, the relevant background relating to volume-averaging and closure of volume-averaged equations will be reviewed. The constitutive equation-based approach of DeGroot and Straatman [20] will be the primary focus for the closure methods, although comparisons will be made to methods based on direct simulation of the pore-level thermal fields [15–17, 21]. Next, numerical results for the thermal dispersion conductivity tensor, obtained using the method of DeGroot and Straatman [20], will be presented for various pore geometries. Results for arrays of square and circular cylinders in various arrangements will be analyzed first and the effects of the solid constituent shape and arrangement will be discussed. Next, dispersion in a complicated three-dimensional pore geometry representing an idealized metallic foam [22] will be examined and compared to the simple two-dimensional geometries considered previously. Finally, volume-averaged simulations of heat transfer in a graphite foam block will be examined and the impact of the choice of dispersion model on the overall heat transfer prediction is discussed.

2 Volume-Averaging and Closure

The equations governing heat and fluid flow at the pore level of a general porous medium are the standard mass, momentum, and energy equations. Since there are typically many thousands of pores within a given porous medium, it is not practical to directly compute the flow and thermal fields with such fine resolution. Instead, the relevant governing equations are integral averaged using the method of volume-averaging [9, 10] to obtain new sets of equations that can be solved for quantities averaged over many pores. The method of volume-averaging will be discussed in Sect. 1, followed by the derivation of the volume-averaged energy equations under local thermal non-equilibrium conditions in Sect. 2. The difficulty with any spatial averaging technique is that terms involving unresolved pore-level details remain in the averaged governing equations. In Sect. 3 closure methods based on both the direct pore-level approach and the constitutive equation-based approach are presented and discussed.

2.1 The Method of Volume-Averaging

The method of volume-averaging is a procedure by which the partial differential equations describing the flow and thermal fields in a general medium are integral averaged over many pores of a porous medium to obtain a new set of governing equations in terms of quantities averaged over many pores. In the paragraphs to follow, the basics of volume-averaging are briefly described. Further details are readily available in the literature [9, 10, 23–25].

Consider the flow in an arbitrary porous medium composed of both fluid and solid constituents, as depicted in Fig. 1. Let V_p denote the space occupied by the porous medium with length scale L and let $V \subset V_p$ define the averaging volume having length scale ℓ. In order to have statistically-meaningful averages, V must be taken large enough that the volume-average at a point is relatively insensitive to the particular choice of V. Also, V must be taken small enough that $\ell \ll L$, ensuring that the volume-average of a quantity does not vary significantly within V and can be considered as a constant in integral terms [26].

The definition of the extrinsic volume-average of a quantity ϕ_k, defined at the centroid of V, is given by

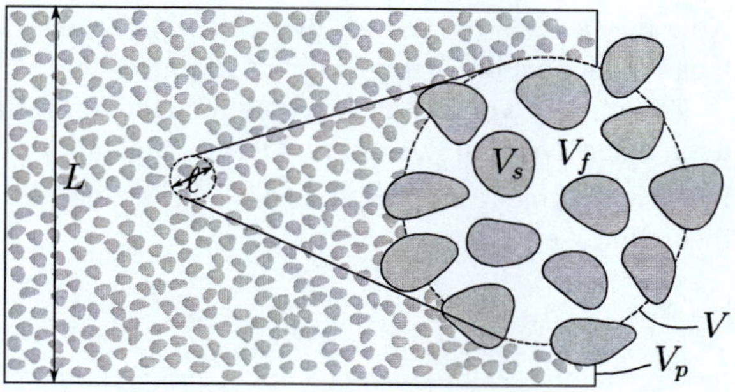

Fig. 1 An illustration of a typical averaging volume, V, for an arbitrary porous medium containing fluid and solid volumes V_f and V_s, respectively [20]

$$\langle \phi_k \rangle = \frac{1}{V} \int_{V_k} \phi_k dV, \tag{1}$$

where $k \in \{f, s\}$ denotes the phase in which the quantity ϕ_k is defined with 'f' and 's' indicating the fluid and solid phases, respectively. Alternatively, the intrinsic average is defined as

$$\langle \phi_k \rangle^k = \frac{1}{V_k} \int_{V_k} \phi_k dV, \tag{2}$$

which is an average over a single phase only. The intrinsic average may be more convenient for quantities such as pressure because it is the physical quantity that would be measured experimentally. The two types of averages are related through the porosity, $\varepsilon = V_f/V$, according to

$$\langle \phi_k \rangle = \begin{cases} \varepsilon \langle \phi_f \rangle^f & \text{if} \quad k = f \\ (1-\varepsilon)\langle \phi_s \rangle^s & \text{if} \quad k = s \end{cases}. \tag{3}$$

Simply applying the integral operator given in Eq. (1) to the governing partial differential equations is not sufficient to derive volume-averaged equations that can be solved in practice. The resulting equations would contain averages of derivatives, while derivatives of averages are required for ease of solution. Terms involving spatial derivatives are thus simplified using the 'spatial averaging theorem' [9, 24, 25]. For the gradient operator, this theorem is stated as

$$\langle \nabla \phi_k \rangle = \nabla \langle \phi_k \rangle + \frac{1}{V} \int_{A_{kl}} \phi_k \mathbf{n}_{kl} dA, \tag{4}$$

where $k, l \in \{f, s\}$, $k \neq l$, and the unit normal vector $\mathbf{n}_{kl}$ is oriented from the k-phase to the l-phase. A_{kl} is the area contained within V forming the intersection of V_k and V_l. Analogous to Eq. (4), the spatial averaging theorem for the divergence of a vector or second-rank tensor, $\mathbf{a}_k$, is given as

$$\langle \nabla \cdot \mathbf{a}_k \rangle = \nabla \cdot \langle \mathbf{a}_k \rangle + \frac{1}{V} \int_{A_{kl}} \mathbf{a}_k \cdot \mathbf{n}_{kl} dA. \tag{5}$$

Note that in Eqs. (4) and (5) the subscripts denote either the fluid or solid phase, thus no summation is implied over repeated indices.

To simplify volume-averages of products of variables, such as those encountered when averaging convection terms, a quantity ϕ_k may be decomposed into its intrinsic volume-average and a pore-level spatial deviation $\tilde{\phi}_k$. The quantity ϕ_k is then expressed as

$$\phi_k = \langle \phi_k \rangle^k + \tilde{\phi}_k, \tag{6}$$

leading to the volume-average of a product of variables, $\phi_{k,1}$ and $\phi_{k,2}$, being defined as

$$\langle \phi_{k,1} \phi_{k,2} \rangle = \frac{1}{\varepsilon_k} \langle \phi_{k,1} \rangle \langle \phi_{k,2} \rangle + \langle \tilde{\phi}_{k,1} \tilde{\phi}_{k,2} \rangle, \tag{7}$$

where all terms are expressed using extrinsic averages and the k-phase porosity, $\varepsilon_k = V_k / V$, which is introduced so that results are general to both fluid- and solid-phase averages. Note that in arriving at Eq. (7) it has been assumed that volume-averaged quantities are constant within a particular averaging volume, which can be shown to be valid provided $\ell \ll L$. Additionally, it is assumed that the volume-average of spatial deviations are zero [25].

2.2 Volume-Averaged Energy Equations

To derive the volume-averaged energy equations, we must first consider the equations governing the conservation of energy in the fluid and solid phases, which are given as

$$\rho_f c_{p,f} \left(\frac{\partial T_f}{\partial t} + \mathbf{u} \cdot \nabla T_f \right) = k_f \nabla^2 T_f \tag{8}$$

and

$$\rho_s c_s \frac{\partial T_s}{\partial t} = k_s \nabla^2 T_s, \tag{9}$$

respectively. Note that ρ, c, and k represent density, specific heat capacity (in the case of the fluid it is the specific heat capacity at constant pressure, c_p), and thermal conductivity, respectively, where the subscripts f and s indicate quantities associated with either the fluid or solid phase. Also note that T represents temperature, $\mathbf{u}$ the fluid velocity vector, and t time.

Volume-averaging Eqs. (8) and (9) results in

$$\rho_f c_{p,f}\left[\varepsilon\frac{\partial\langle T_f\rangle^f}{\partial t}+\langle\mathbf{u}\rangle\cdot\nabla\langle T_f\rangle^f\right]=\varepsilon k_f\nabla^2\langle T_f\rangle^f+\nabla\cdot\left(\frac{1}{V}\int_{A_{fs}}k_f\tilde{T}_f\mathbf{n}_{fs}dA\right)$$
$$+\frac{1}{V}\int_{A_{fs}}k_f\nabla\tilde{T}_f\cdot\mathbf{n}_{fs}dA-\varepsilon\rho_f c_{p,f}\nabla\cdot\langle\tilde{\mathbf{u}}\tilde{T}_f\rangle^f \qquad (10)$$

and

$$(1-\varepsilon)\rho_s c_s\frac{\partial\langle T_s\rangle^s}{\partial t}=(1-\varepsilon)k_s\nabla^2\langle T_s\rangle^s+\nabla\cdot\left(\frac{1}{V}\int_{A_{fs}}k_s\tilde{T}_s\mathbf{n}_{sf}dA\right)$$
$$+\frac{1}{V}\int_{A_{fs}}k_s\nabla\tilde{T}_s\cdot\mathbf{n}_{sf}dA, \qquad (11)$$

respectively. Equations 10 and 11 are the unsimplified result of volume-averaging Eqs. (8) and (9), however, when considering highly conductive solid matrices, there are simplifications that can be made. When the solid-to-fluid phase conductivity ratio is high and convection is the dominant heat transfer mechanism, it is common to neglect the so-called tortuosity term in the fluid energy equation (the second term on the right of Eq. (10)) [20, 27]. With this simplification, the volume-averaged fluid energy equation becomes

$$\rho_f c_{p,f}\left[\varepsilon\frac{\partial\langle T_f\rangle^f}{\partial t}+\langle\mathbf{u}\rangle\cdot\nabla\langle T_f\rangle^f\right]=\varepsilon k_f\nabla^2\langle T_f\rangle^f+\frac{1}{V}\int_{A_{fs}}k_f\nabla\tilde{T}_f\cdot\mathbf{n}_{fs}dA$$
$$-\varepsilon\rho_f c_{p,f}\nabla\cdot\langle\tilde{\mathbf{u}}\tilde{T}_f\rangle^f. \qquad (12)$$

Note that the final term in Eq. (12) represents thermal dispersion, while the second last term in Eq. (12) and the last term in Eq. (11) represent heat exchange between phases. In the solid energy equation, the tortuosity term is retained [as the second last term in Eq. (11)] since this term can be significant in forming the effective solid conductivity [28]. Since pore-level spatial deviations remain in Eqs. (11) and (12), they cannot be solved in general without suitable closure models to characterize these terms. The topic of closure will be discussed next.

2.3 Closure

In this section, two different approaches for obtaining a closed form of Eq. (12) are discussed. The first is based on directly simulating the flow and heat transfer at the pore level, such that the terms involving velocity and temperature deviations may be evaluated directly. The second involves the use of constitutive equations to express the pore-level deviations in terms of volume-averaged quantities. Substituting these constitutive equations into the governing equations for the pore-level deviations results in closure problems that are solved and integrated at the pore-level

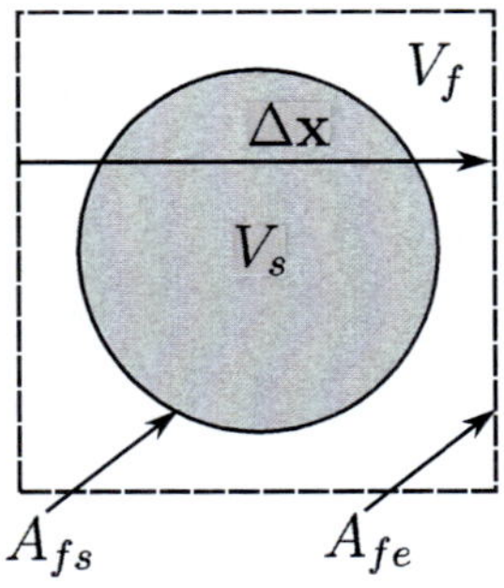

Fig. 2 An illustration of a periodic unit-cell for an array of cylinders with the relevant geometric parameters indicated [20]

to obtain the relevant effective properties. Although the direct pore-level approach is simpler to implement and solve, there are certain advantages to the constitutive equation-based approach in terms of accuracy and its firm theoretical basis.

Before discussing the two approaches for closing the volume-averaged energy equation, it is important to discuss the concept of periodic unit cells, which simplify the process of obtaining the closure parameters. In closure modelling, it is assumed that there exists a periodic unit-cell that repeats itself throughout the porous domain. This corresponds to a homogeneous porous medium in which the region of interest is far removed from any boundaries so that the flow is periodic from cell-to-cell. Figure 2 gives an example of a periodic unit-cell for an array of circular cylinders, which is used to define the relevant geometric parameters for the periodic boundary conditions to be imposed. The displacement vector from the inlet to the corresponding location on the periodic outlet is indicated as $\Delta\mathbf{x}$ on the areas bounding the unit-cell, indicated by A_{fe}. With these definitions in place, the two closure methods are to be outlined next.

2.3.1 Direct Pore-Level Approach

In this section, the approach of Kuwahara et al. [15, 16] for determining the dispersion behaviour in porous media is discussed. Using this approach, the dispersion term in Eq. (12) is closed quite simply by employing the gradient-diffusion hypothesis, given by

$$-\rho_f c_{p,f} \langle \tilde{\mathbf{u}} \tilde{T}_f \rangle^f = \mathbf{k}_d \cdot \nabla \langle T_f \rangle^f \tag{13}$$

in terms of intrinsic averages. Note that Kuwahara et al. [15, 16] applied their analysis for the case of local thermal equilibrium conditions, so all averages used were extrinsic. Here, we are modifying their analysis slightly so that it is applicable for the non-equilibrium energy equations. The dispersion term can then be written by taking the divergence of Eq. (13) and multiplying by ε, resulting in

$$-\varepsilon \rho_f c_{p,f} \nabla \cdot \langle \tilde{\mathbf{u}} \tilde{T}_f \rangle^f = \nabla \cdot \left(\varepsilon \mathbf{k}_d \cdot \nabla \langle T_f \rangle^f \right). \tag{14}$$

Once $\mathbf{k}_d$ is determined, Eq. (14) can be substituted directly into Eq. (12), which closes the dispersion term. Note that Moghari [21] extended the method proposed by Kuwahara et al. [15, 16] in order to close the remaining terms which arise due to the thermal non-equilibrium assumption, however, the focus here is on thermal dispersion so these additional terms will not be discussed. To determine $\mathbf{k}_d$, Kuwahara et al. [15, 16] impose a constant temperature gradient across the unit-cell such that Eq. (13) may be rewritten as

$$\frac{-\rho_f c_{p,f}}{V_f}\int_{V_f}\left(T_f - \langle T_f\rangle^f\right)\left(\mathbf{u} - \langle \mathbf{u}\rangle^f\right)dV = \mathbf{k}_d \cdot \mathbf{n}_T \frac{\Delta T_f}{|\Delta \mathbf{x}|} \tag{15}$$

where ΔT_f is the imposed temperature difference across the unit-cell of size $|\Delta \mathbf{x}|$ and $\mathbf{n}_T$ is a unit-normal vector in the direction of the temperature gradient. Typically, one is interested in the axial and transverse components of the dispersion tensor, so it is common to select the x-direction to coincide with the flow direction and the y- and z-directions to be perpendicular to the flow. In this case, for an isotropic medium, the dispersion tensor is diagonal, and it is simple to extract the components of the dispersion tensor from Eq. (15) after computing the pore-level velocity and temperature fields on a given unit-cell. However, there are some theoretical difficulties encountered using this approach, outlined below.

It is well documented that the volume-average of a quantity is associated with the *centroid* of the averaging volume and that it may, in general, vary within that volume. Thus, evaluation of the term on the left side of Eq. (15) is somewhat questionable if one simply averages the temperature over the unit-cell and claims this represents $\langle T_f\rangle^f$ everywhere due to the fact that the fluid is heated or cooled as it moves through the cell. If the imposed volume-averaged temperature gradient is assumed constant as in Eq. (15) and the fluid temperature varies linearly across the unit-cell this is not an issue, however, the generality of the results with respect to the thermal boundary conditions must be examined. To this end, Pedras and de Lemos [17] studied the impact of imposing a temperature difference directly as opposed to applying a boundary heat flux to generate a temperature difference. They concluded that the differences were small, but indeed there are some differences attributed to applying different boundary conditions. This naturally leads to questions regarding the impact of the magnitude of the imposed heat flux, which could generate temperature variations which are far from being linear.

Although direct pore-level closure methods are considered as useful approximations for determining thermal dispersion behaviour, there are some questions about their theoretical basis, particularly in the case of thermal non-equilibrium cases where the heated solid constituent may cause significant variations in the volume-averaged temperature within the unit-cell. In the next section, the constitutive equation-based approach for seeking closure to the volume-averaged energy equations will be discussed. Such methods are firmly based on the theory of volume-averaging and do not require a specific temperature boundary condition as in direct pore-level approaches. On this basis, it is believed that constitutive equation-based methods can provide better accuracy than direct pore-level approaches.

2.3.2 Constitutive-Equation and Closure Problem Approach

In this section, a closure method based on the use of constitutive equations for the temperature deviations and solutions of the resulting closure problems is discussed. Although there have been several studies employing this approach for thermal equilibrium conditions, there have been few that have considered general thermal non-equilibrium conditions. Quintard et al. [19] proposed a comprehensive set of closure problems for closure of general non-equilibrium energy equations. Although their approach was elegant, it was also complicated to solve numerically. Recently, DeGroot and Straatman [20] modified the method proposed by Quintard et al. [19] for the case of porous media where the ratio of solid-to-fluid conductivity is very high. The resulting model was simpler to solve and was shown to approach the results of Quinard et al. [19] for high solid-to-fluid conductivity ratios [20].

Since this work is focused on thermal dispersion, only closure of the fluid energy equation is considered, although the solid energy equation can be closed in a similar manner [22]. To obtain a transport equation for the temperature deviations in the fluid phase, the volume-averaged energy equation for the fluid phase, divided by ε, is subtracted from the energy equation for a fluid continuum. This results in

$$\rho_f c_{p,f}\left[\frac{\partial \tilde{T}_f}{\partial t} + \mathbf{u}\cdot\nabla\tilde{T}_f + \tilde{\mathbf{u}}\cdot\nabla\langle T_f\rangle^f\right] = k_f\nabla^2\tilde{T}_f - \frac{1}{V_f}\int_{A_{fs}} k_f\nabla\tilde{T}_f\cdot\mathbf{n}_{fs}dA \\ + \rho_f c_{p,f}\nabla\cdot\langle\tilde{\mathbf{u}}\tilde{T}_f\rangle^f. \tag{16}$$

By order-of-magnitude arguments, outlined by DeGroot and Straatman [20], Eq. (16) may be reduced to

$$\rho_f c_{p,f}\left(\mathbf{u}\cdot\nabla\tilde{T}_f + \tilde{\mathbf{u}}\cdot\nabla\langle T_f\rangle^f\right) = k_f\nabla^2\tilde{T}_f - \frac{1}{V_f}\int_{A_{fs}} k_f\nabla\tilde{T}_f\cdot\mathbf{n}_{fs}dA, \tag{17}$$

subject to

$$\tilde{T}_f = T_w - \langle T_f\rangle^f, \text{on } A_{fs}, \tag{18a}$$

$$\tilde{T}_f(\mathbf{x}_{in} + \Delta\mathbf{x}) = \tilde{T}_f(\mathbf{x}_{in}), \text{on } A_{fe}, \tag{18b}$$

$$\langle\tilde{T}_f\rangle^f = 0, \tag{18c}$$

where T_w is the constant wall temperature on A_{fs}. Note that the boundary condition listed in Eq. (18b) expresses the periodicity from a position on the inflow face, $\mathbf{x}_{in}$, to the corresponding location on the outflow face.

It can be seen that the temperature deviation field is driven by the fluid-to-solid temperature difference and the fluid temperature gradient, as indicated by the boundary condition listed in Eq. (18a) and the integral source term in Eq. (17), respectively. Thus, the closure constitutive equation proposed by DeGroot and Straatman [20] for the fluid temperature deviations was

$$\tilde{T}_f = \mathbf{b} \cdot \nabla \langle T_f \rangle^f + \psi \left(T_w - \langle T_f \rangle^f \right), \tag{19}$$

where $\mathbf{b}$ and ψ are coefficients used to express the local temperature deviation in terms of the gradient of the volume-averaged temperature and the wall to volume-averaged temperature difference.

In the work of Quintard et al. [19] a different closure constitutive equation of the form

$$\tilde{T}_f = \mathbf{b}_{ff} \cdot \nabla \langle T_f \rangle^f + \mathbf{b}_{fs} \cdot \nabla \langle T_s \rangle^s + \psi \left(\langle T_s \rangle^s - \langle T_f \rangle^f \right) \tag{20}$$

was used in the fluid region, which is a result of the influence of the solid region on the fluid temperature field. In their work, no assumption was made regarding the ratio of the solid and fluid conductivities, since their approach was not specific to any particular type of porous media. Since this work is focused on high-conductivity porous media, Eq. (21) is valid and indeed the results have been shown to approach the results of Quintard et al. [19] as $k_s/k_f \to \infty$ [20].

Substitution of Eq. (21) into (17) leads to two closure problems, since the temperature deviations are represented by a linear superposition of two separate effects. Thus, if both the $\mathbf{b}$ and ψ fields satisfy their respective partial differential equations, the superposition given by Eq. (21) will satisfy the original equation in terms of $\tilde{T}_f$. The two closure problems for $\mathbf{b}$ and ψ are then

$$\rho_f c_{p,f} (\tilde{\mathbf{u}} + \mathbf{u} \cdot \nabla \mathbf{b}) = k_f \nabla^2 \mathbf{b} - \frac{1}{V_f} \int_{A_{fs}} k_f \nabla \mathbf{b} \cdot \mathbf{n}_{fs} dA, \tag{21}$$

subject to

$$\mathbf{b} = \mathbf{0}, \text{on } A_{fs}, \tag{22a}$$

$$\mathbf{b}(\mathbf{x}_{in} + \Delta \mathbf{x}) = \mathbf{b}(\mathbf{x}_{in}), \text{on } A_{fe}, \tag{22b}$$

$$\langle \mathbf{b} \rangle^f = \mathbf{0}, \tag{22c}$$

as well as

$$\rho_f c_{p,f} \mathbf{u} \cdot \nabla \psi = k_f \nabla^2 \psi - \frac{1}{V_f} \int_{A_{fs}} k_f \nabla \psi \cdot \mathbf{n}_{fs} dA, \tag{23}$$

subject to

$$\psi = 1, \text{ on } A_{fs}, \tag{24a}$$

$$\psi(\mathbf{x}_{in} + \Delta \mathbf{x}) = \psi(\mathbf{x}_{in}), \text{on } A_{fe}, \tag{24b}$$

$$\langle \psi \rangle^f = 0. \tag{24c}$$

The boundary conditions on A_{fs} for the two closure problems listed above are derived based on the fact that the temperature deviation at the wall is $T_w - \langle T_f \rangle^f$, since the wall temperature is assumed to be constant within the periodic unit-cell. In terms of the constitutive equation for the temperature deviation, we have

$$T_w - \langle T_f \rangle^f = \mathbf{b} \cdot \nabla \langle T_f \rangle^f + \psi \left(T_w - \langle T_f \rangle^f \right). \tag{25}$$

For an arbitrary volume-averaged temperature field, it is clear that Eq. (25) can only be satisfied when $\mathbf{b} = \mathbf{0}$ and $\psi = 1$ on the fluid-solid interface, leading to the conditions given in Eqs. (22a) and (24a).

One may also note that the solutions of Eqs. (21) and (23) are not unique without the constraints listed in Eqs. (22c) and (24c). For every possible integral source term in Eqs. (21) and (23), there is a unique solution, however the solution being sought is the one that satisfies the constraints given in Eqs. (22c) and (24c). As such, the proper integral source terms are found using an iterative algorithm.

Substituting the constitutive equation for the temperature deviation, given in Eq. (21), into the volume-averaged energy equation for the fluid phase results in an interfacial exchange term given by

$$\begin{aligned} \frac{1}{V} \int_{A_{fs}} k_f \nabla \tilde{T}_f \cdot \mathbf{n}_{fs} dA = & \left(\frac{1}{V} \int_{A_{fs}} k_f \nabla \mathbf{b} \cdot \mathbf{n}_{fs} dA \right) \cdot \nabla \langle T_f \rangle^f \\ & + \left(\frac{1}{V} \int_{A_{fs}} k_f \nabla \psi \cdot \mathbf{n}_{fs} dA \right) \left(T_w - \langle T_f \rangle^f \right) \end{aligned} \tag{26}$$

while the dispersion term is given by

$$-\varepsilon \rho_f c_{p,f} \nabla \cdot \langle \tilde{\mathbf{u}} \tilde{T}_f \rangle^f = -\varepsilon \rho_f c_{p,f} \nabla \cdot \left(\langle \tilde{\mathbf{u}} \mathbf{b} \rangle^f \cdot \nabla \langle T_f \rangle^f \right) + \varepsilon \rho_f c_{p,f} \langle \tilde{\mathbf{u}} \psi \rangle^f \cdot \nabla \langle T_f \rangle^f \tag{27}$$

Noting that $T_w = \langle T_s \rangle^s$ within any averaging volume, the closed volume-averaged energy equation in the fluid phase is given by

$$\rho_f c_{p,f} \left(\varepsilon \frac{\partial \langle T_f \rangle^f}{\partial t} + \bar{\mathbf{u}}_f \cdot \nabla \langle T_f \rangle^f \right) = \nabla \cdot \left(\mathbf{k}_{fe} \cdot \nabla \langle T_f \rangle^f \right) + a_{fs} h_{fs} \left(\langle T_s \rangle^s - \langle T_f \rangle^f \right) \tag{28}$$

where the convecting velocity, which is different from the volume-averaged velocity, is defined by

$$\bar{\mathbf{u}}_f = \langle \mathbf{u} \rangle - \varepsilon \langle \tilde{\mathbf{u}} \psi \rangle^f - \frac{\varepsilon}{V_f} \int_{A_{fs}} \frac{k_f}{\rho_f c_{p,f}} \nabla \mathbf{b} \cdot \mathbf{n}_{fs} dA \tag{29}$$

and the effective conductivity tensor is defined by

$$\mathbf{k}_{fe} = \varepsilon k_f \mathbf{I} + \varepsilon \mathbf{k}_d \tag{30}$$

where the dispersion conductivity tensor is given by

$$\mathbf{k}_d = -\rho_f c_{p,f} \langle \tilde{\mathbf{u}} \mathbf{b} \rangle^f \tag{31}$$

Finally, the interfacial heat transfer coefficient is defined by

$$a_{fs} h_{fs} = \frac{\varepsilon}{V_f} \int_{A_{fs}} k_f \nabla \psi \cdot \mathbf{n}_{fs} dA \tag{32}$$

where a_{fs} is the specific surface area of the porous material (*i.e.* the surface area per unit volume of porous media). Note the definition of the dispersion tensor in Eq. (31) in this case is completely independent of any computed temperature field and that the vector $\mathbf{b}$ is periodic. Thus, the difficulties discussed in the previous section regarding the definition of the volume-averaged temperature and temperature gradient are removed. Also, Eq. (27) indicates that the dispersion term is actually split into two parts: one constituting the typical dispersion term and one contributing to the convecting velocity term. Thus, it would seem that for thermal non-equilibrium conditions that the gradient-diffusion hypothesis underlying direct pore-level approaches is not entirely obeyed according the the derivation above.

3 Dispersion in High-Conductivity Cylinder Arrays

In this section, thermal dispersion in arrays of high-conductivity square and circular cylinders will be discussed. Sahraoui and Kaviany [29] undertook one of the earliest numerical studies into dispersion in cylinder arrays by solving the relevant closure problem for local thermal equilibrium conditions. Kuwahara et al. [15, 16] took a different approach, studying dispersion in arrays of square cylinders under local thermal equilibrium conditions by integrating computed velocity and temperature fields to determine the dispersion term directly, assuming dispersion is a gradient-diffusion process, as described in Sect. 2. Quintard et al. [19] presented the first results for all relevant closure parameters under general thermal non-equilibrium conditions. Moghari [21] also computed the relevant closure parameters for cylinder arrays under local thermal non-equilibrium conditions using a methodology similar to Kuwahara et al. [15, 16]. DeGroot and Straatman [20] studied the same configuration as Quintard et al. [19] to show that their simplified model produced the same results for high ratios of k_s/k_f.

3.1 Parameters of Study

Geometries under consideration in this section are periodic arrays of square and circular cylinders in various arrangements with a porosity $\varepsilon = 0.38$ as in Quintard et al. [19] and DeGroot and Straatman [20]. Figure 3 gives a schematic diagram of

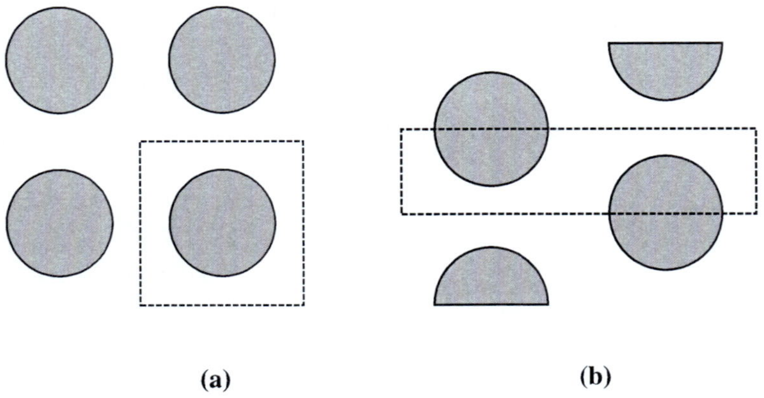

Fig. 3 Schematic diagrams of the **a** inline and **b** staggered arrangements of circular cylinders. The periodic computational domain is denoted with *dotted lines*. In both cases, the primary flow direction is horizontal

both inline and staggered circular cylinders with the associated unit-cell considered. Reynolds numbers $Re_D \in \{1, 10, 50, 100\}$ are considered and the Prandtl number is varied to give Péclet numbers $Pe_D \in [0.1, 1000]$, where D is the cylinder diameter in the case of a circular cylinder and the side length in the case of a square cylinder. Results will first be presented for inline arrangements of square and circular cylinders to evaluate the effect of the solid constituent shape. Subsequently, results are given for circular cylinders in both inline and staggered arrangements to determine the effect of the solid constituent arrangement.

3.2 Effect of Solid Constituent Shape

Results for the dimensionless dispersion conductivity in the axial direction, $(\mathbf{k}_d)_{xx}/k_f$ for inline arrays of square and circular cylinders are shown in Fig. 4 as a function of the Péclet number. These results show that increasing flow inertia, represented by increasing Reynolds number, serves to increase dispersion for arrays of circular cylinders, while for square cylinders there is no significant difference observed between the various Reynolds numbers considered. This observation is confirmed using least-squares fits of the data, which reveal the following models:

$$\frac{(\mathbf{k}_d)_{xx}}{k_f} = 0.0027 Re_D^{2.1} Pr^{1.9} \tag{33}$$

for circular cylinders and

$$\frac{(\mathbf{k}_d)_{xx}}{k_f} = 0.0060 Re_D^{2.0} Pr^{2.0} = 0.0060 Pe_D^2 \tag{34}$$

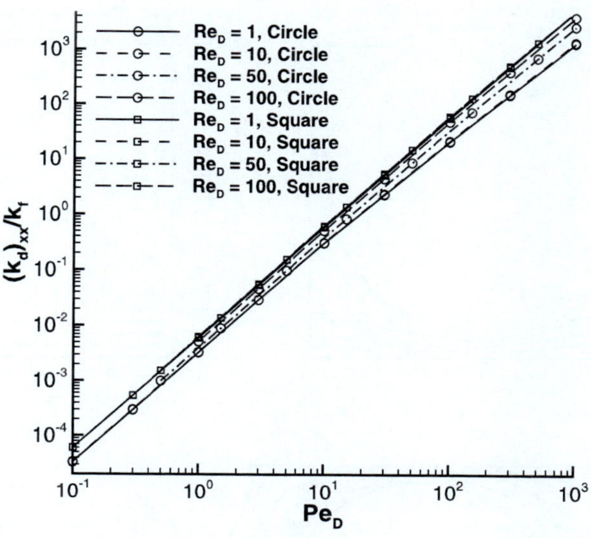

Fig. 4 A plot of the dimensionless axial thermal dispersion conductivity as a function of the Péclet number for arrays of inline square and circular cylinders

for square cylinders. It is interesting to note the elegance of the above models. In these cases, a single expression is able to fully capture the dispersion behaviour over the full range of Péclet numbers considered, whereas many existing models require multiple Péclet number regimes [15, 16, 21]. It will be interesting to determine if this is a property of the different closure model employed or more fundamentally related to the specific porosity or k_s/k_f ratio under consideration. This will be an interesting topic for future research.

In light of the preceding results, it is concluded that for square cylinders under the conditions considered herein, that axial dispersion is a function of Pe_D alone and that the relationship resembles the classical Pe_D^2 relationship derived by Taylor for circular tubes [11]. Note that the expression for Taylor dispersion is $(\mathbf{k}_d)_{xx}/k_f = Pe_D^2/192 \approx 0.0052 Pe_D^2$, where in this case D is the tube diameter. This indicates that the mechanism of Taylor dispersion is similar to that for arrays of square cylinders, although the magnitude is different. The similarity can be rationalized on the basis that between the rows of square cylinders there exists a largely parabolic velocity profile as for flow in a circular tube. Thus, the resulting spatial variations in velocity tend to drive similar dispersion behaviour, although with a higher magnitude due to recirculations behind the cylinders.

For arrays of circular cylinders it is found that there is a slightly stronger dependence on the Reynolds number than the Prandtl number. This indicates that, in general, one should not simply lump their effects into the Péclet number as is commonly done; instead, they should be considered as separate influences. From a physical perspective, it is logical that dispersion in circular cylinders is sensitive

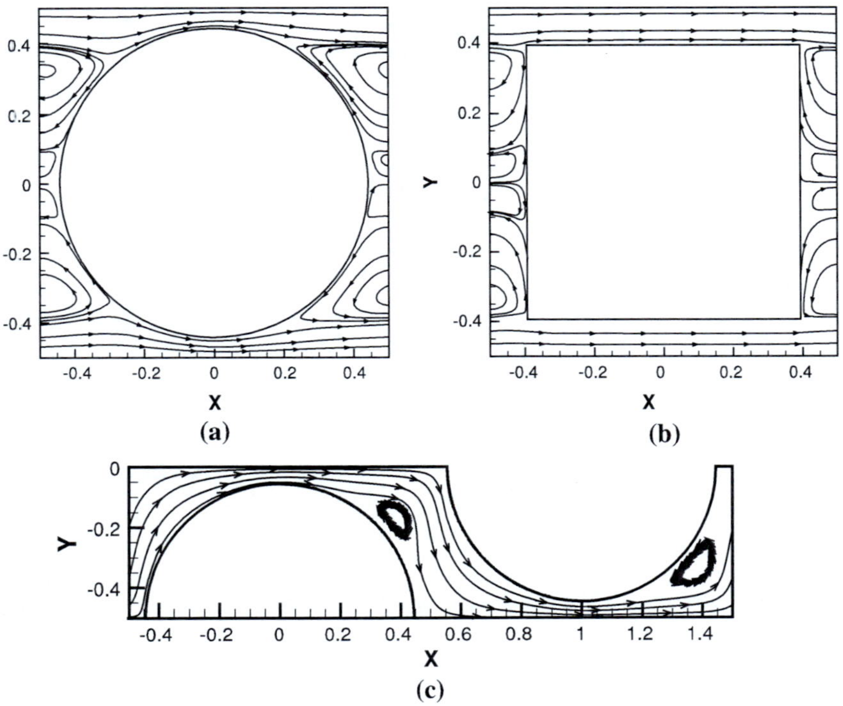

Fig. 5 Plots of streamlines for **a** inline circular cylinders, **b** inline square cylinders, and **c** staggered circular cylinders for $Re_D = 100$. Note that X and Y represent spatial coordinates non-dimensionalized by the unit-cell size

to inertial effects, while in square cylinders there is no significant effect because of the nature of the separation points in these flows, shown in Fig. 5. For flow past square cylinders, the separation points are well-defined at the trailing edge of the cylinder, so the velocity deviations that drive dispersion are similar, although different in magnitude, regardless of the flow speed. For circular cylinders, however, the separation point is strongly affected by inertia, leading to substantially different velocity deviations at different Reynolds numbers. Such inertial effects have been previously observed by Sahraoui and Kaviany [29] under local thermal equilibrium conditions.

3.3 *Effect of Solid Constituent Arrangement*

In this section, the effect of solid constituent arrangement is examined by comparing the dispersion behaviour of both inline and staggered arrays of circular cylinders. In the previous section, it was determined that the effect of inertia on the dispersion behaviour was minimal for an array of square cylinders. Thus, to study

Fig. 6 A plot of the dimensionless axial thermal dispersion conductivity as a function of the Péclet number for staggered arrays of circular cylinders

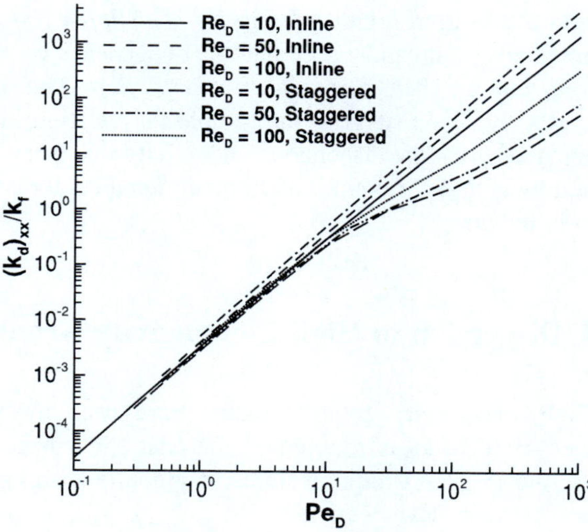

the effects of the solid constituent arrangement we will focus only circular cylinders since the results are more general. It was also found previously for an inline array of cylinders that there was little difference between the results obtained at $Re_D = 1$ and $Re_D = 10$. Thus, to study the effect of particle arrangement the Reynolds number will be considered to be $Re_D \in \{10, 50, 100\}$.

Results for the axial dispersion conductivity for a staggered array of circular cylinders are plotted along with those for an inline array in Fig. 6. These results indicate that similar dispersion behaviour is observed for low Pe_D, while as Pe_D increases beyond approximately 10 it is found that the dispersion conductivity is much lower for staggered arrays of cylinders. This result can be explained on the basis that while there are minimal recirculation regions contributing to dispersion in the case of the staggered arrays (at low Reynolds numbers there are no recirculations), the fluid follows a much more tortuous path as it passes through the medium as compared to inline arrays. As concluded by Sahraoui and Kaviany [29] for creeping flows under conditions of local thermal equilibrium, it appears that the net effect of these differences serves to decrease the overall dispersion, particularly at high Pe_D.

For $Pe_D \leq 10$, the axial dispersion conductivity can be modelled as

$$\frac{(\mathbf{k}_d)_{xx}}{k_f} = 0.0054 Re_D^{1.8} Pr^{1.7} \tag{35}$$

whereas for $Pe_D \geq 100$ it can be expressed as

$$\frac{(\mathbf{k}_d)_{xx}}{k_f} = 0.016 Re_D^{1.5} Pr^{1.1}, \tag{36}$$

with a transition region present for $10 < Pe_D < 100$. Therefore, as in the case of inline arrays, it can be seen that the dependence on Re_D and Pr are quite different from one another, particularly at high Pe_D. This highlights the importance of inertia and again suggests that simple Pe_D relationship is unsuitable. It can also be concluded that the dispersion conductivity is substantially lower and increases less rapidly with the Reynolds and Prandtl numbers for staggered arrays in comparison to inline arrays.

4 Dispersion in High-Conductivity Graphite Foams

High-conductivity graphite foams have been the subject much research over the past decade as a potential material for enhanced heat transfer applications [2, 4–6, 8, 30]. Unique features of graphite foams that make them attractive for heat transfer devices include:

1. A high solid phase thermal conductivity (800–1,900 W/m K) leading to a very high effective solid conductivity (40–160 W/m K [5]).
2. A large amount of exposed internal surface area per unit volume available for convective heat exchange (5,000–50,000 m^2/m^3).

While the heat transfer characteristics of graphite foams have been investigated experimentally [4–6, 30], there have been relatively few attempts to determine the relevant effective properties such that the performance of designs incorporating graphite foams can be accurately simulated numerically by solving volume-averaged governing equations. Although there exist numerical methods to solve these equations [31–33], the accuracy of such simulations is limited by the accuracy of the effective properties specified for the foam region. Further complicating matters is the fact that the extremely high solid phase conductivity necessitates the consideration of local thermal non-equilibrium between the fluid and solid phases.

Many of the required geometric parameters of graphite foams can be obtained from the idealized geometric model of the pore structure proposed by Yu et al. [34], shown in Fig. 7 in comparison to images of actual foam. From this geometric model, many important parameters such as the effective thermal conductivity of the solid matrix can be obtained. Using the model of Yu et al. [34] and experimental data Straatman et al. [5] calibrated an interstitial exchange model to match experimental results. Note that in this work, thermal dispersion was modelled simply using the correlation of Calmidi and Mahajan [35] for aluminum foams and was not part of the calibration procedure. Recent work by Karimian and Straatman [36] used direct simulations of flow and heat transfer at the pore-level to investigate axial dispersion using a simplified model. More recently, DeGroot and Straatman [22] performed a comprehensive analysis of the effective properties of graphite foam, although calibration with experimental results is still required [36].

In this section, results for both the axial and transverse dispersion behaviour in graphite foams is reviewed and discussed. Subsequently, volume-averaged

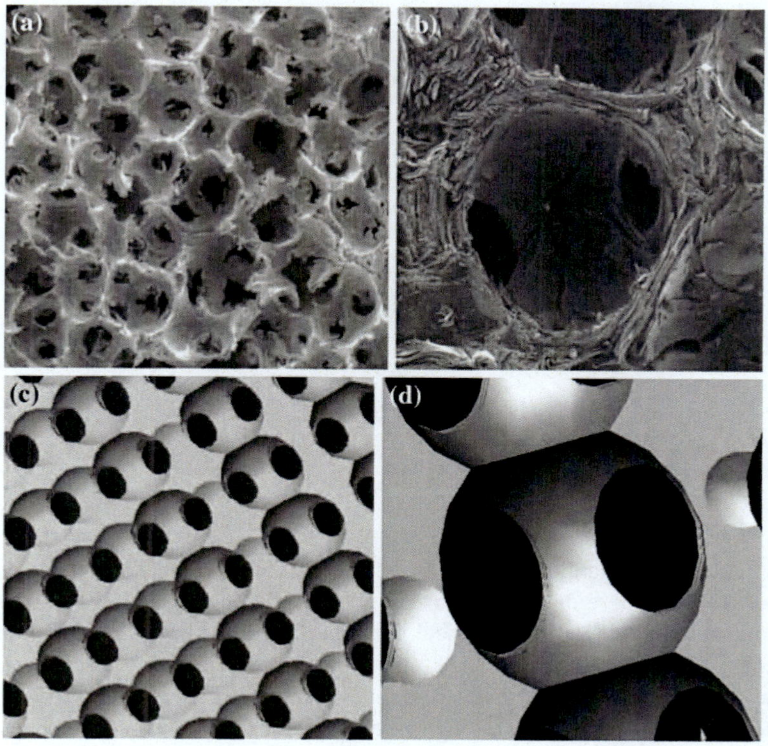

Fig. 7 Comparison between the actual graphite foam pore geometry, shown in the electron micrograph images (**a**) and (**b**), to the geometric idealization proposed by Yu et al. [34], shown in the rendered images (**c**) and (**d**). Used with permission from Yu et al. [34] and from the ASME Journal of Heat Transfer

simulations are presented to demonstrate the differences in the overall heat transfer prediction when employing different dispersion models.

4.1 Parameters of Study

Graphite foams are typically manufactured with porosities in the range $\varepsilon \in [0.8, 0.9]$. Below $\varepsilon = 0.8$, the smaller pore windows lead to exceedingly high pressure drops when passing fluid through the foam, making such foams inefficient for use in heat transfer applications. Above $\varepsilon = 0.9$, foams are difficult to manufacture as the interconnecting ligaments become too small. Thus, the numerical results that are presented in this section are restricted to $\varepsilon \in \{0.80, 0.85, 0.90\}$ which covers the primary region of interest for use in heat transfer applications.

The Reynolds numbers considered are confined to the laminar flow regime, since it is more difficult to account for turbulence effects in closure modelling. Additionally, due to the relatively low hydraulic permeability of graphite foams, it

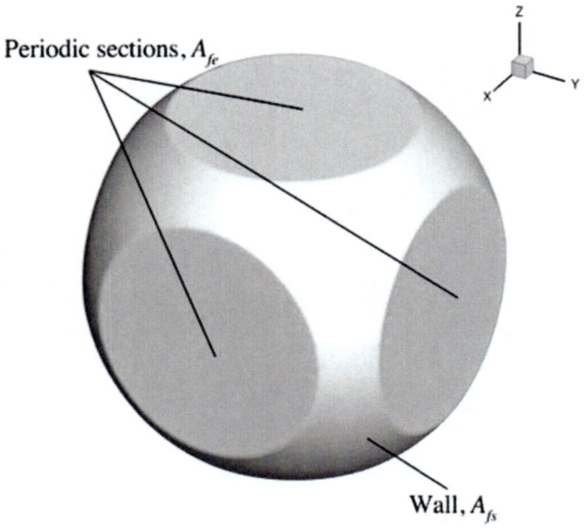

Fig. 8 Schematic diagram of the computational domain for the graphite foam pore geometry, in this case for $\varepsilon = 0.85$ [22]. Used with permission from the ASME Journal of Heat Transfer

would require a substantial pressure drop to sustain turbulence. Thus, the Reynolds number based on pore diameter, d, is confined to the interval $Re_d \in [1, 100]$ since creeping flows are of limited practical relevance in heat transfer. Prandtl numbers of the fluid are taken to be $Pr \in \{1, 5, 10\}$ which cover the approximate range of typical fluids used in heat transfer (i.e. air, water, and refrigerants).

The results presented in the following sections come from the comprehensive study conducted by DeGroot and Straatman [22] which involved 198 numerical simulations. The computational domain used in this work for the solution of the flow field and the associated closure problems consists of a single periodic unit-cell, shown in Fig. 8. The flow is taken to enter the pore in such a way that the average velocity vector makes equal angles with all of the coordinate axes (X, Y, Z) shown in Fig. 8. Karimian and Straatman [36, 37] found that this flow condition is of greatest practical relevance as it represents somewhat of an average flow direction for the randomly oriented pores of a real foam geometry. While this coordinate system is convenient for computations, it is more customary to discuss axial and transverse dispersion where the axial direction is the direction of the average velocity. Thus, all values reported herein are transformed to a flow-oriented system (x, y, z) by the appropriate change of basis.

4.2 Axial Dispersion

Similar to the dispersion behaviour in cylinder arrays, it has been found that dispersion in idealized graphite foams does not obey a simple relationship with respect to the Péclet number [22]. Instead, it has been found that the dependence on the Reynolds and Prandtl numbers is different from one another. DeGroot and Straatman [22] found that the dimensionless axial dispersion conductivity for idealized graphite foams can accurately be represented by the equation

Fig. 9 A plot of the dimensionless axial thermal dispersion conductivity as a function of the dimensionless group $Re_d^{1.91}Pr^{1.72}(1-\varepsilon)^{1.05}$ along with a least squares fit of the data [22]. Used with permission from the ASME Journal of Heat Transfer

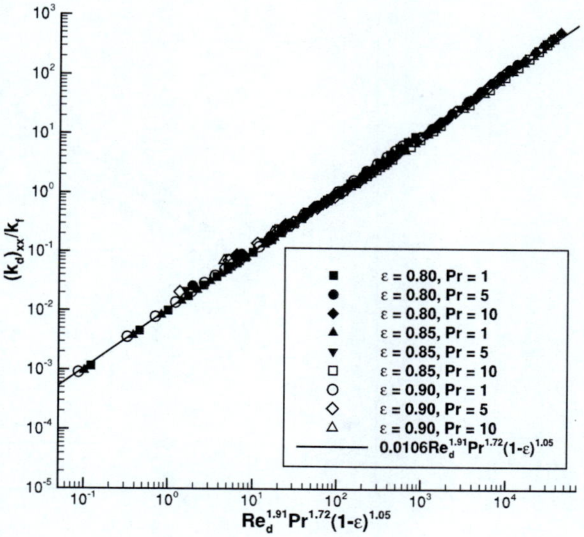

$$\frac{(\mathbf{k}_d)_{xx}}{k_f} = 0.0106 Re_d^{1.91} Pr^{1.72} (1-\varepsilon)^{1.05}. \tag{37}$$

Numerical results are plotted in Fig. 9 along with the least squares fit given by Eq. (37). Plotted against the dimensionless group $Re_d^{1.91}Pr^{1.72}(1-\varepsilon)^{1.05}$, this figure shows excellent collapse of the data onto the given line and clearly shows that the dispersion conductivity cannot be simply considered as a function of the Péclet number. It is found that the dispersion conductivity has a sightly stronger dependance on the Reynolds number than the Prandtl number. There is a roughly linear relationship between the dispersion conductivity and the solid fraction, $1-\varepsilon$.

4.3 Transverse Dispersion

The transverse component of the dimensionless dispersion tensor is plotted in Fig. 10 as a function of the Péclet number, Pe_d. It is clear from this figure that performing a least-squares fit over the full range of Pe_d would not give a good fit, since there exist different behaviours for different Péclet number regimes. For $Pe_d < 10$, it has been found that the dimensionless transverse dispersion conductivity is adequately described in terms of the Péclet number and the solid fraction according to the model

$$\frac{(\mathbf{k}_d)_{yy}}{k_f} = 0.0296 Pe_d^{1.78} (1-\varepsilon)^{1.38} \tag{38}$$

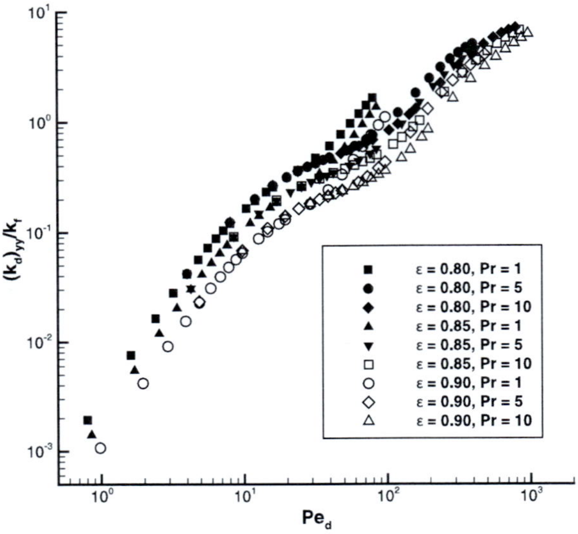

Fig. 10 A plot of the dimensionless transverse thermal dispersion conductivity as a function of Pe_d [22]. Used with permission from the ASME Journal of Heat Transfer

for $Pe_d < 10$. With this model established, the dimensionless transverse dispersion conductivity is plotted as a function of the dimensionless group $Pe_d^{1.78}(1-\varepsilon)^{1.37}$ in Fig. 11a. This figure demonstrates that the data collapses very well onto the model given by Eq. (38) within its range of applicability.

For higher Péclet numbers, it has been found that the Reynolds and Prandtl numbers should be considered as separate influences, as was the case for the axial dispersion conductivity. A least-squares fit indicated that the following model was appropriate for the higher Péclet number regime:

$$\frac{(\mathbf{k}_d)_{yy}}{k_f} = 0.0269 Re_d^{1.21} Pr^{0.776} (1-\varepsilon)^{0.786} \tag{39}$$

for $Re_d^{1.21} Pr^{0.776}(1-\varepsilon)^{0.786} > 30$. The transverse dispersion data is plotted in Fig. 11b as a function of the dimensionless group $Re_d^{1.21} Pr^{0.776}(1-\varepsilon)^{0.786}$. This figure shows reasonable agreement with the model of Eq. (39), although there is some departure at the highest Péclet numbers.

4.4 Volume-Averaged Calculations

With the dispersion models given in the previous sections, along with appropriate models for the remaining effecting properties, calculations at the volume-averaged level can be undertaken. In this section, results are presented for a

Fig. 11 Plots of the dimensionless transverse thermal dispersion conductivity as a function of the dimensionless groups **a** $Pe_d^{1.78}(1-\varepsilon)^{1.37}$ and **b** $Re_d^{1.21}Pr^{0.776}(1-\varepsilon)^{0.786}$ along with the corresponding least squares fits [22]. Used with permission from the ASME Journal of Heat Transfer

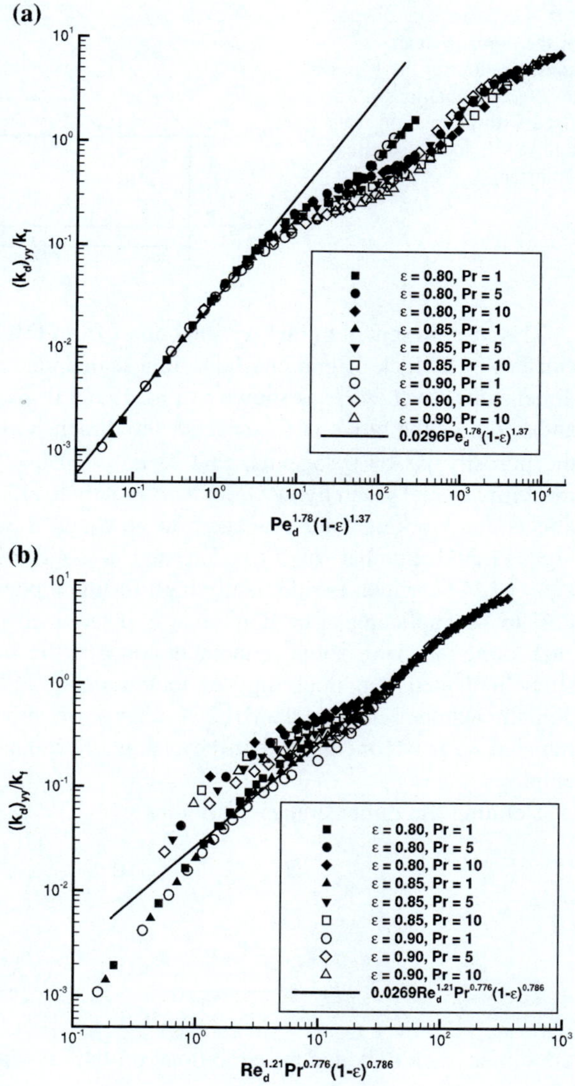

simple test case to compare the results obtained using the dispersion models presented in the previous sections with equivalent results obtained using the dispersion model of Calmidi and Mahajan [35], which is linear with respect to the Péclet number. Although this model was developed for aluminum foams, it has also been employed by Straatman et al. [5] in the study of graphite foams since at that time there were no models developed specifically for such materials. To isolate the differences due changing the dispersion model, all other effective properties are identical for both cases and are the values used by DeGroot and Straatman [22].

Fig. 12 Schematic diagram of the domain under consideration for the volume-averaged simulations [22]. Used with permission from the ASME Journal of Heat Transfer

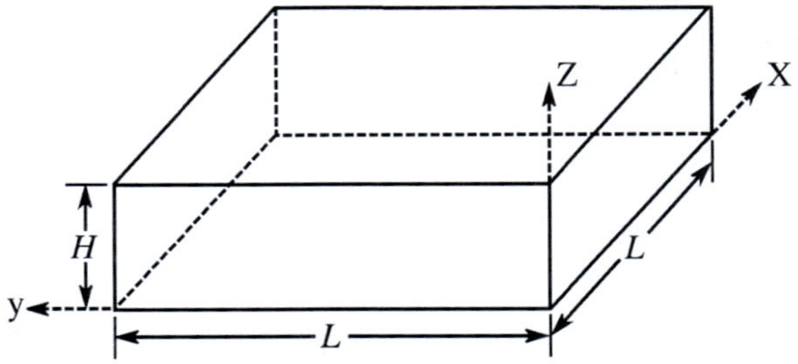

The case presented in this section comes from DeGroot and Straatman [22] and consists of a block of graphite foam in a sealed channel, heated from below, with dimensions $L \times L \times H$, as shown in Fig. 12. In this case the dimensionless length and width of the block is $L/d = 100$, the dimensionless height is $H/d = 12$, and the porosity is $\varepsilon = 0.85$, such that $K/d^2 = 1.24 \times 10^{-2}$ according the the permeability model given by DeGroot and Straatman [22]. To highlight the dispersion effects, the working fluid was taken to be water at 300 K, which has a relatively high Prandtl number of $Pr = 5.8$ and a solid-to-fluid conductivity ratio of $k_s/k_f = 2447$, which is sufficiently high for the models of DeGroot and Straatman [22] to be applicable. The domain was discretized using 30 control volumes in each direction, with some refinement towards the boundaries. A grid resolution study indicated that doubling the total number of control volumes within the domain resulted in less than 0.2 % change in pressure drop and overall heat transfer, so it is concluded that the results are independent of the particular grid employed.

Defining the dimensionless variables

$$\mathbf{x}^* = \frac{\mathbf{x}}{d} \qquad \langle \mathbf{u} \rangle^* = \frac{\langle \mathbf{u} \rangle}{U} \qquad \langle p \rangle^{f*} = \frac{\langle p \rangle^f}{\rho_f U^2} \tag{40a, b, c}$$

$$\theta_f = \frac{\langle T_f \rangle^f - T_{in}}{T_w - T_{in}} \qquad \theta_s = \frac{\langle T_s \rangle^s - T_{in}}{T_w - T_{in}}, \tag{40d, e}$$

the dimensionless boundary conditions on the velocity field are summarized as

$$\langle \mathbf{u} \rangle^* |_{x^*=0} = \mathbf{i} \quad \nabla \langle \mathbf{u} \rangle^* |_{x^*=L/d} = \mathbf{0} \quad \langle \mathbf{u} \rangle^* |_{y^*=0} = \mathbf{0} \tag{41a, b, c}$$

$$\langle \mathbf{u} \rangle^* |_{y^*=L/d} = \mathbf{0} \quad \langle \mathbf{u} \rangle^* |_{z^*=0} = \mathbf{0} \quad \langle \mathbf{u} \rangle^* |_{z^*=H/d} = \mathbf{0}. \tag{41d, e, f}$$

The dimensionless pressure is simply set to zero at $x^* = H/d$ and is extrapolated at all other boundaries. The dimensionless boundary conditions on the temperature fields are summarized as

$$\theta_f |_{x^*=0} = 0 \quad \left. \frac{\partial \theta_f}{\partial x^*} \right|_{x^*=L/d} = 0 \quad \left. \frac{\partial \theta_f}{\partial y^*} \right|_{y^*=0} = 0 \tag{42a, b, c}$$

Table 1 Summary of the average Nusselt number computed using the two dispersion models for $Re_d \in \{50, 100\}$ and the percent difference between them

Re_d	$\overline{Nu}$ (Calmidi and Mahajan model [35])	$\overline{Nu}$ (DeGroot and Straatman model [22])	Percent difference
50	35.7	36.8	3.4
100	55.0	61.8	12.4

$$\left.\frac{\partial \theta_f}{\partial y^*}\right|_{y^*=L/d} = 0 \quad \theta_f\big|_{z^*=0} = 1 \quad \left.\frac{\partial \theta_f}{\partial z^*}\right|_{z^*=H/d} = 0 \tag{42d, e, f}$$

$$\left.\frac{\partial \theta_s}{\partial x^*}\right|_{x^*=0} = 0 \quad \left.\frac{\partial \theta_s}{\partial x^*}\right|_{x^*=L/d} = 0 \quad \left.\frac{\partial \theta_s}{\partial y^*}\right|_{y^*=0} = 0 \tag{42g, h, i}$$

$$\left.\frac{\partial \theta_s}{\partial y^*}\right|_{y^*=L/d} = 0 \quad \theta_s\big|_{z^*=0} = 1 \quad \left.\frac{\partial \theta_s}{\partial z^*}\right|_{z^*=H/d} = 0. \tag{42j, k, l}$$

Results were obtained for two Reynolds numbers, $Re_d \in \{50, 100\}$, using the two different dispersion models and are presented in terms of the average Nusselt number, defined as

$$\begin{aligned}\overline{Nu} &= \frac{q_{tot} d}{k_f A_b (T_w - T_{in})} \\ &= \left(\frac{d}{L}\right)^2 \int_{x^*=0}^{x^*=L/d} \int_{y^*=0}^{y^*=L/d} \left(\frac{(\mathbf{k}_{fe})_{zz}}{k_f} \frac{\partial \theta_f}{\partial z^*} + \frac{(\mathbf{k}_{se})_{zz}}{k_f} \frac{\partial \theta_s}{\partial z^*} \right)_{z^*=0} dx^* dy^*,\end{aligned} \tag{43}$$

where q_{tot} is the total heat transfer from the heated base of area A_b for a temperature difference between the heated wall to inlet fluid of $T_w - T_{in}$. The average Nusselt number results, summarized in Table 1, indicate that at $Re_d = 50$ the difference between the two dispersion models is relatively small, whereas at $Re_d = 100$ the difference is much more significant. This is also illustrated in the contour plots of the dimensionless fluid temperature in each case, given in Figs. 13 and 14, which show that the dispersion model of DeGroot and Straatman [22] results in more conduction within the fluid in comparison to the model of Calmidi and Mahajan [35], with this effect being stronger in the $Re_d = 100$ case. DeGroot and Straatman [22] also reported that the difference in $\overline{Nu}$ between the two dispersion models was further increased for larger values of H/d, corresponding to thicker foam blocks.

These volume-averaged simulation results indicate that there is a significant difference between the dispersion models of Calmidi and Mahajan [35] and DeGroot and Straatman [22] when applied to graphite foams. At lower Reynolds numbers, the difference between the two models is much less, which explains why the Calmidi and Mahajan [35] model gave good results for the cases considered by Straatman et al. [5]. However, it has been shown that as the Reynolds number and

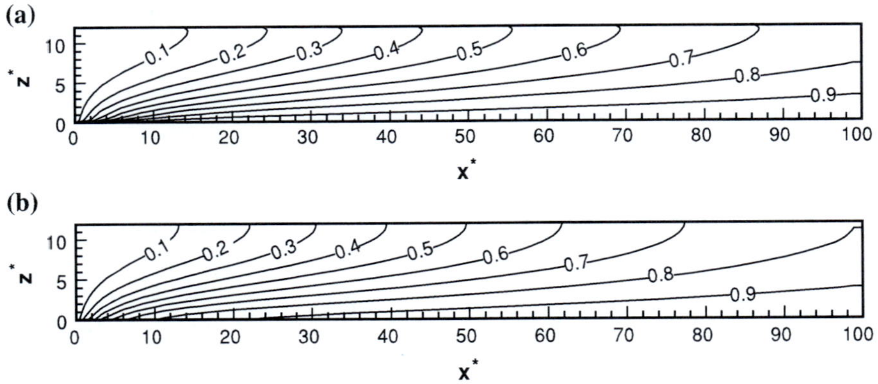

Fig. 13 Contour plots of the dimensionless fluid temperature, θ_f, along the center plane defined by $y^* = \frac{1}{2}\frac{L}{d}$ for $Re_d = 50$ using **a** the dispersion model of Calmidi and Mahajan [35] and **b** the present dispersion model [22]. Used with permission from the ASME Journal of Heat Transfer

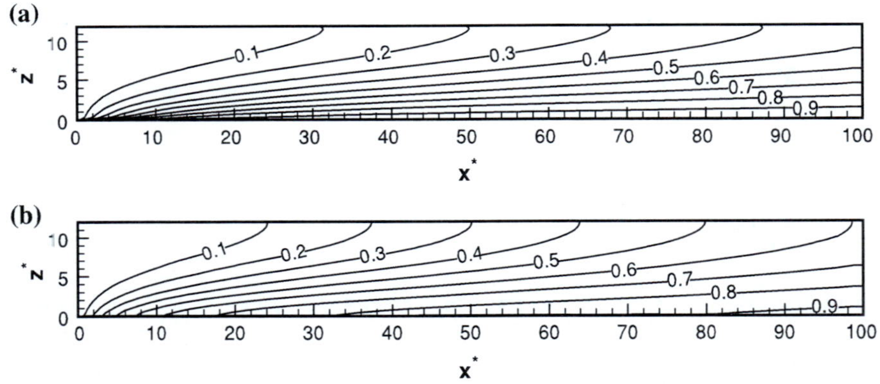

Fig. 14 Contour plots of the dimensionless fluid temperature, θ_f, along the center plane defined by $y^* = \frac{1}{2}\frac{L}{d}$ for $Re_d = 100$ using **a** the dispersion model of Calmidi and Mahajan [35] and **b** the present dispersion model [22]. Used with permission from the ASME Journal of Heat Transfer

block height are increased, it becomes important to use a model developed specifically for graphite foams [22].

5 Summary

In this chapter, thermal dispersion in high-conductivity porous materials was reviewed and discussed. It is clear through the review of the literature that there have been a limited number of studies focusing on thermal dispersion under local thermal non-equilibrium conditions which are of great practical importance for highly conductive media. The two major techniques that exist for obtaining closed forms of the non-equilibrium energy equations, namely direct pore-level and constitutive equation-based techniques, were discussed in terms of their relative advantages and disadvantages. In this chapter, the major focus was placed on results obtained using the newly proposed constitutive equation-based technique of DeGroot and Straatman [20], which is specific to high-conductivity porous media.

The overall message of this work is that thermal dispersion is a complicated phenomenon that is not easily described using traditional models posed only in terms of the Péclet number. This is particularly true when considering complicated three-dimensional pore shapes such as a graphite foam pore. It must be remembered that thermal dispersion is a result of pore-level variations in both velocity and temperature, which are highly dependent on the geometry and Reynolds number under consideration. Thus, it should come as no surprise that in all but the simplest cases the dispersion conductivity was found to be a function of both Reynolds and Prandtl numbers, rather than a simple function of the Péclet number.

With this in mind, it is clear that much work needs to be done to properly characterize the effective properties of newly developed porous materials such as metal foams in order to simulate their performance in a volume-averaged framework. In addition to solving the appropriate closure models on an idealized geometric model, calibration to experimental results may be necessary depending on the extent to which the geometry model is idealized [36]. This is likely to be an active area of research over the coming years as metal foams become a more mainstream material and find new applications in the field of heat transfer.

References

1. Bhattacharya, A., Mahajan, R.L.: J. Electron. Packag. **124**, 155 (2002)
2. Gallego, N.C., Klett, J.W.: Carbon **41**, 1461 (2003)
3. Boomsma, K., Poulikakos, D., Zwick, F.: Mech. Mater. **35**, 1161 (2003)
4. Straatman, A.G., Gallego, N.C., Thompson, B.E., Hangan, H.: Int. J. Heat Mass Transfer **49**, 1991 (2006)
5. Straatman, A.G., Gallego, N.C., Yu, Q., Betchen, L.J., Thompson, B.: J. Heat Transfer **129**, 1237 (2007)
6. Straatman, A.G., Gallego, N.C., Yu, Q., Thompson, B.E.: J. Eng. Gas Turbines Power Trans. ASME **129**, 326 (2007)
7. DeGroot, C.T., Straatman, A.G., Betchen, L.J.: J. Electron. Packag. **131**, 021001 (2009)
8. Leong, K.C., Li, H.Y., Jin, L.W., Chai, J.C.: Appl. Therm. Eng. **30**, 520 (2010)
9. Whitaker, S.: A.I.Ch.E. J. **13**, 420 (1967)
10. Slattery, J.C.: A.I.Ch.E. J. **13**, 1066 (1967)
11. Taylor, G.: Proc. R. Soc. Lond. A **219**, 186 (1953)
12. Aris, R.: Proc. R. Soc. Lond. A **235**, 67 (1955)
13. Yagi, S., Kunii, D., Wakao, N.: A.I.Ch.E. J. **6**, 543 (1960)
14. Hunt, M.L., Tien, C.L.: Int. J. Heat Mass Transfer **31**, 301 (1988)
15. Kuwahara, F., Nakayama, A., Koyama, H.: J. Heat Transfer **118**, 756 (1996)
16. Kuwahara, F., Nakayama, A.: J. Heat Transfer **121**, 160 (1999)

17. Pedras, M.H.J., de Lemos, M.J.S.: Int. J. Heat Mass Transfer **51**, 5359 (2008)
18. Amiri, A., Vafai, K., Kuzay, T.M.: Numer. Heat Transfer A-Appl. **27**, 651 (1955)
19. Quintard, M., Kaviany, M., Whitaker, S.: Adv. Water Resour. **20**, 77 (1997)
20. DeGroot, C.T., Straatman, A.G.: Int. J. Heat Mass Transfer **54**, 5039 (2011)
21. Moghari, M.: J. Enhanc. Heat Transfer **15**, 81 (2008)
22. DeGroot, C.T., Straatman, A.G.: J. Heat Transfer **134**, 042603 (2012)
23. Gray, W.G.: Chem. Eng. Sci **30**(1), 229 (1975)
24. Kaviany, M.: Principles of Heat Transfer in Porous Media, 2nd edn. Springer, New York (1955)
25. Whitaker, S.: Fluid Transport in Porous Media, Chap. 1, pp. 1–59. Springer, Southampton (1997)
26. Whitaker, S.: Ind. Eng. Chem. **61**, 14 (1969)
27. Nakayama, A., Kuwahara, F., Kodama, Y.: J. Fluid Mech. **563**, 81 (2006)
28. Krishnan, S., Murthy, J.Y., Garimella, S.V.: J. Heat Transfer **128**, 793 (2006)
29. Amiri, A., Vafai, K., Kuzay, T.M.: Int. J. Heat Mass Transfer **37**, 1029 (1994)
30. Sultan, K., DeGroot, C.T., Straatman, A.G., Gallego, N.C., Hangan, H.: Int. J. Heat Mass Transfer **52**, 4296 (2009)
31. Costa, V.A.F., Oliveira, L.A., Baliga, B.R., Sousa, A.C.M.: Numer. Heat Transfer, Part A **45**, 675 (2004)
32. Betchen, L.J., Straatman, A.G., Thompson, B.E.: Numer. Heat Transfer, Part A **49**, 543 (2006)
33. DeGroot, C.T., Straatman, A.G.: Numer. Heat Transfer, Part B **60**, 252 (2011)
34. Yu, Q., Thompson, B.E., Straatman, A.G.: J. Heat Transfer **128**, 352 (2006)
35. Calmidi, V.V., Mahajan, R.L.: J. Heat Transfer **122**, 557 (2000)
36. Karimian, S.A.M., Straatman, A.G.: J. Heat Transfer **131**, 052602 (2009)
37. Karimian, S.A.M., Straatman, A.G.: Int. J. Heat Fluid Flow **29**, 292 (2008)

Heat Transfer Enhancement in Short Corrugated Mini-Tubes

P. Kumar, F. Topin, M. Miscevic, P. Lavieille and L. Tadrist

Abstract Heat transfer phenomena are studied with standing waves inside the tubes for static and moving sinusoidal corrugated walls. The past studies have been done on big-size (dimensions in m) and micro-sized circular tubes (dimensions in μm). We are focusing on intermediate size tubes (dimensions in mm). Numerical simulations, using finite volume commercial software, were performed to study the effects of spatial wavelengths on heat transfer enhancement and associated pressure drop. We imposed 5, 10, 15 and 20 3D sinusoidal radial sine waves along the length of the tube. Heat transfer characteristics of static corrugated wavy walls were calculated for various imposed Reynolds numbers ($1 < Re < 120$) and amplitude of the wave was varied from 1 to 20 % of the diameter of the tube. For static wall case, upon increasing the number of sine waves, the Nusselt number starts to decrease; the associated pressure drop and friction factor increases very rapidly at the highest values of amplitude. On the other hand, in comparison to the static corrugated wall tube, the pressure drop is reduced by 20–80 % and heat transfer is enhanced by 35–70 % for highest amplitude when frequencies in the range $0 < f < 60$ Hz are imposed on tube wall to make the corrugated tube moving in transverse direction.

P. Kumar (✉) · F. Topin · L. Tadrist
IUSTI, CNRS UMR 7343, Aix Marseille University, Marseille, France
e-mail: prashant.kumar@polytech.univ-mrs.fr

F. Topin
e-mail: frederic.topin@polytech.univ-mrs.fr

L. Tadrist
e-mail: Lounes.Tadrist@polytech.univ-mrs.fr

M. Miscevic · P. Lavieille
LAPLACE, UMR 5213, Université Paul Sabatier, Toulouse, France
e-mail: marc.miscevic@laplace.univ-tlse.fr

P. Lavieille
e-mail: pascal.lavieille@laplace.univ-tlse.fr

J. M. P. Q. Delgado et al. (eds.), *Numerical Analysis of Heat and Mass Transfer in Porous Media*, Advanced Structured Materials 27,
DOI: 10.1007/978-3-642-30532-0_7,

Keywords Heat transfer · Pressure drop · Static wall · Moving wall · Merit factor

1 Introduction

The enhancement of heat transfer phenomena for laminar flow has been considered a significant research area in many fields such as cooling of electronic devices, HVAC systems, onboard compact heat exchangers (aerospace, automobiles) etc. Corrugated channels with a periodically converging–diverging cross section is one of the devices that are widely used in order to achieve heat transfer enhancement because such corrugated channel breaks the thermal boundary layer, enhance mixing, and thereby increase the convection heat transfer. It has been found experimentally that both flow and heat transfer becomes fully developed after 3–5 cycles [1] for moderate Reynolds numbers.

During the last decades, studies mainly focused on hydraulic diameter reduction in order to increase surface over volume ratio. For small diameters in usual operating conditions, the Reynolds number is relatively low and the flow is laminar rather than turbulent. Thus, the heat transfer characteristics are fully different from those obtained in classical devices. Further enhancement could not be obtained by increasing fluid velocity due to pressure drop increase. So, other ways to enhance heat transfer need to be explored, such as the use of corrugations similar to those used in classical size channels.

Goldstein and Sparrow [2] compared heat transfer in a corrugated channel and in a parallel plate channel. They found an enhancement in the average convective heat transfer coefficient by a factor of 3 in turbulent flow regimes obtained for relatively large tube diameter. The analyses of laminar forced convection in a wavy channel by Wang [3], Zhang [4] and Metwally [5] shown that heat transfer increases with the Reynolds number due to increase in size of vortices which enhance mixing. Niceno and Nobile [6] numerically investigated fluid flow and heat transfer through converging–diverging sinusoidal and arc shaped wavy channel. They pointed out little or no heat transfer enhancement compared to parallel-plate channel in steady flow regimes at low Reynolds number. In unsteady regime, an enhancement of heat transfer up to three times is obtained for $175 < Re < 200$ in sinusoidal channel and $60 < Re < 80$ in arc-shaped channel as a result of self-sustained oscillations in the fluid flow. They also reported that friction factor for both the geometries are higher than for parallel-plate channel. Mohamed et al. [7] numerically studied the effect of corrugation amplitudes for different flow rates ($100 < Re < 1{,}500$) in symmetric channels for water and air. They found that the heat transfer is slightly enhanced for low amplitudes while it decreases for amplitude above a critical value. They also reported that the local Nusselt number is characterized by a very sharp decrease near the entry of the channel and tends towards a constant value like in the study of Yuan et al. [1]. They did not discuss about the friction factor variation with respect to Reynolds number and corrugation amplitudes.

Studies were also conducted to determine heat and mass transfer characteristics of pulsatile flow in corrugated or grooved channels.

First study on pulsatile flow was carried by Bellhouse et al. [8] who reported improvement in heat and mass transfer in a furrowed channel due to large fluid oscillations. Subsequently, Macley and Stonestreet [9] analyzed the heat transfer in baffled tubes by varying frequency and amplitude of fluid oscillation. Little effect of fluid oscillation on heat and mass transfer was found. The authors explained this small effect by the laminar behavior of the flow; there is no mixing as described by Howes et al. [10]. Various other numerical and experimental studies [11–15] had been done in the past which explained the periodic redevelopment of the thermal and compositional boundary layers from each wave of grooved channel and its contribution on heat transfer enhancement. Wang and Vanka [16] explained that self-sustained oscillations lead to the destabilization of laminar thermal boundary layers, which replenish the near-wall fluid with the fluid in the core region, and thus provide a natural mechanism of heat transfer enhancement. In steady-flow regime, average Nusselt numbers for the wavy wall channel were only slightly larger than those for a parallel-plate channel but in the transitional-flow regime, the enhancement of heat transfer was about a factor of 2.5. Friction factors for the wavy channel were about twice those for the parallel-plate channel in the steady flow region, and remained almost constant in the transitional regime. Nishimura and Matsune [17] numerically studied pulsatile flow and performed flow visualizations to study the dynamical behavior of vortices generated in channels. Vortices expand in each furrow during the deceleration phase and shrink during the acceleration phase. An increment of the frequency leads to promote the vortex strength, but above a certain frequency the vortex strength reversely decreases.

To our knowledge, very few studies are available concerning low diameter circular corrugated tubes. Depending on Reynolds number range, geometry (symmetric or asymmetric) as well as corrugation amplitude, enhancement of heat transfer up to three times or decrease of heat transfer coefficient was obtained. No clear statement was found on the heat transfer enhancement produced by using corrugations in low diameter channel at low Reynolds number.

In this work, the heat transfer and pressure drop in a short circular corrugated tube crossed by a uniformly heated permanent fluid flow are investigated. Dynamically morphed sinusoidal wall tubes with imposed frequency and amplitude are not studied so far unlike wavy channel for industrial purposes [17, 18].

It is however important to find a solution that minimizes pressure drop and maximizes heat transfer rate. So, both static and moving wavy wall tube are considered in this work. Firstly, the model, mesh and numerical procedure are detailed. Then local analyses of flow and heat transfer are presented. Finally, effects of variation of main parameters on global behavior are determined. Conclusions about the use of static or moving corrugations in small circular channels in order to enhance heat transfer are derived.

2 Numerical Model

The numerical simulations were realized using commercial software, StarCCM+ . Constant fluid properties were assumed and the flow was considered three-dimensional and laminar. Flow and conjugate heat transfers problems were solved simultaneously.

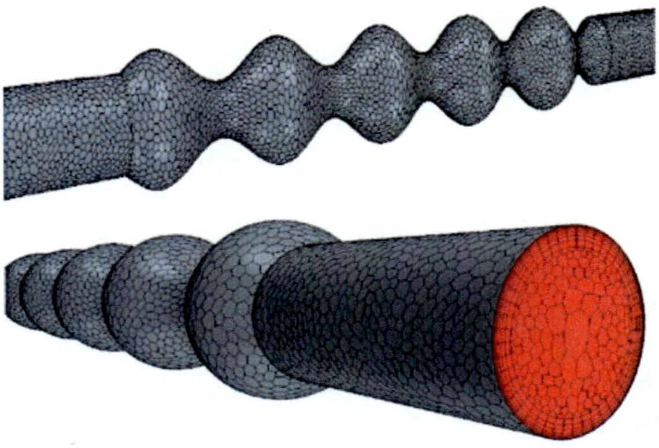

Fig. 1 View of the deformed tube, and of the mesh

Both Navier–Stokes and energy equations were solved using a segregated approach. The test section as shown in the Fig. 1 is composed of an adiabatic zone of length $2D_0$ at inlet, a heated zone of length $10D_0$ and an outlet adiabatic zone of length $4D_0$. The lengths of the adiabatic zones were chosen to limit the mesh size as well as the impact of the boundary conditions on the flow pattern and heat transfer.

Boundary conditions were: imposed uniform velocity and temperature at inlet, uniform heat flux (10,000 W/m^2) on length $10D_0$ and uniform pressure at the exit of the tube.

The tube diameter and amplitude of the corrugated wall are related as below:

$$A = nD_0 \tag{1}$$

where n is varying from 1 to 20 %, A and D_0 are deformation amplitude and initial diameter of tube respectively. Note that for $n = 25$ %, the contraction of the tube close the section which means that the walls are in contact at all the contractions.

2.1 Numerical Validation on Static Wall

In order to validate the numerical procedure, several calculations were conducted for fully developed flow in the deformed tube of 1 mm diameter. The mesh is composed of polyhedral cells core (approximately 10,000 up to 150,000 cells) with additional prism layers mesh in the vicinity of solid surfaces. The numerical analyses were made first on static wall (stationary regime) and then, subsequently on moving wall (transient regime).

Heat transfer as a function of the number of cells for different Reynolds numbers on a deformed corrugated wall is represented in Fig. 2. For all studied cases, convergence is obtained for more than 110,000 cells approximately, corresponding to an equivalent mesh size of 0.2 mm.

Moreover, meshing introduces small geometrical errors that decrease when wavelength increases and amplitude decreases. These geometrical errors change a

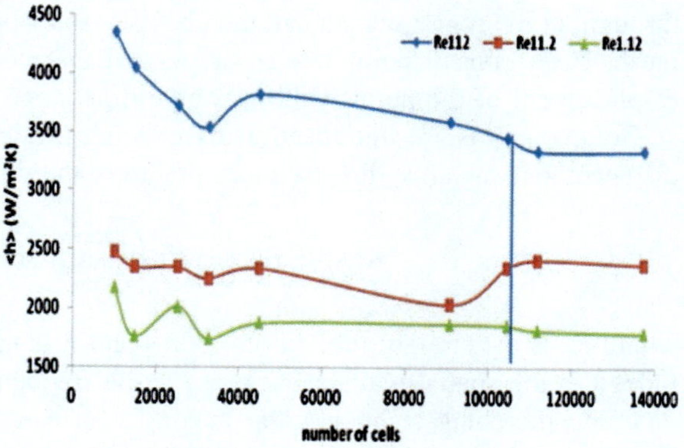

Fig. 2 Variations of heat transfer coefficient as a function of the number of cells at for several Reynolds numbers for the static corrugated wall case ($n = 20\ \%$)

Table 1 Influence of mesh size on global heat transfer coefficient and pressure drop at for Reynolds numbers 56 and 112 (Wave amplitude = 20 %, f = 10 Hz)

Re	f(Hz)	Mesh size (mm)	h (W/m^2K)	ΔP
56	10	0.4	4202	29
56	10	0.2	3759	30
56	10	0.08	3729	30
112	10	0.4	5173	71
112	10	0.02	5058	74
112	10	0.08	4926	74

little bit hydraulic diameter (see Sect. 4 and Table 2). The chosen mesh size leads to a maximum error of 2 % on hydraulic diameter for the worst case i.e. for the smallest wavelength and highest amplitude considered in this chapter.

2.2 *Numerical Validation on Moving Wall*

In the case of moving wall, three different mesh sizes of 0.4, 0.2 and 0.08 mm have been tested. Simulations were performed considering two Reynolds numbers (56 and 112), the highest amplitude (20 %) and 10 Hz frequency imposed on wavy surface as presented in Table 1.

The results start to converge at lower mesh size as shown in Table 1. For 0.08 mm, the computational time is far longer and the results are varying less than 2 % compared to results obtained with mesh size of 0.2 mm. So, like for the static wall case, we have retained 0.2 mm mesh size for all the calculations carried on moving wall case.

3 Morphing of the Tube

Morpher of the commercial software StarCCM + enables the movement of the wall with an imposed frequency. The wall motion redistributes mesh vertices in response to the movement of control points. Control points and their associated displacements are used by the mesh morpher to generate an interpolation field throughout the region which can then be used to displace the actual vertices of the mesh. Each control point has an associated distance vector which specifies the displacement of the point within a single time-step.

Control points are initialised using existing mesh vertices on a boundary. The displacement specifies directly more precisely the position of the wall defined by:

$$R(x) = R_0 + A\sin\left(\frac{2\pi}{\lambda}x\right).\sin(2\pi ft) \tag{2}$$

where R_0 is the straight tube initial radius and A is the amplitude of the deformed tube, λ is the spatial wavelength and f is the frequency imposed on the wall.

Using morphing technique, the moving wall case has been studied for various frequencies ($0 < f < 60$ Hz) to determine heat transfer and pressure drop characterstics.

4 Useful Parameters

We extract, for moving wall case, only time averaged quantity over a period, the term $\cos(2\pi ft)$ vanishes. For sake of clarity in all definitions below, the time integration has been omitted. Consequently, definitions are similar both static and moving wall cases.

Parametric studies of the effect of wall displacement and flow characteristics on heat transfer and pressure drop were made, involving the knowledge of the various parameters defined below. First, the local Reynolds number is defined according to the hydraulic diameter by :

$$\mathrm{Re} = \frac{\rho\,\bar{u}\,D_h}{\mu} \tag{3}$$

where,

$$D_h = \frac{1}{\lambda}\int_0^{\lambda}\frac{4S}{P}d\lambda = \frac{2}{\lambda}\int_0^{\lambda}R(x)dx \tag{4}$$

where S is the cross-section area of deformed tube, P is wetted perimeter, λ is spatial wavelength and R(x) is defined in Eq. 2. On substituting the value of R(x) in Eq. 4, we get

$$D_h = 2R_0 \tag{5}$$

We calculated D_h from the actual volume and surface of the deformed region for various amplitudes. Because of deformation, the meshing introduced some discrepancies. A maximum error of 2 % was obtained between analytical and numerical D_h for relative amplitude between 0 and 20 % (Table 2). Analytical (see Appendix) and actual wall surface area were compared and the error was found to be less than 1 %.

Table 2 Numerical hydraulic diameter and wall surface area compared to analytical ones with associated errors for a wavelength of 1 mm

Amplitude (%)	D_h(Analytical) $\times 10^4$(m)	D_h(Numerical) $\times 10^4$(m)	Error (%)	Analytical area ($\times 10^5 m^2$)	Numerical area ($\times 10^5 m^2$)	Error (%)
0	10	9.91	0.9	3.14	3.13	0.45
1	10	9.95	0.5	3.14	3.13	0.46
2	10	9.95	0.5	3.15	3.14	0.47
2.5	10	9.95	0.5	3.16	3.15	0.47
5	10	9.95	0.5	3.22	3.20	0.52
7.5	10	9.94	0.6	3.31	3.29	0.59
10	10	9.87	1.3	3.43	3.41	0.67
15	10	9.83	1.7	3.75	3.72	0.86
20	10	9.80	2.0	4.15	4.11	1.03

Local heat transfer coefficient, h(x) was calculated as follows :

$$h(x) = \frac{q}{\overline{T_w}(x) - \overline{T_f}(x)} \tag{6}$$

where, q is the imposed wall heat flux on the deformed wall, and $\bar{T}_f$ and $\bar{T}_w$ are the average local mean temperature of the fluid inside a cross-section and the average wall temperature over the perimeter of a cross-section, respectively, defined by:

$$\overline{T_f}(x) = T_{in} + \frac{qS(x)}{\dot{m}C_p} \tag{7}$$

$$\overline{T_w}(x) = \frac{1}{2\pi}\int_0^{2\pi} T_w(x,\theta)d\theta \tag{8}$$

where T_{in} is the fluid temperature at inlet and S(x) is the wall surface area of the deformed tube between the inlet and the position x.

Local friction factor f was evaluated as:

$$f(x) = \frac{\frac{dP(x)}{dx} D_h(x)}{\frac{1}{2}\rho\,\bar{u}(x)^2} \tag{9}$$

where $\frac{dP(x)}{dx}$ is pressure gradient, ρ is density and $\bar{u}(x)$ is mean velocity in a cross-section of the tube.

Several quantities were also defined at global averaged scale, i.e. averaged quantities between two abscissa (x_0 and x_1) or on the whole deformed region. So, the averaged wall temperature was calculated as:

$$\langle T\rangle_w = \frac{1}{S(x_1) - S(x_0)}\int_{S(x_0)}^{S(x_1)} T_w dS \tag{10}$$

Similarly, the averaged fluid temperature was calculated as:

$$\langle T \rangle_f = T_{in} + \frac{q\{S(x_0) + S(x_1)\}}{2\dot{m}C_p} \tag{11}$$

We adopted the following convention in this chapter: when x_0 and x_1 are omitted, values are averaged along the whole length ($0 \rightarrow L$) of the deformed region.

Averaged global heat transfer coefficient $\langle h \rangle$ and Nusselt number, $\langle Nu \rangle$ of deformed wall were calculated as below:

$$\langle h \rangle = \frac{q}{\langle T_w \rangle - \langle T_f \rangle} \tag{12}$$

$$\langle Nu \rangle = \frac{\langle h \rangle D_h}{k} \tag{13}$$

The Colburn factor j is defined as:

$$j = \frac{\langle Nu \rangle}{\langle \mathrm{Re} \rangle \, \mathrm{Pr}^{1/2}} \tag{14}$$

where, $\langle \mathrm{Re} \rangle = \frac{\rho \langle \bar{u} \rangle D_h}{\mu}$ is the average Reynolds number in the tube.

The performance comparisons were made considering flow area goodness factor defined as the ratio of the Colburn factor (*j*) to friction factor (*f*). A surface having a higher *j/f* factor is "good" because it requires lower free flow area and hence a lower frontal area for heat exchanger [19].

5 Results and Discussion

5.1 Local Analysis of Static Corrugated Wall Case

We study the influence of amplitudes and Reynolds number variations on local heat transfer, instantaneous fluid and wall temperature and velocity profiles. The analyses were presented on reduced scale (given by Eq. 15) for all the parameters. Local fluid and wall temperature, velocity and heat transfer, T_f^*,T_w^*, V^* and h^* with respect to x-direction (main flow axis) can be rewritten in the reduced form as:

$$\psi^* = \frac{\psi - \psi_{\min}}{\psi_{\max} - \psi_{\min}} \tag{15}$$

where, $\psi_{\min}(x)$ and $\psi_{\max}(x)$represents any thermal or flow properties.

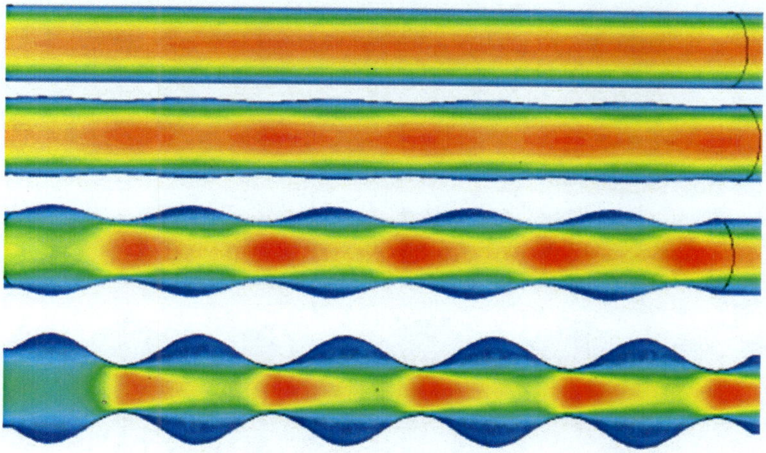

Fig. 3 Flow field in the straight tube and corrugated tube for A = 2, 10 and 20 % for a Reynolds number Re = 112

5.1.1 Impact of Amplitude

The changes in the flow fields according to an increase of the amplitude of the corrugation are illustrated in Fig. 3 for a constant Reynolds number of 112. Four cases are represented corresponding to A = 0 (straight tube), 2, 10 and 20 %. Major differences between the corrugated channel and the straight channel are obtained for A = 10 and 20 %. The fluid velocity is sharply increased in the shrinkage zone (see Fig. 3). Despite the low value of the Reynolds number, maximum velocity is reached downstream the neck of the corrugation, which is characteristic of a boundary layer detachment. As a consequence, the velocity is nearly zero close to the channel wall in the enlargement region. The flow appears to be periodically developed after the second corrugation.

Regarding the fluid temperature fields (see Fig. 4), one can notice hot zones near the wall in the diverging region of the corrugations whatever is the amplitude. These high temperature zones are related to the low recirculation velocity of the fluid is these regions, which doesn't contribute significantly to heat transfer. In the converging zones, the wall temperature is less due to the impact of the jet formed upstream the corrugation neck.

Heat transfer coefficient profiles were derived from these fluid temperature fields according to Eq. 6. These profiles present a oscillating behavior when A is increased (see Fig. 5).

For A = 0, 2 and 10 %, h* presents a decreasing trend according to the position in the channel. For higher amplitudes, the flow appears to be periodically thermally fully developed at approximately the end of the first corrugation location. The heat transfer coefficient is less in the diverging region of the corrugation, where the recirculation velocity is low. It reaches its higher value in the converging zones. The maximum value is reached upstream of the contraction where

Fig. 4 Temperature field in the straight tube and corrugated tube for A = 2, 10 and 20 % for a Reynolds number Re = 112

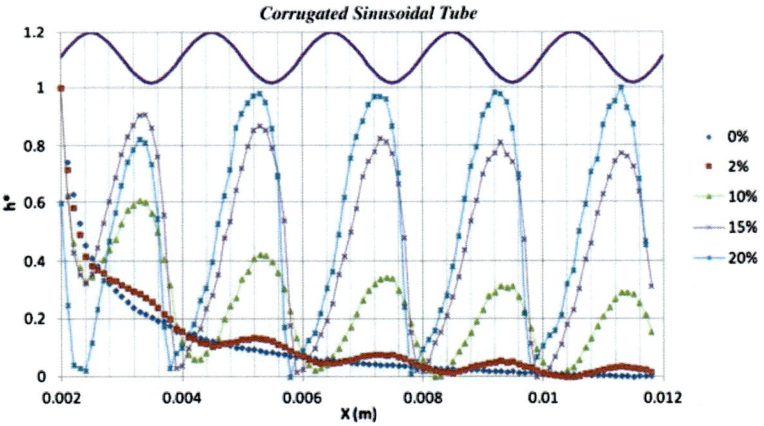

Fig. 5 Impact of corrugation amplitude on local heat transfer coefficient profiles for Re = 112. We superimposed the tube shape above the heat transfer curves

the core high velocity fluid goes near the wall. The minimum values are obtained downstream the contraction where the core flow separates from the wall by the recirculation zone.

5.1.2 Impact of Reynolds Number

Simulations have been performed for various Reynolds number at 20 % amplitude. The reduced form of centerline velocity as a function of longitudinal abscissa is plotted in Fig. 6. It is clear that for low Reynolds number (Re $\approx$ 10), the velocity reaches a

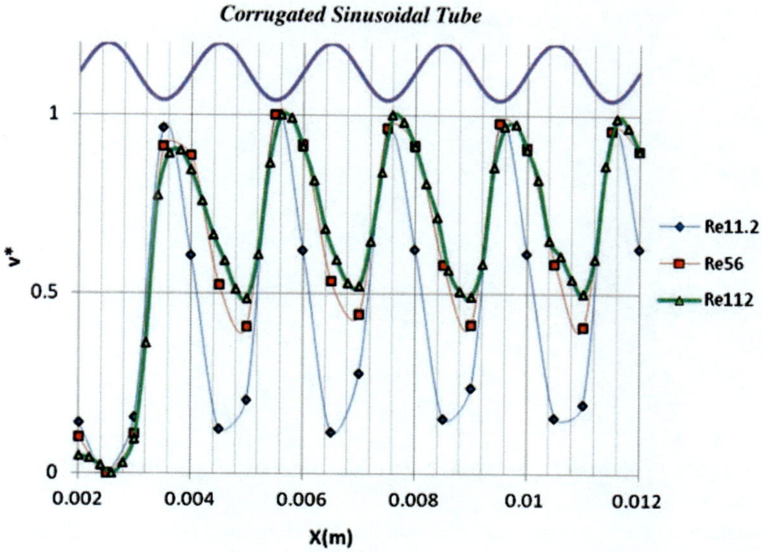

Fig. 6 Centerline velocity profiles at 20 % amplitude for several Re. We superimposed the tube shape above the velocity curves

maximum at the contraction of the tube. But as the Reynolds number increases, there is a shift in the location of maximum velocity position. The maximum velocity is obtained just after wall contraction and indicates boundary layer detachment. The fluid flow doesn't follow the corrugated wall at high Reynolds number.

For higher Reynolds number at highest amplitude, the flow appears to be periodically thermally fully developed at approximately the end of the first corrugation location. The local heat transfer characteristic doesn't follow the corrugated wall at high Reynolds number as explained in the case of reduced velocity profiles (see Fig. 6).

The location of maximum local heat transfer is obtained just before the contraction of the tube. Upon increasing Reynolds number, the position of maximum local heat transfer shifts upstream of the actual minimum contraction of the tube as shown in Fig. 7.

The streamlines pattern characteristics of recirculation at different Reynolds number (at 20 % amplitude) are shown in the Fig. 8 where the recirculation zone is almost stationary in the furrows at higher Reynolds number.

5.2 Global Analysis of Static Corrugated Wall Case

For lower Reynolds number i.e. Re = 1, the wavy wall heat transfer coefficient compared to straight tube one increases for amplitudes up to 5 % and then starts to decrease as shown in Fig. 9. For higher Reynolds numbers, (Re > 10), it has been seen that heat transfer increases for amplitudes up to only 1–2 % and then starts to decrease.

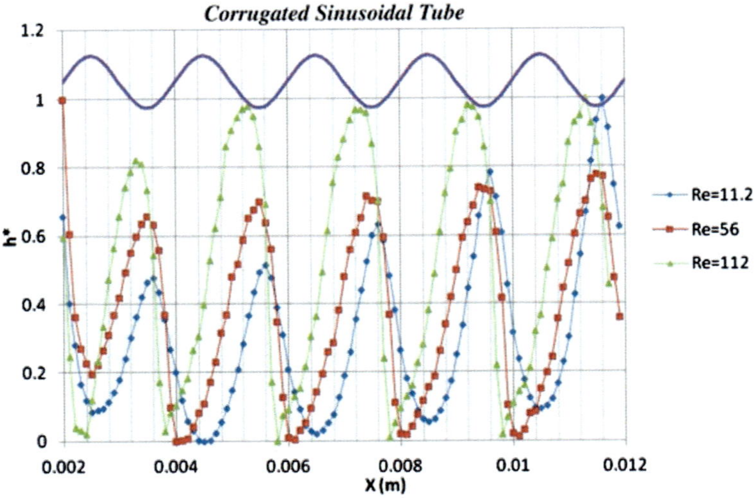

Fig. 7 Impact of Reynolds on local heat transfer coefficient profiles at 20 % amplitude. We superimposed the tube shape above the heat transfer curves

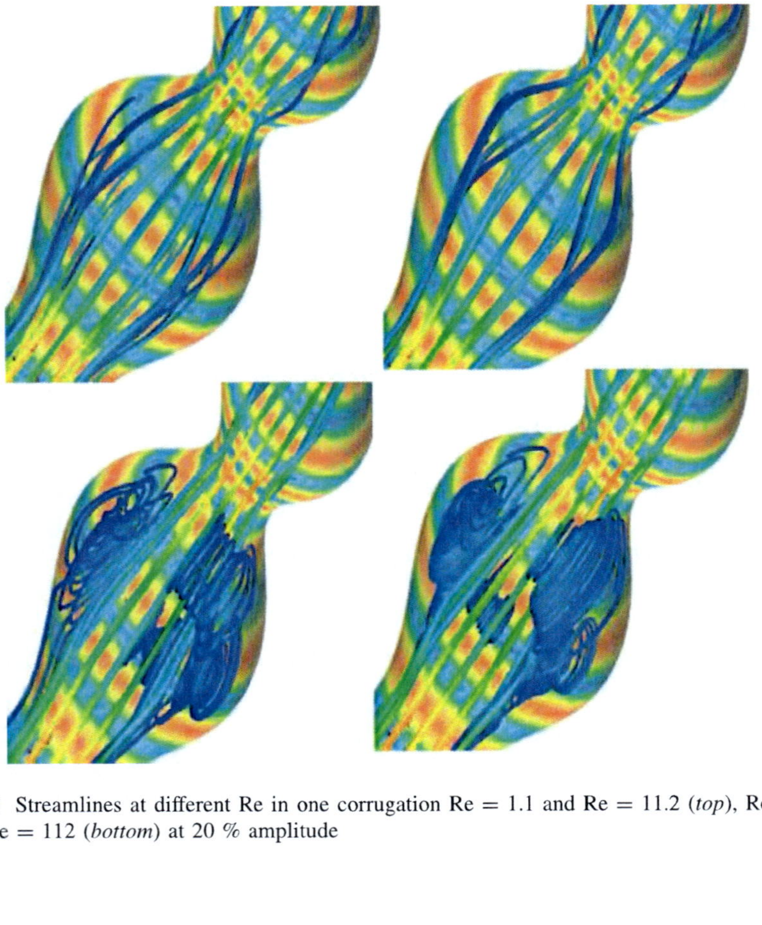

Fig. 8 Streamlines at different Re in one corrugation Re = 1.1 and Re = 11.2 (*top*), Re = 56 and Re = 112 (*bottom*) at 20 % amplitude

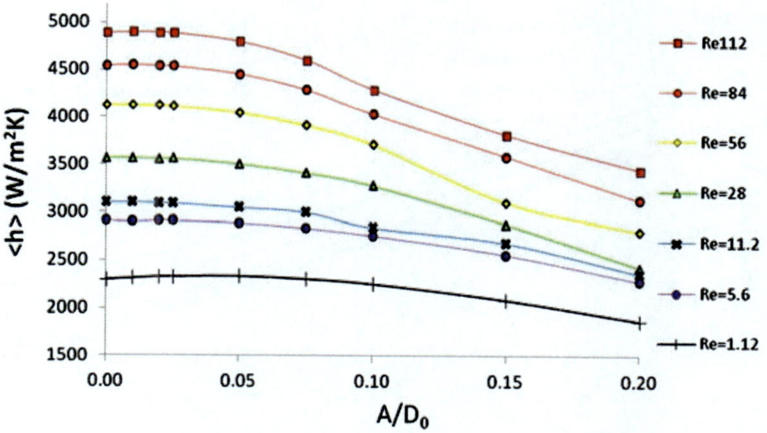

Fig. 9 Variation of heat transfer coefficient with A/D_0 (reduced form) at different Re

Small deformation on a straight tube enhances heat transfer due to recirculation zone which has enough velocity for better fluid mixing and increases the heat transfer rate but on the other hand, when the deformed structure is too big compared to 5 % of amplitude, the recirculation zone becomes bigger which reduces the flow velocity in that area and so nullifies the effect of mixing; heat transfer enhancement starts to fall down below the value obtained with straight wall circular tubes. For very small amplitudes (1–2.5 %), no real recirculation zones exist and no clear detachment is observed. The global heat transfer enhancement is mainly due to slight increase in velocity in contraction zone without occurrence of a buffer zone in the furrow.

Pressure drop across the wavy wall section is found to increase with Reynolds number. The velocity is continously increasing with the increase of amplitude as presented in Fig. 3 which has direct impact on the pressure drop which reaches higher value for higher Reynolds number. Friction factor increases with amplitude as a huge amount of stationary buffer fluid stays inside the furrows implying very high velocity at the centre of tube. Also with increase in the Reynolds number, friction factor decreases having the same behavior of slope of -1 like the case of straight tube as shown in Fig. 10. This indicates that the flow regime is always laminar even in presence of recirculation zone (detached boundary layers) for tested cases.

Figure 11 indicates that for short tubes (i.e. when flow is not fully developed), the highest performances are obtained with straight tubes compared to symmetric wavy ones. These results are true only for low Re. On the other hand, for long tubes, as flow becomes established in straight tube, better performance will be obtained with moderate amplitude wavy tubes. It is probable that for better heat transfer characteristics, amplitude of 5 % is sufficient. At high amplitudes, the j/f ratio starts to decrease and high amplitude corrugation is not appropriate for heat exchanger design due to high friction factor and decreasing heat transfer characteristics.

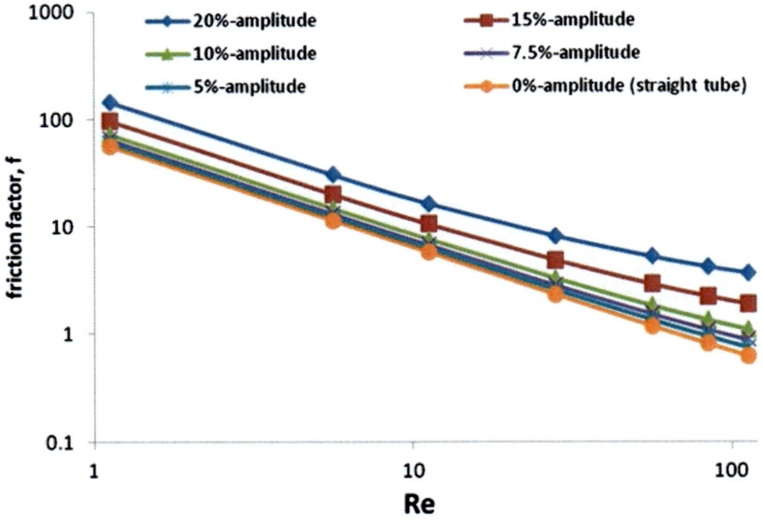

Fig. 10 Variation of friction factor 'f' with different Re at different amplitudes

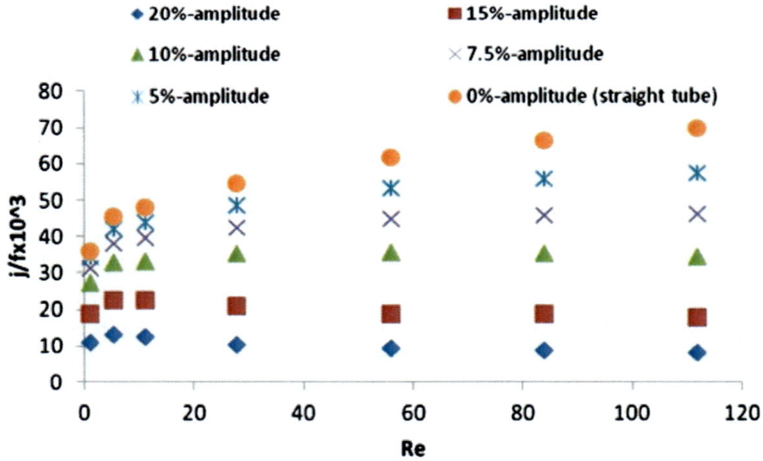

Fig. 11 Effect of Reynolds number on heat transfer performance factor for several amplitudes. All deformations lead to a decrease in performance for this short tube at low Reynolds number

5.3 Spatial Wavelength Analysis Across Wavy Wall

Other wavelengths such as $\lambda = 1/2$, 2/3 and 1 mm have been chosen i.e. 10, 15 and 20 spatial wavelengths along the length of the channel and compared with $\lambda = 2$ mm (5 spatial wavelengths) to see its impact on heat transfer and pressure drop charateristics.

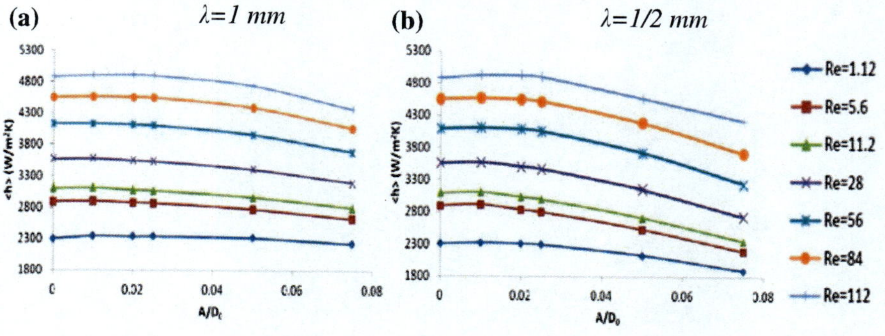

Fig. 12 Variation of heat transfer coefficient with A/D_0 at different Re for **a** $\lambda = 1$ mm **b** $\lambda = 1/2$ mm

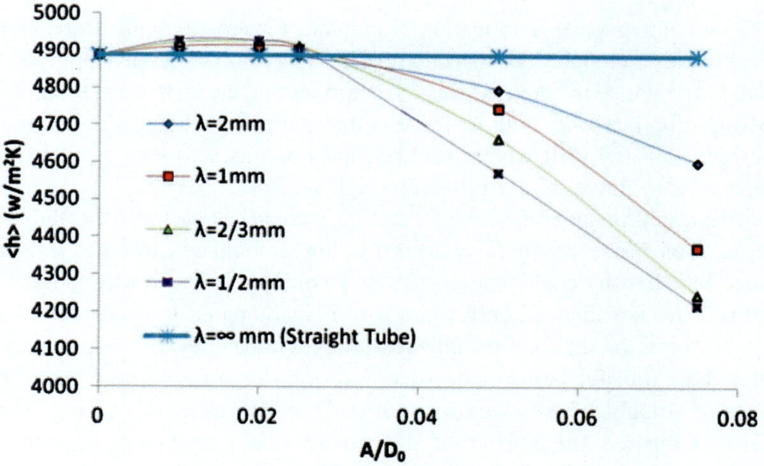

Fig. 13 Variation of heat transfer versus reduced amplitude for several wavelengths at Re 112

Meshing introduced geometrical error, these being more important for small spatial wavelengths case. These discretization errors introduce small variations (about 2 %) for amplitudes less than 10 % on hydraulic diameter. Consequently, we present here only the results obtained for amplitudes less than 10 % of amplitude in order to compare results at constant hydraulic diameter. From Fig. 12, heat transfer at different spatial wavelengths for all Re follows the same trend i.e. increase for 1–2 % of amplitude and decreases with increase in the amplitude like the case of 5 sine waves.

The behavior of heat transfer for different wavelengths at Re = 112 is shown in Fig. 13. One could see that the impact of wavelength is smaller than the amplitudes ones. The maximal variations on heat transfer are of about 10 % at high amplitudes and negligible for smaller ones. On the other hand, wavelength impacts strongly on pressure drop. For a given tube length, pressure drop increases when wavelength decreases whilst pressure drop per unit pattern (sine wave) decreases when wavelength decreases.

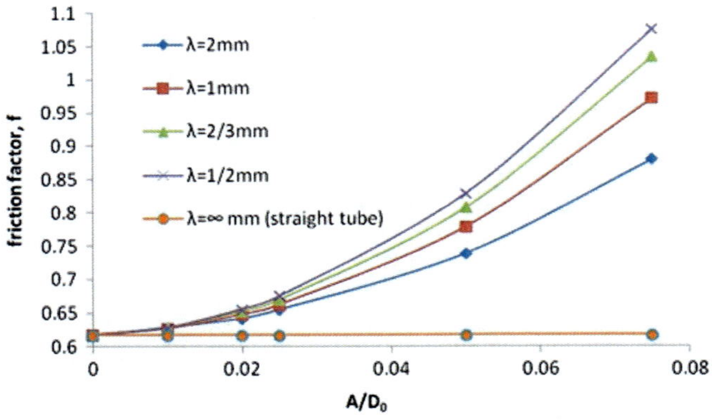

Fig. 14 Variation of 'f' versus relative amplitude for several spatial wavelengths at Re = 112

In Fig. 14, for a given corrugation amplitude, f increases with decrease in the wavelength. This is probably due to increase of furrows number per unit length and hence repeated obstruction to the fluid flow which contributes to the increase in friction factor.

Pressure drop increases with increase in the spatial wavelengths due to repeated obstructions of wavy wall where a lot of fluid remains stationary which imposes the mean flow to move with a relatively high velocity.

From the results obtained in Sects. 5.2 and 5.3, it could be concluded that the pressure drop in case of 5 sine waves ($\lambda = 2$ mm) is lower compared to other sine waves. Moreover, heat transfer coefficient doesn't vary considerably with wavelengths.

Heat transfer is enhanced only when $n < 5$ % compared to established straight tube. j/f factor is better than established straight tube flow but when amplitude increases, heat transfer performance decreases and is always lower than the performance of straight tube of small length (L/D $\approx$ 10) as shown in Fig. 15a. We have also compared the impact of decreasing spatial wavelengths (increasing number of sine waves) in Fig. 15b. Increase in spatial wavelength increases j/f factor. This latter is always better than established straight tube case for Re > 30 but never reaches j/f values of small length straight tube.

An optimal design for low Reynolds heat exchange devices will be obtained using a composite pattern such as composed of straight part followed by few corrugations. Such a pattern will lead to moderate pressure drop because major portion is a straight tube and will give good heat transfer performance until the thermal boundary layer is established. More in-depth study should be performed to reach full optimisation of the pattern for a given Reynolds range.

5.4 Local Analysis of Moving Wall Case

In case of short mini tube with static wavy wall, we have shown that the decrease in wavelength decreases heat transfer and increases pressure drop. In order to compare with this reference case, a wavelength, $\lambda = 2$ mm (5 sinusoidal waves) is

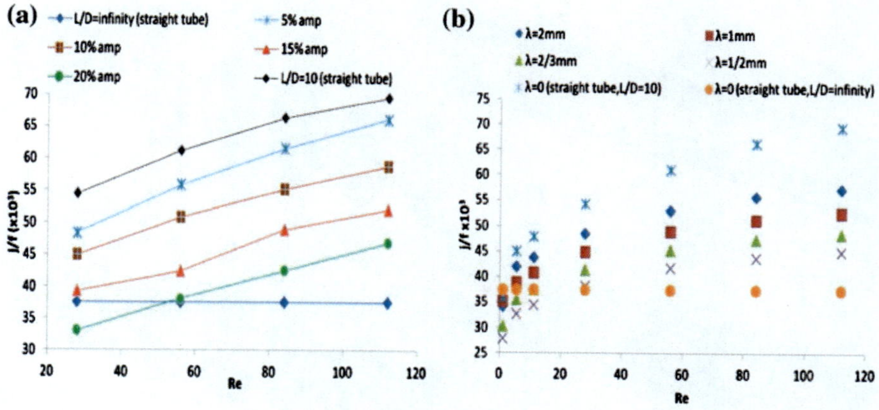

Fig. 15 Effect of **a** amplitudes for $\lambda = 2$ mm [static wall] **b** spatial wavelengths (λ = 1/2, 2/3 and 1 mm) at 5 % amplitude [static wall] on j/f ratio versus Reynolds number

chosen to perform analysis on moving corrugated wall. Various simulations for frequencies varying from 0 up to 60 Hz have been performed on same amplitude range like static wavy wall case.

In order to produce moving wall results, we have used the following procedure. We initialized both tube shape and fluid flow. Then, calculation (including wall movement) is carried out until a periodic stationary regime is reached. This latter point is checked by comparing global heat transfer: calculations are performed until the heat transfer coefficient value between two consecutive time periods differs by less than 1 %. Consequently, an additional time period is calculated to extract all instantaneous and time averaged values of all physical quantities.

The local results during one period are presented in Figs. 16, 17 and 18. In static wall case, the core flow has high velocity and stationary buffer fluid stays near the wall. This latter separates the cold core fluid and the wall and thus, low heat transfer is obtained. In the case of moving wall, the temperature fields and velocity fields are completely different compared to stationary wall case. Velocity pattern is complex and depends strongly on time and expansion of the radius of tube.

During expansion of the tube, it moves the fluid in both radial and longitudinal direction. Wall movement induces radial displacement of the fluid. Volume is not conserved and produces a velocity fluctuation along the main flow axis (pulsating flow) that increases along the main axis and is superimposed on average fluid velocity. Due to radial movement, cold fluid is introduced near the wall which gets heated. Mixing then takes place with the core flow and thus enhancing heat transfer (see Figs. 16, 17 and 18). The heat transfer in this case strongly depends on frequency and Reynolds number but no clear dependence has yet been found.

The pressure drop is expected to change because wall movement gives mechanical energy to the fluid. The flow pattern is different and hence mechanical energy dissipation is also different and produces lower pressure drop compared to

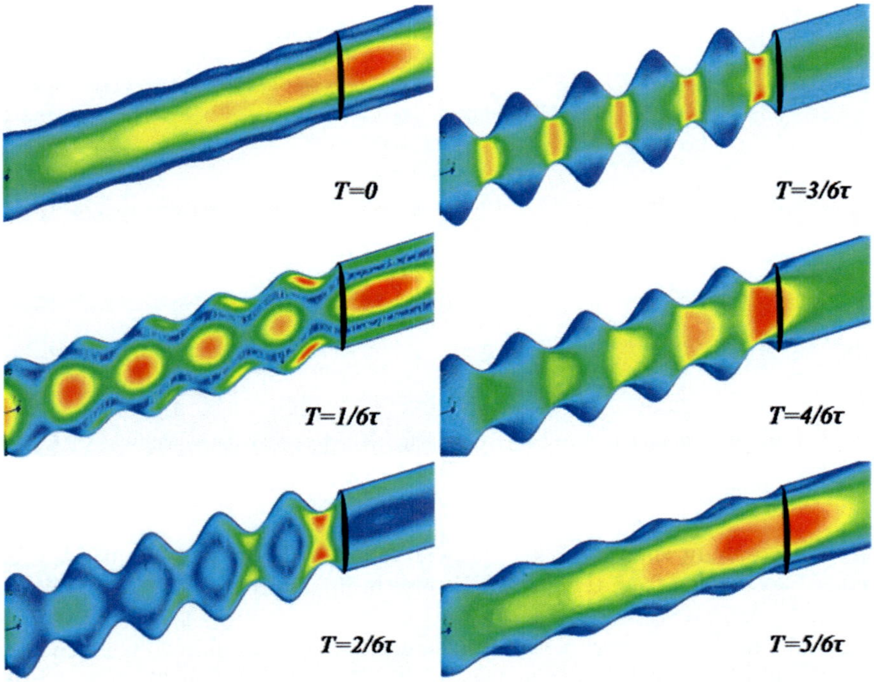

Fig. 16 Dynamics of velocity field during one period

static wall case. The core flow does not remain all the time at high velocity contrary to the static wall case. We even observed minimum velocity in the smallest section of tube at some point of time.

5.5 Global Analysis of Moving Wall Case

Numerical simulations on moving wavy walls were performed at Reynolds number ranging from 25 to 120 but only detailed analysis of Re = 112 are presented. The analyses are classified into parametric studies of influence of frequency, amplitude and Reynolds number on heat transfer and pressure drop characteristics.

5.5.1 Influence of Frequency

From Fig. 19, heat transfer coefficients at different amplitudes are following a common trend at all the frequencies (increasing and decreasing characteristics) except at 20 % amplitude. For this amplitude, heat transfer coefficient is lower at

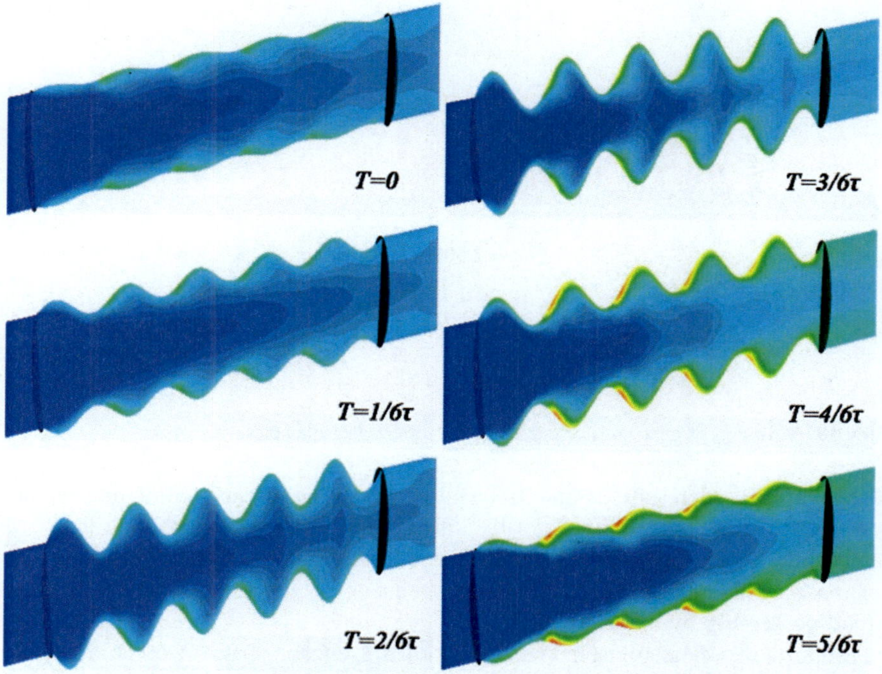

Fig. 17 Dynamics of temperature field during one period

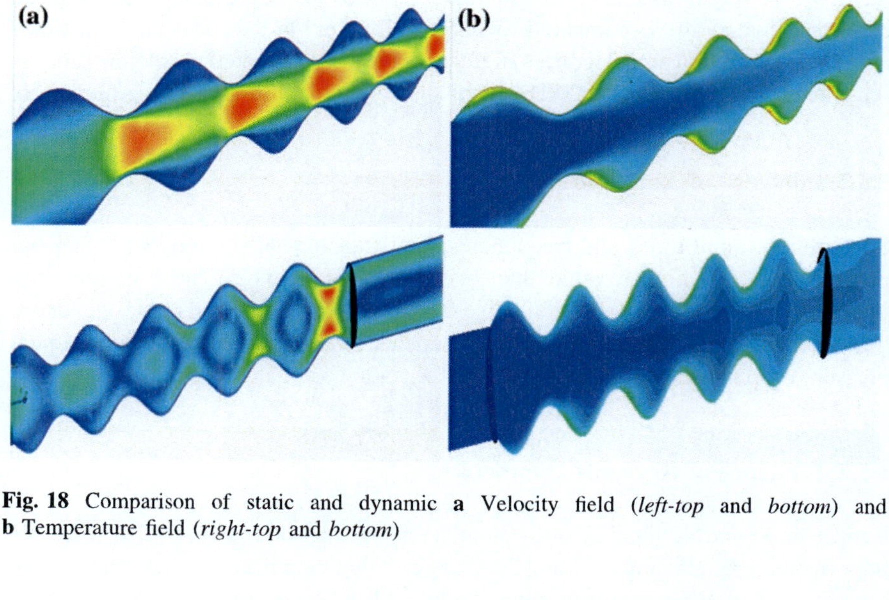

Fig. 18 Comparison of static and dynamic **a** Velocity field (*left-top* and *bottom*) and **b** Temperature field (*right-top* and *bottom*)

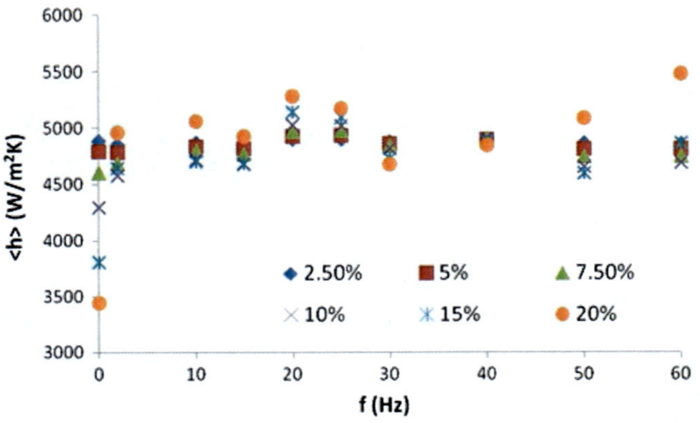

Fig. 19 Variation of heat transfer coefficient at different frequencies

0 Hz, it then increases as the frequency reaches 2 Hz and dominates values obtained for other amplitudes till 25 Hz. Heat transfer coefficient at 20 % amplitude then possess lower value than other amplitudes at 30 Hz. It starts to increase again monotonically at higher frequencies, deviating from the general trend possessing by other amplitudes.

Importantly, heat transfer characteristics are enhanced up to 60 % at higher amplitudes (20 %) unlike the case of static wall case. For intermediate amplitudes, enhancement in heat transfer reaches only 20 %. This trend of heat transfer enhancement as a function of amplitude leads us to think that for very high amplitude (20–25 %), extremely high heat transfer performance is expected.

Figure 20 reveals that at higher amplitudes (10–20 %), the pressure drop is reduced by approximately 40 % compared to static wall case. On the other hand, at lower amplitudes, wall movement reduces slightly pressure drop as detailed latter in Table 3. It has been clearly seen that at 20 Hz, for all the amplitudes pressure drop is minimum.

5.5.2 Influence of Amplitude

For different amplitudes and frequencies, heat transfer has no apparent order but possesses always higher value than static corrugated tube. The pressure drop increases with increase in amplitude like the case of static wall and posses quadratic behavior with amplitude. Again, no clear dependency on frequency is visualized but pressure drop values are always less than for static corrugated tube case.

5.5.3 Influence of Reynolds Number

Impact of Reynolds number on heat transfer and pressure drop is also of prime importance. From the Figs. 21 and 22, it appears that both heat transfer and pressure drop increase with Reynolds number at a given frequency. Small variations in heat

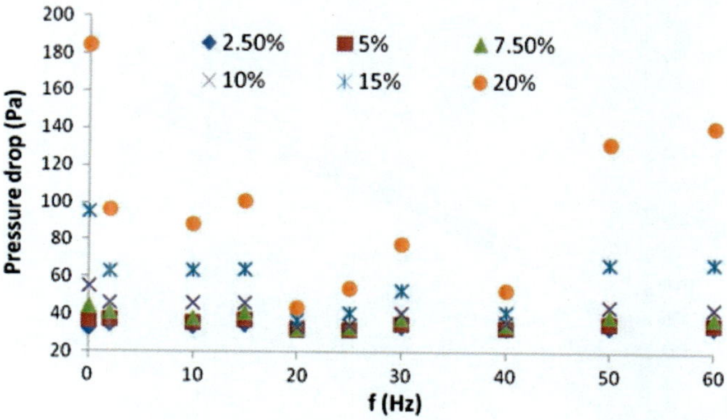

Fig. 20 Variation of ΔP at different frequencies

transfer coefficient and pressure drop are noticed till 10 % of amplitude. The heat transfer possess cubic behavior while the pressure drop possesses slightly quadratic behavior with Reynolds number (at a given frequency, here the results are shown for 20 Hz). The quadratic behavior is produced by inertia effects that are induced by wall movement. These effects become significant only for high amplitudes and frequencies. More precisely, these effects are induced by the pulsating longitudinal component due to the volume variations of the tube.

5.5.4 Synopsis of Global Analysis

We made the studies on 20 % amplitude case in order to gain a better understanding of impact of frequencies on heat transfer and pressure drop. The range of frequencies (0–60 Hz) is divided into low frequency range (0–30 Hz) and high frequency range (30–60 Hz) for better analysis of the phenomena. The global results are presented for Re = 112 at 20 % amplitude within the entire range of frequency.

From the Fig. 23, it appears that there is an enhancement in heat transfer and reduction in pressure drop compared to static wall case. The low frequencies points form a cloud of size of height of about 0.2 and width of about 0.3 around the position (0.3, 1.4). Most of the high frequencies points gather along a curve going from (0.6, 1.4) up to (0.8, 1.7). Several high frequencies points lie also in the same area than low frequencies ones. Note that we studied only about 20 data points in high frequency range. Considering, the behavior obtained for low frequency, it is possible that some not studied frequencies may lead to dispersed results anywhere in the frame presented in the Fig. 23. Globally for the majority of the points, heat transfer is increased by a factor of about 1.5. In worst case, the enhancement factor is about of 1.3 and the best enhancement obtained is 1.7.

For the majority of points, the pressure drop is 3 times smaller than static wall case and in the worst case, the reduction is only 20 %. For low frequency range, pressure drop

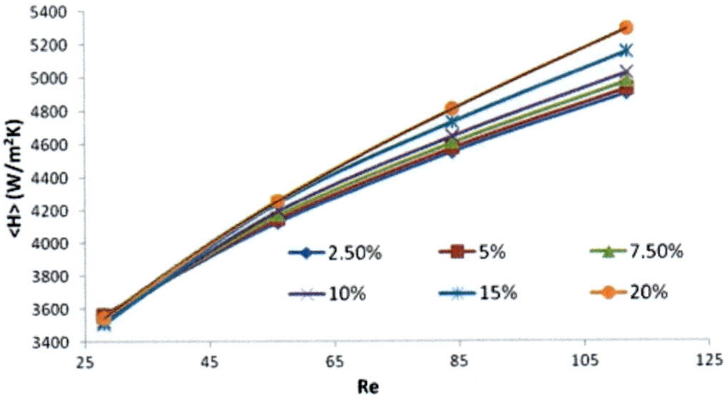

Fig. 21 Variation of heat transfer coefficient with Re at f = 20 Hz

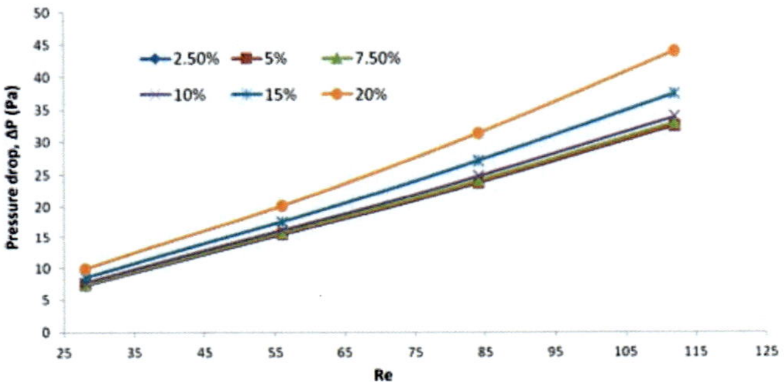

Fig. 22 Variation of ΔP with Re at f = 20 Hz

is always lesser than half of the static wall case value. On the other hand, for high frequency range, we obtained two different results: half of the points lie relatively high pressure drop range while the other half are close to the minimum value of pressure drop.

No clear dependence with studied parameter is obtained. Apparently, frequency does not govern directly pressure drop nor heat transfer coefficient values.

In order to quantify the cloud and dispersed behavior of values shown in Fig. 23, we have plotted the results in a form that compares reduced heat transfer enhancement to reduced pressure drop. The scattering disappears when considering the performance factor $[\langle h\rangle/\langle h_0\rangle]/[\Delta P/\Delta P_0]$ instead of$\langle h\rangle/\langle h_0\rangle$. Indeed, in that case, the points merge on a single curve as illustrated in Fig. 24. The performance factor varies roughly like the inverse of reduced pressure drop with a small scattering of values around this trend.

Nevertheless, no order is obtained regarding the frequency. Two close frequencies could lead to very different enhancement ratio. Reciprocally, two neighbour points on this curve could correspond to completely different frequencies.

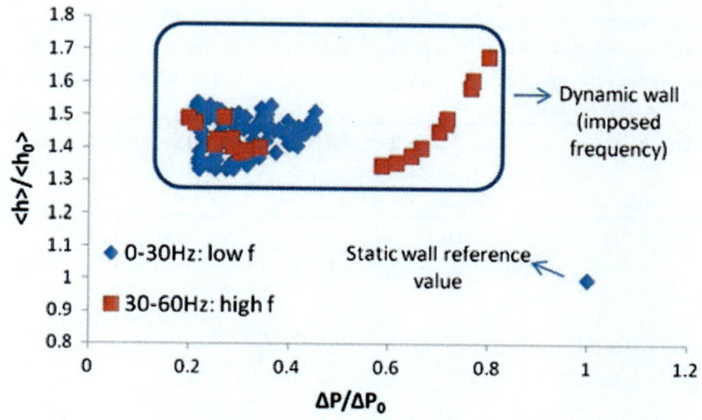

Fig. 23 Reduced heat transfer coefficient versus reduced ΔP. Relative amplitude 20 % and Re = 112

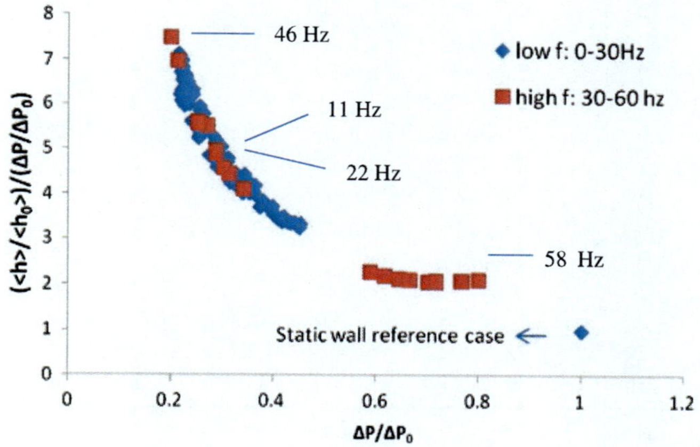

Fig. 24 h/ΔP versus pressure drop ΔP (reduced form)

Table 3 presents the minimum and maximum heat transfer enhancement and pressure drop decrease as well as their associated frequencies. One important point to highlight is that higher amplitudes (10–20 %) conduct to higher heat transfer performance and lower pressure drop which is completely contrary to results obtained in static wall case. On the other hand, lower amplitudes (2.5–7.5 %) have not shown significant advantages in terms of heat transfer enhancement and reduction in pressure drop. In order to achieve high heat transfer and lower pressure drop, one must utilize high amplitude. It is evident from Table 3 and Figs. 23 and 24 that no clear dependency of performance factor (heat transfer coefficient or pressure drop) with frequency, Reynolds number and amplitude could be derived.

Table 3 Comparison of minimum and maximum average heat transfer increase and average pressure drop decrease (in %) for different Reynolds numbers and amplitudes. The variations in heat transfer and pressure drop are calculated with respect to the values obtained on static wall case at their respective amplitudes

Re	Amplitude (%)	Min "h" % increase	f(Hz)	Max "h" % increase	f(Hz)	Min "ΔP" % decrease	f(Hz)	Max "ΔP" % decrease	f(Hz)
28	2.5	0	30	0	10	0	40	4	30
	5	1	30	2	10	4	40	11	25
	7.5	2	30	4	10	12	2	21	25
	10	5	30	8	10	20	2	32	25
	15	17	2	26	40	36	15	53	25
	20	32	2	46	20	40	30	71	25
56	2.5	0	2	0	30	-4	2	1	30
	5	0	2	3	20	-1	2	9	30
	7.5	1	2	7	20	5	2	19	30
	10	5	2	13	20	13	2	31	30
	15	19	2	37	20	26	25	52	20
	20	30	25	52	20	31	25	69	20
84	2.5	0	2	0	30	-5	2	2	30
	5	0	2	3	30	-1	2	11	25
	7.5	2	2	7	25	6	2	23	20
	10	6	2	15	20	15	2	35	20
	15	15	2	32	20	26	60	57	20
	20	31	40	67	60	15	60	74	20
112	2.5	0	2	0	25	-5	2	2	25
	5	0	2	3	25	-1	2	13	20
	7.5	2	2	8	25	7	2	26	20
	10	7	2	18	20	16	10	39	20
	15	20	22	35	20	29	60	62	17
	20	36	30	59	60	24	60	76	20

At high Re ($> 10^3$), it is known that corrugated tubes gives better heat transfer performance than straight tubes. The behavior is completely different at low Re. As discussed in static wall case, heat transfer is enhanced only when $n < 5$ % compared to established static straight tube for low range of Reynolds number. j/f factor is better than established straight tube flow but when amplitude increases, heat transfer performance decreases and remains always lower than straight tube of small length (L/D $\approx$ 10). We have also compared j/f factor (Fig. 15b) of decreasing spatial wavelengths (increasing number of sine waves).

In moving wall case, we have calculated best and worse j/f factors by utilizing maximum and minimum heat transfer (20 % amplitude results are presented) at different frequencies. We have then compared best and worst j/f factors with j/f factors of static wall, straight tube of short length (L/D $\approx$ 10) and established straight tube (L/D $\approx \infty$). We have shown in static wall case that low amplitude play an important role in design of heat exchanger through the j/f factor. In moving wall case, higher amplitudes conducted to higher heat transfer and lower pressure

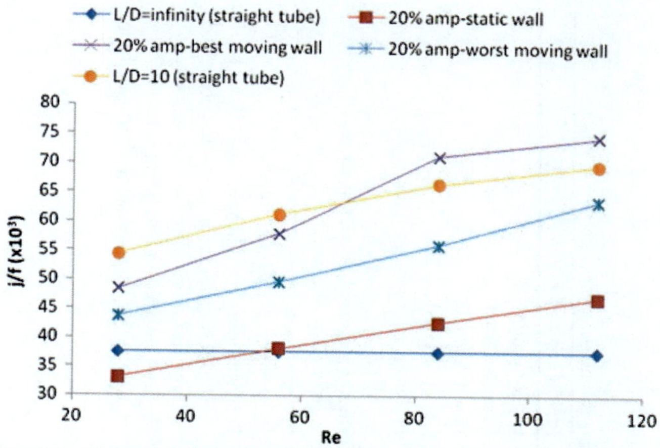

Fig. 25 Comparison of static and moving wall transfer performances for $\lambda = 2$ mm : j/f ratio Vs Re at 20 % amplitude

drop characteristics compared to static wavy wall case and moving wall with low amplitudes.

From Fig. 25, it is seen that moving wall always gives better heat transfer performance than static wall and established straight tube. Moreover, moving wall also gives better heat exchanger performance over short straight tube (L/D $\approx$ 10) for Re $>$ 75. The main reason is due to the enhanced mixing effect induced by increase in fluctuating velocity in longitudinal direction with average flow velocity (pulsating flow) along main axis at high Reynolds. Moreover, Table 3 clearly indicates that pressure drop is reduced by a factor of 1.2–5 for moving wall case compared to static straight and corrugated tube. Table 4 indicates the preference of different structures of tubes for their use in different engineering applications.

6 Conclusions

Numerical analysis were performed to study the effect of spatial wavelengths (λ = 1/2, 2/3, 1/2 mm), Reynolds number (1–120) and amplitude (1–20 % of tube diameter, D_0) on heat transfer and pressure drop in static and moving sinusodial corrugated wavy walls. It has been seen that for static wall, heat transfer increases with increase in the Reynolds number. There is an enhancement of heat transfer at 1–2 % amplitude of deformation of wall (compared to straight tube) but as the amplitude of the wall increases, the heat transfer starts to decrease. Moreover, pressure drop increases due to obstruction produced within fluid flow by increasing number of sine waves (λ decreases) on wavy wall but heat transfer coefficient remains almost unchanged. The heat transfer coefficient, h and thus Nusselt number, Nu for moving wavy wall has a higher value for all frequencies compared to static wall case. In moving wall case, at higher amplitudes, there is a sharp

Table 4 Guidelines for tube design according to tube lengths for straight or corrugated tube wall at different Reynolds

Tube	Type	Re	*J/F* VALUE	Preference
L/D $\approx$ 10, Straight tube	Static tube	Low Re	High	Highly preferred
L/D $\approx$ 10, Straight tube	Static tube	High Re	Low	Less preferred
L/D $\approx$ ∞, Straight tube	Static tube	Low Re	Low	Less preferred
L/D $\approx$ ∞, Straight tube	Static tube	High Re	Low	Less preferred
L/D $\approx$ 10, Corruagted tube	Static tube	Low Re	High	Preferred over straight tube of L/D $\approx$ ∞
L/D $\approx$ 10 or ∞, Corruagted tube	Static tube	High Re	High	Highly preferred over straight tube of L/D $\approx$ 10 or ∞
L/D $\approx$ 10 or ∞, Corruagted tube	Moving tube (best and worst case)	Low Re (Re > 75)	Low	Preferred over straight tube of L/D $\approx$ ∞ or static corrugated tube
L/D $\approx$ 10 or ∞, Corruagted tube	Moving tube (best and worst case)	Low Re (Re > 75)	High	Preferred over straight tube of L/D $\approx$ 10 (for best case) or ∞ or static corrugated tube (for worst case)
L/D $\approx$ 10 or ∞, Corruagted tube	Moving tube (best and worst case)	High Re	High	Highly Preferred over straight tube of L/D $\approx$ 10 or ∞ or static corrugated tube

decrease in pressure drop by a factor of 1.2–5. Moreover, heat transfer increases up to 35–70 % is observed. Mainly, heat transfer and pressure drop values change apparently erratically with wall frequency and no general trend in respect with operating parameters has been found.

Acknowledgments Financial support from CNRS Energy CITAMPE PR09-3.1.3-2 and FNRAE SYRTIPE are gratefully acknowledged.

A.1 7 Appendix

Local tube radius is given by:

$$R(x) = R_0 + A\sin\left(\frac{2\pi x}{\lambda}\right) \tag{16}$$

The surface between 0 and x is given by:

$$S(x) = 2\pi\int_0^x \left(R_0 + A\sin\left(\frac{2\pi x}{\lambda}\right)\sqrt{1+\frac{4A^2\pi^2\cos\left(\frac{2\pi x}{\lambda}\right)^2}{\lambda^2}}\right)dx \tag{17}$$

The result is given by the following expression:

$$S(x) = \frac{1}{4A\pi\lambda}\left(\frac{8A^4\pi^3 + 2A^2\pi\lambda^2 + \sqrt{A^2}\lambda^2\sqrt{4A^2\pi^2+\lambda^2}Log\left[2\sqrt{A^2}\pi + \sqrt{4A^2\pi^2+\lambda^2}\right]}{\sqrt{1+\frac{4A^2\pi^2}{\lambda^2}}} + \right.$$

$$\left(\sqrt{2A^2\pi^2+\lambda^2+2A^2\pi^2\cos\left(\frac{4\pi x}{\lambda}\right)}\left(4A\pi R_0\sqrt{2A^2\pi^2+\lambda^2+2A^2\pi^2\cos\left(\frac{4\pi x}{\lambda}\right)}\right.\right.$$

$$EllipticE\left[\frac{2\pi x}{\lambda}, \frac{4A^2\pi^2}{4A^2\pi^2+\lambda^2}\right] - \sqrt{\frac{2A^2\pi^2+\lambda^2+2A^2\pi^2\cos\left(\frac{4\pi x}{\lambda}\right)}{4A^2\pi^2+\lambda^2}}$$

$$\left(2A^2\pi\cos\left(\frac{2\pi x}{\lambda}\right)\sqrt{2A^2\pi^2+\lambda^2+2A^2\pi^2\cos\left(\frac{4\pi x}{\lambda}\right)} + \sqrt{A^2}\lambda^2\right.$$

$$\left.\left.\left.Log\left[2\sqrt{A^2}\pi\cos\left(\frac{2\pi x}{\lambda}\right) + \sqrt{2A^2\pi^2+\lambda^2+2A^2\pi^2\cos\left(\frac{4\pi x}{\lambda}\right)}\right]\right)\right)\right) \Bigg/$$

$$\left.\left(\sqrt{\frac{2A^2\pi^2+\lambda^2+2A^2\pi^2\cos\left(\frac{4\pi x}{\lambda}\right)}{\lambda^2}}\sqrt{\frac{2A^2\pi^2+\lambda^2+2A^2\pi^2\cos\left(\frac{4\pi x}{\lambda}\right)}{4A^2\pi^2+\lambda^2}}\right)\right) \tag{18}$$

References

1. Yuan, Z., Tao, W., Wang, Q.: Numerical prediction for laminar forced convection heat transfer in parallel-plate channels with streamwise-periodic rod disturbances. Int. J. Numer. Method Fluids **28**, 1371–1388 (1998)
2. Goldstein, L., Sparrow, E.M.: Heat and Mass transfer characterstics for flow in a corrugated wall channel. Trans. ASME J. Heat Transf.**99**, 187–195 (1997)
3. Wang, C.C., Chen, C.K.: Forced Convection in a wavy-wall channel. Int. J. Heat Mass Transf. **45**, 2587–2795 (2002)
4. Zhang, J., Kundu, J., Manglik, R.M.: Effect of waviness and spacing on lateral vortex structure and laminar heat transfer in wavy-plate-fin cores. Int. J. Heat Mass Transf. **47**, 1719–1730 (2004)
5. Metwally, H.M., Manglik, R.M.: Enhanced heat transfer due to curvature-induced lateral vortices in laminar flows in sinusoidal corrugated plate channels. Int. J. Heat Mass Transf. **47**, 2283–2292 (2004)
6. Niceno, B., Nobile, E.: Numerical analysis of fluid flow and heat transfer in periodic wavy channels. Int. J. Heat Fluid Flow **22**, 156–167 (2001)
7. Mohamed, N., Khedidja, B., Abdelkader, S., Belkacem, Z.: Heat Transfer and Flow field in the entrance region of a symmetric wavy-channel with constant wall heat flux density. Int. J. Dyn. Fluid **3**, 63–79 (2007)
8. Bellhouse, B.J., Bellhouse, F.H., Curl, C.M., MacMillan, T.I., Gunning, A.J., Spratt, E.H., MacMurray, S.B., Nelems, J.M.: A high efficiency membrane oxygenator and pulsatile pumping system and its application to animal trials. Trans A. Soc. Artif. Internal Organs **19**, 72–79 (1973)
9. Mackley, M.R., Stonestreet, P.: Heat transfer and associated energy dissipation for oscillatory flow in baffled tubes. Chem. Eng. Sci. **50**, 2211–2224 (1995)
10. Howes, T., Mackley, M.R., Roberts, E.P.L.: The simulaton of choatic mixing and dispersion for periodic flows in baffled channels. Chem. Eng. Sci. **46**, 1669–1677 (1991)
11. Farhanieh, B., Herman, C., Sunden, B.: Numerical enhancement and experimental analysis of laminar fluid flow and forced convection heat transfer in a grooved duct. Int. J. Heat Mass Transf. **36**, 1609–1617 (1993)
12. Fusegi, T.: Numerical study of mixed convection in a channel with periodic cavities. Proceedings of 9th International heat Transfer conference, 5 (1994) 471-476
13. Nigen, J.S., Amon, C.H.: Time-dependent conjugate heat transfer characterstics of self-sustained oscillatory flows in a grooved channel. ASME J. Fluid Engg. **116**, 499–507 (1994)
14. Herman, C., Kang, E., Huang, H., Puranik, B.: Experimental visualization of unsteady temperature fields in electronic cooling applications, ASME HTD vol. 319/EEP-vol. 15 (1995) 33–40
15. Nishimura, T., Kunitsugu, K., Nakagiri, H.: Fluid Mixing and local mass transfer characterstics in a grooved channel for self-sustained oscillatory flow. Heat Transf. Jpn Res **27**, 522–534 (1998)
16. Wang, G., Vanka, S.P.: 1995 Convective heat transfer in periodic wavy passages. Int. J. Heat Mass Transf. **38**(17), 3219–3230 (1995)
17. Nishimura, T., Matsune, S.: Mass transfer enhancement in a sinusoidal wavy channel for pulsatile flow. Warme und Stoffuberagung **32**, 65–72 (1996)
18. Nishimura, T., Matsune, S.: Vortices and wall shear stresses in asymmetric and symmetric channels with sinusoidal wavy walls for pulsatile fow at low Reynolds numbers. Int. J. Heat Fluid Flow **19**(1998), 583–593 (1998)
19. Kuppan, T.: Heat Exchanger Design Book. Marcel Dekker Inc, New York (2000)

Assessment of Heat Affected Zone of Submerged Arc Welding Process through Digital Image Processing

A. Ghosh, S. Chattopadhyaya and N. K. Singh

Abstract Submerged arc welding (SAW) is a high quality, high deposition rate welding process commonly used to join plates of higher thickness in load bearing components. This process of arc welding provides a purer and cleaner high volume weldment that has relatively a higher material deposition rate compared to the tra-ditional welding methods. A common issue in the application of SAW process raises a concern about the uncertainties involved with the heat affected zone (HAZ) in and around the weldment. The most intriguing issue is about HAZ softening that imparts some uncertainties in the welded quality. It increases the probability of fatigue failures at the weakest zones caused by the heating and cooling cycle of the weld zone. An attempt has been made in this paper to assess the heat affected zone of submerged arc welding of structural steel plates through the analysis of the grain structure by means of digital image processing techniques.

Keywords Digital image processing · HAZ · Fatigue failure · Grain structure · Heating and cooling cycle

A. Ghosh (✉)
Department of Mechanical Engineering, Government College of Engineering and Textile Technology, Berhampore, WB Pin-742101, India
e-mail: agmech74@gmail.com

S. Chattopadhyaya
Department of ME and MME, ISM, Dhanbad, India
e-mail: somuismu@gmail.com

N. K. Singh
Department of ME and MME, ISM, Dhanbad, India
e-mail: nks_221@yahoo.co.in

J. M. P. Q. Delgado et al. (eds.), *Numerical Analysis of Heat and Mass Transfer in Porous Media*, Advanced Structured Materials 27,
DOI: 10.1007/978-3-642-30532-0_8,

1 Introduction

Submerged arc welding (SAW) is a high quality, high deposition rate welding process commonly used to join plates of higher thickness in load bearing components. SAW is a very important process for fabricating structures, bridges, ships, boilers, etc. This process of arc welding provides a purer and cleaner high volume weldment that has relatively a higher material deposition rate compared to the traditional welding methods. In a country like India, in the context of infrastructural development, the SAW process has much more useful applications in welding of critical components and equipment. Use of this technology has huge economic and social implications in the national perspective. Lot of critical set of input variables i.e. current, voltage, electrode diameter, travel speed, wire feed rate, stick out etc. are involved in submerged arc welding process [1]. Temperature distribution patterns depend on these welding process parameters because these are the functions of heat input [2]. So, shape of heat distribution is changed with the change of input parameters of SAW process [3]. Study of temperature distribution of welded plates is very essential for designing submerged arc welding joint [4]. As key parameter for change of HAZ width is heat input. Generally in welded plates prominent grain growths have found. Existence of prominent grain growth provides the confirmatory evidence of the HAZ softening phenomenon. In the welded portion, grain refinement occurs in most of the region due to the heating and cooling cycle of SAW method. Predominant direction of the grain growth is clearly observed from the photograph of the microstructure. Hall–Petch equation states the strength of the metal is to vary reciprocally with size of subgrain. The similar phenomenon is also revealed in case of hardness. In the context of this equation one can say that the hardness of the grain growth portion will also manifest lower values related to higher grain sizes. In the grain growth portion of the welded region longer grains have been found depicting the chances of dislocation, slip, low yield strength and low hardness values. That is why it is very essential to find out accurate transient temperature distribution on welded plates.

An attempt to develop mathematical model of travelling heat source was made more than fifty years ago [5]. After that, lot of research works have been continuing on this area. Initially two dimensional surface Gaussian heat source with effective arc radius has been adopted to find out the temperature distribution on welded plates and weld pool geometry [6]. This solution is an improved version of estimation temperature distribution in the near heat source area in (x,y) plane but this solution is failed to find out temperature distribution along Z direction. However, this attempt has indicated new direction for finding out temperature distribution on welded plates. Then an analytical solution for transient temperature distribution for welded joint based on similar Gaussian heat distribution is proposed. Different distribution parameters have also been introduced [7]. These [6, 7] parameters are still limited to 2D heat source. But these solutions failed to describe the effect of penetration. Recently this short coming has been overcome by considering 3D heat source. Goldak et al [8] first described 3D double ellipsoidal mobile heat source and three dimensionally temperature distribution has been calculated through Finite Element modeling. This could overcome the shortcoming of the two dimensional Gaussian model to estimate the temperature distribution of the welded plates with deeper weld bead penetration. Nguyen et al. [9] described an analytical solution of transient temperature distribution of a body (semi-infinite) subjected to three dimensional heat density of semi ellipsoidal and double ellipsoidal mobile heat source. Very good agreement between predicted and measured temperature distribution data achieved when double ellipsoidal heat source has been assumed. But this analytical solution has some limitations. It is correct when both semi ellipsoids are equal in size and this solution is valid only when $c_r = c_f$ and $f_r = f_f = 1$ where c_r, c_f are radii of rear and front ellipsoid respectively and f_r (resp. f_f) is the portion of the heat deposited in the rare (resp. front) ellipsoid. Victor et al. [10] described a semi-analytical solution which has ability to overcome the aforesaid limitations. Nguyen et al. [11] again described an analytical approximate solution for double ellipsoidal heat source in finite thick plates. In this work, they established that approximate solution can be directly used for simulation of welding of finite thick plate without applying the mirror method which is required in a semi-infinite body. This analytical approximate solution is an effective tool for finding thermal stress, microstructure modeling etc. Many problems of welding are involved with complicated geometries with complex boundary conditions or variable properties, and cannot be solved analytically. In such cases, sufficiently accurate approximate solutions can be obtained by computers using a numerical methods. Ravichandran et al. [12] developed a model of temperature distribution during circumferential arc welding of spherical and cylindrical components using the finite element method and got very good results. Chandra [13] described the extension of Rosenthal's method for the estimation of thermal field in a pipe with a mobile heat source by taking many simplifying assumptions. Ahmet Akkus [14] described the effect of sheet thickness and current density on cooling rate and distribution of temperature in resistance spot welding through his experimental results. He found thicker sheets have higher temperatures in weld zone. Heat generation in welding zone is affected by current density and sheet thickness. Veensra et al. [15] measured temperature between electrodes for RSW by applying thermal method and Bentley et al. [16] investigated temperature distribution in RSW by applying metallographic method. Kermanpur et al. [17] investigated experimentally distribution of temperature of GTAW process for butt weld joint. Maheshwari et al. [18] used thermocouples to measure temperature at different locations of GMA welded plates. Temperature readings have been taken in every 10 s interval and temperature profile has been generated for GMAW process which may be helpful to optimize the GMAW. Inspite of more laborious and time consuming method, experimentally measured temperature distribution has some advantages. One may guess the change of output parameters with the change of temperature distribution accurately. Many researchers investigated analytical solution of thermal field [9–11] only considering heat conduction, without considering heat lost through convection and radiation. But practically three modes of heat transfer occur in welding process. Araya [19] has carried out the numerical simulation of the temperature distribution generated by mobile heat source through control volume approach. Here convection and radiation effect have been considered. Bianaco et al. [20, 21] carried out two numerical methods for two and three dimensional models for evaluating transient conductive fields due to moving heat source. Bianaco et al. [22] also investigated transient three dimensional temperature distribution numerically by COMSOL Multiphysics 3.2. Radiation and convection modes of heat transfer from work piece surfaces as well as variable thermo physical properties has been considered here. They found that surface heat transfer strongly affected the temperature distributions in the worpiece. Ohring et al. [23] considered radiative heat transfer, evaporation, and viscous stress to find out temperature distribution but he assumed material properties are constant. Mundra et al. [24] considered specific heat and thermal conductivity values for solid and liquid metal only for analysis of temperature distribution. Heat transfer mechanism in a molten pool is extremely complex and its physics is not well understood till now [25]. Biswas et al. [25] considered a numerical model based the finite element package ANSYS for single pass single sides submerged arc welding of square butt joints. Using it, they determined the distortion of welded joint. In their study, heat lost through natural heat convection has been considred. Many researchers tried to find out stress distribution during welding process using different tools. One of them [26] carried out numerical simulation of arc welding heat transfer and stress distribution for thin plates. For predicting temperature and stress distribution of thin plates during arc welding, computational fluid dynamics and computational solid mechanics have been utilized [26]. Postacioghu et al. [27] investigated thermal stress, strain during welding using linear theory of elasticity. They considered in their study that the weld pool is approximated by an elliptic region which is constant in cross sectional shape and depth. Few researchers investigated thermal analysis of welding on Aluminum plates. For this study, they considered three modes of heat transfer (i.e. conduction, convection, radiation). They solved the equation of steady state temperature profiles of welded plates by finite difference method. They compared calculated and measured temperature near the heat source and obtained good agreement. In the year 2005, Ali et al. [28, 29] described theory of relativistic heat conduction and relativistic moving heat source which have given new direction to researchers for finding out temperature distribution, thermal stress etc. due to moving heat source. In a research paper [30] an improved version of shape of moving heat source i.e. double central Conicoidal is considered and temperature distribution during submerged arc welding HAZ width and weld pool dimensions are calculated and compared with experimental values.

It is observed that a refined microstructure of the HAZ, imparts largely the intended properties of the welded joint [31]. In order to bring out an appropriate combination of SAW parameters and a methodology to control such parameters an in depth investigations and characterizations of HAZ softening zone are necessary to enrich this Submerged Arc Welding technology. In Submerged Arc welding process, major process control parameters are current, arc voltage and travel speed. They all affect the bead shape, depth of penetration and chemical composition of the deposited weld metal. Another very critical issue in the understanding of the joint performance obtained from SAW process rests on the analysis of heat affected zone. It is difficult for the operator to observe the weld pool during the process. So, better control comes from SAW process parameter settings than dependence on the operator's expertise. It is shown by the researchers [31] that the HAZ has various regions those influence the ability of the joint to provide crack resistance and uniform strength in both the direction of the weld. An estimation of bead width and depth of penetration obtained from infrared thermal imaging technique (IRTI technique) is also found to influence the quality of SAW process [32]. Very little information is available in the literature on the aspects of HAZ softening during SAW process. A three dimensional analysis to predict the zones of microstructure of SAW process is indicated in [33]. A model to predict HAZ in case of SAW is addressed by [31] attempts to predict HAZ in case of SAW. A combined effect of chemical composition of flux and welding parameters on the mechanical properties of SAW process is shown to be of utmost significance [34, 35]. It is apparent from these references [31–35] that SAW process has drawn much of attention, in recent time, for characterization of its various aspects. It is also very clear that a systematic study to bring out a correlation based performance characterization through identification of control parameters in conjunction with quality assessment is missing. An identification of the contribution of each of its process control parameters on the quality or performance of a SAW joint, possess a challenge to the researchers in this area and demands a very systematic study of the problem.

Livier and Lazzarin [36] discussed the issue of notch stress intensity applied from fatigue life prediction of submerged arc welded joints. They emphasized on crack initiation life in relation to the welded toe geometry. The other parameters related to welding like voltage current, diameter of the electrode are not considered and their interactions are not studied. Bechet and Kahraman [37] investigated on wear behavior of bulldozer rollers joined using a SAW process. They considered only wear as a yield parameter of SAW where the other yield parameters like strength of welds, spread and form of heat affected zones, influence of deposition rate etc. are not addressed. Trindade et al. [38], argued about the influence of zirconium on microstructure and toughness of low alloy steel weld metals. They stressed on microstructure and toughness as the weld parameter in comparison to the other yield parameters, like material deposition rate, heat affected zone and longitudinal strength of the weld. Tuek et al. [39], provided the weld cost analysis to indicate a cost saving accomplished by replacing single wire submerged arc welding with triple wire welding. Kim et al. [40] discussed about the prediction of welding parameters for pipeline welding using an intelligent system. In this study the authors mainly stressed on pipeline welding based on one database and a finite element method (FEM) model. They considered two back-propagation (BP) neural network model and a corrective neural network (CNN) model for validation of their results. Zhao et al. [41] investigated about the numerical simulation of the dynamic characteristics of weld pool geometry with step changes of welding parameters. From their study they developed a three-dimensional numerical model to investigate the dynamic characteristics of the weld pool geometry with a step input change in the welding current and the welding speed. Menaka et al. [32] brought forth an estimate of bead width and depth of penetration during welding by the use of infrared thermal imaging technique. Bang and Kim [42] addressed the issue of estimation and prediction of HAZ softening in thermodynamically controlled and accelerated—cooled steel material.

It is imperative from all above that characterization of HAZ softening is of critical importance in case of SAW process. However, no such work provides a knowledge base for the prediction of the HAZ hardness or the classification of HAZ in SAW. In view of the above information it becomes very clear that the critical investigation to assess the grain structure through some reliable digital image processing were not reported for structural steel plates. Therefore, it is evident that not much work has been reported related to the analysis of the HAZ softening and hardening phenomena of SAW process to the best of knowledge of the authors. Few of the works as mentioned above are there to create an appropriate prelude to go for such in depth analysis of the characterization of HAZ hardness in SAW method of coalescence.

2 Experimental Procedure

The experiment was conducted with the MEMCO semiautomatic welding equipment (see Fig. 1) with constant voltage rectifier. Flux used is ADOR Auto melt Gr II AWS/SFA 5.17 (Granular flux), Electrode selected is ADOR 3.15 diameter copper coated wire, Test Piece is 400 × 75 × 10 mm square butt joint, and weld position is flat with electrode positive and positioned perpendicular to the plate. The job 400 × 25 × 10 mm was firmly fixed to a base plate by means of tack welding and then the welding was carried. The slag was removed and the job was allowed to cool down. The welded sections were cut by saw cutter and they were ground by surface grinding machine and polished with leather cloth. The jointed sections were put into the microscope and the images were captured through image grabber of a TV tuner card. The images processed through digital image processing functions of MATLAB 6.0 software.

3 Heat Affected Zone

Heat-affected zone (HAZ): The HAZ is the area of base material, either a metal or a thermoplastic, which has had its microstructure and properties altered by welding or heat intensive cutting operations. The heat from the welding process and subsequent re-cooling causes this change in the area surrounding the weld. The extent and magnitude of property change depends primarily on the base material, the weld filler metal, and the amount and concentration of heat input by the welding process. The thermal diffusivity of the base material plays a large role—if the diffusivity is high, the material cooling rate is high and the HAZ is relatively small. Alternatively, a low diffusivity leads to slower cooling and a larger HAZ.

Fig. 1 **a** SAW process taking place at the welding lab. **b** Pictorial view of formation of HAZ due to welding. **c** Hardness values are recorded at different grid junctions of the welded specimen (distance between two grid points is equal to 2 mm) **d** Hardness variation on different grid points as shown in (**c**) for B4 Job (heat input = 2.84 kj/mm) of Table 1. **e** Portion1-bead geometry, 3, 4 just end of the welded zone (fusion zone), 5, 8 heat affected zone of Submerged Arc Welded plates. **f** Microstructure at the welded zone of 100 Magnification for 2.84 kJ/mm Heat Input Few points are prominent in this microstructure, these are caused due to grain growth and these are the softening portion of welded zone. (portion-3 of e) **g** Microstructure at the non welded zone of 100 Magnification for 2.84 kJ/mm Heat Input Here more softening portion(portion-4 of e) **h** Microstructure at the welded zone of 50 Magnification for 2.84 kJ/mm Heat Input. The alignment of the grain formation confirms the directional component of the grain growth for the sub merged arc welding process of 2.84 kJ/mm heat input. (portion-5 of e) **i** Microstructure at the non-welded zone of 50 Magnification for 2.84 kJ/mm Heat Input Here more softening portion (non welded zone) with respect to welded portion (portion-8 of e)

The amount of heat inputted by the welding process plays an important role as well, as processes like oxyfuel welding use high heat input and increase the size of the HAZ. Processes like laser beam welding and electron beam welding give a highly concentrated, limited amount of heat, resulting in a small HAZ. Arc welding falls between these two extremes, with the individual processes varying somewhat in heat input. To calculate the heat input for arc welding procedures, the following formula is used:

$$Q = \left(\frac{V \times I \times 60}{S \times 1000}\right) \times \text{Efficiency} \tag{1}$$

where Q = heat input (kJ/mm), V = voltage (V), I = current (A), and S = welding speed (mm/min). The efficiency is dependent on the welding process used, with

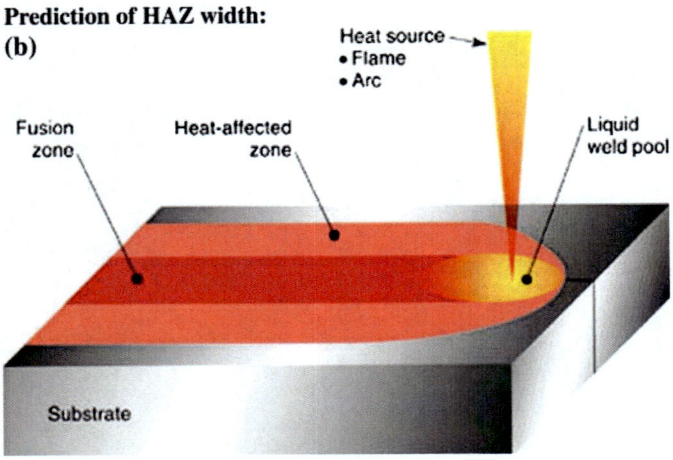

Fig. 1 (continued)

Table 1 Observed values for bead parameters for HAZ analysis

Sl. No.	Job No.	Voltage (V)	Current (A)	Travel Speed (cm/min)	Penetration (mm)	Reinforcement Height (mm)	Bead Width (mm)
1	A1	25	350	17	6.70	2.38	17.96
2	A2	35	350	17	3.72	2.34	21.90
3	A3	25	450	17	6.69	3.16	21.00
4	A4	35	450	17	8.26	2.76	30.92
5	B1	25	350	30	5.28	1.00	13.94
6	B2	35	350	30	4.58	1.78	20.12
7	B3	25	450	30	6.60	2.25	15.90
8	B4	35	450	30	7.78	1.94	22.66

shielded metal arc welding having a value of 0.75, gas metal arc welding and submerged arc welding, 0.9, and gas tungsten arc welding, 0.8.

It has found from the literature [45] that moving heat source shape for submerged arc welding process is double Central Conicoidal heat source and temperature induced by the double Central Conicoidal heat source defined by equation is

$$\Delta T\,(x,\,y,\,z,\,t) = \frac{Q_0\sqrt{a'a''bc}}{\rho c\pi\sqrt{\pi}}\left[\frac{1}{\sqrt{a'}\sqrt{a''}}\right]\left[\frac{1}{[4\alpha\pi]^{\frac{3}{2}}}\right]\int_0^t (I_x^y + I_x^f)I_yI_z(t-t')^{-3/2}\mathrm{d}t'$$

$$\text{or } \Delta T\,(x,\,y,\,z,\,t) = \frac{Q_0\sqrt{a'a''bc}}{8\rho c\pi^3\alpha^{3/2}}\left[\frac{1}{\sqrt{a'}\sqrt{a''}}\right]\int_0^t (I_x^y + I_x^f)I_yI_z(t-\mathrm{t}')^{-3/2}\mathrm{d}t' \qquad (2)$$

Where

$$I_x^r = \int_{-\infty}^{vt'} e^{\left[-a'(x-vt') - \frac{(x-x')^2}{4\alpha(t-t')}\right]} dx'$$

$$= \sqrt{\pi}\sqrt{\frac{\alpha(t-t')}{(1+4\alpha a''(t-t')}} \times e\left[-\frac{a''(x-vt')^2}{1+4\alpha a''(t-t')}\right]$$

$$\times \left[1 - \text{erf}\left\{\frac{(x-vt')}{2\sqrt{\alpha(t-t')\{1+4\alpha a''(t-t')\}}}\right\}\right]$$

$$I_x^f = \int_{vt'}^{\infty} e^{\left[-a'(x-vt') - \frac{(x-x')^2}{4\alpha(t-t')}\right]} dx'$$

$$= \sqrt{\pi}\sqrt{\frac{\alpha(t-t')}{(1+4\alpha a'(t-t')}} \times e\left[-\frac{a'(x-vt')^2}{1+4\alpha a'(t-t')}\right]$$

$$\times \left[1 - \text{erf}\left\{\frac{(x-vt')}{2\sqrt{\alpha(t-t')\{1+4\alpha a'(t-t')\}}}\right\}\right]$$

$$I_y = \int_{-\infty}^{\infty} e^{\left[-b(y) - \frac{(y-y')^2}{4\alpha(t-t')}\right]} dy'$$

$$= 2\sqrt{\pi}\sqrt{\frac{\alpha(t-t')}{(1+4\alpha b(t-t'))}} \times e^{\left[-\frac{by^2}{1+4\alpha b(t-t')}\right]}$$

$$I_z = \int_{-\infty}^{\infty} e^{\left[-c(z- \frac{(z-z')^2}{4\alpha(t-t')}\right]} dz' = 2\sqrt{\pi}\sqrt{\frac{\alpha(t-t')}{(1+4\alpha c(t-t'))}} \times e^{\left[-\frac{cz^2}{1+4\alpha c(t-t')}\right]}$$

Finally, by assuming the body was initially at the homogeneous temperature T_0, the temperature field is defined by

$$T(x, y, z, t) - T_0 = \frac{Q_0\sqrt{a'a''bc}}{8\rho c\pi^3\alpha^{3/2}}\left[\frac{1}{\sqrt{a'}+\sqrt{a''}}\right]\int_0^t (I_x^y + I_x^f) I_y I_z (t-t')^{-3/2} dt'$$

$$\text{or}\, T(x,y,z,t) - T_0 = \frac{Q_0\sqrt{a'a''bc}}{2\rho c\pi\sqrt{\pi}}\left[\frac{1}{\sqrt{a'}+\sqrt{a''}}\right]$$
$$\int_0^t \frac{e^{\left[-\frac{by^2}{1+4\alpha b(t-t')}+\left[\frac{cz^2}{1+4\alpha c(t-t')}\right]\right]}}{\sqrt{((1+4\alpha b(t-t'))(1+4\alpha c(t-t')))}}[A'(1-B')+A''(1+B'')]\mathrm{d}t' \tag{3}$$

$$\text{Where } A^i = A(x,t,t;a^i) = e^{\left[-\frac{a^i(x-vt')^2}{1+4\alpha a^i(t-t')}\right]}$$

$$B^i = B(x,t,t;a^i) = \mathrm{erf}\left\{\frac{x-vt'}{2\sqrt{(\alpha(t-t'))\{1+4\alpha a^i(t-t')\}}}\right\}\Bigg]$$

It is worth noting this double Central Conicoidal distribution heat source is described by four unknown parameters: a′, a″, b, c. Goldak et al. [43] implies equivalence between the source dimensions and those of the weld pool and suggested that appropriate values of a′, a″, b, c could be obtained by measurement of weld pool geometry .

Values of (m′), (m″) are taken in front and rare portion of weld bead along x-axis for the two semi-Central Conicoids having equations

$a'x^2 + by^2 + cz^2 = 1$ [44]
$a''x^2 + by^2 + cz^2 = 1$ [44]

And,
Half of bead width = n, penetration = o.

It has been found from experiment for submerged arc welding process that 4 m′ = m″.

And,
n = 0.87 m′ (for B4 Job of Table 1).

HAZ width of a structural steel is the region heated from recrystalization temperature (i.e. 973 K) to the temperature just below the melting point temperature of welded materials (i.e. 1684 K). Putting these values in the equation No (3) HAZ width(s) have been calculated at z = 0, x = vt′, t = t′ = travel time of moving heat source, which are tabulated below.

It has been found from the Fig. 1d that at just below the fusion zone, the hardness values are low (indicated by red mark) and from Fig. 1f, g, h, i grain growth have been found at just below the fusion zone and that is why the decrease in strength is observed. Around 3 mm in both side at just below the fusion zone the hardness values are comparatively low. This low hardness and prominent grain growth portion is the heat affected zone (HAZ). Above description proves that HAZ width approximately 3 mm for 2.84 kJ/mm heat input in welding process which is having a good agreement with the above predicted HAZ width data (Table 2).

Validation of predicted data of HAZ width:
Hardness values measurement on grid points:

(c)

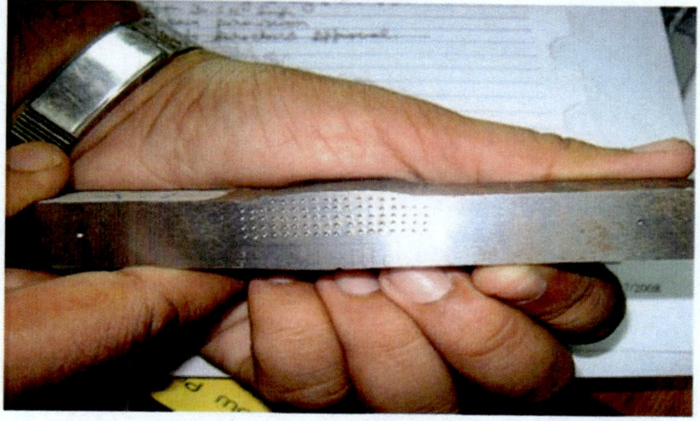

(d)

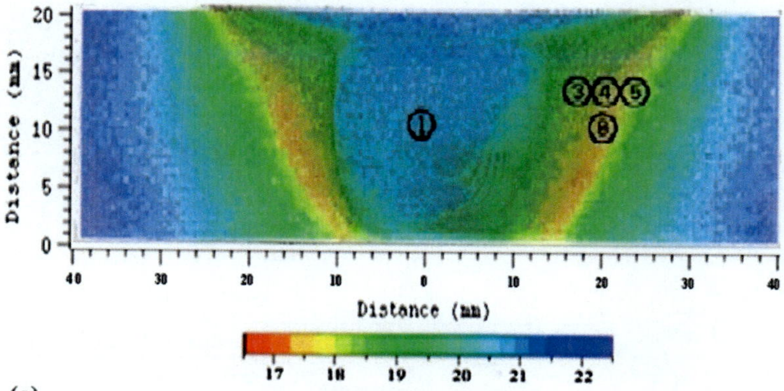

(e)

BEAD GEOMETRY

20	18.5	19	23	18.5	19	23	18	23	21	20	19.5	21	20	20	18.5
17	20	22	12	23	22.5	19	20	19	22.5	20	23	11	21.5	21	21
19	21	21.5	21	16	22.5	21	20	18	22	20	16	20	21	21	21
17	24	21	20	21	13	24	21.5	19.5	23	12	23	22	21	20.5	22.5
21	25	22	23	21.5	23	16	15	14	21.5	21.5	20.5	24	19.5	19	14.5

HEAT AFFECTED ZONE

Fig. 1 (continued)

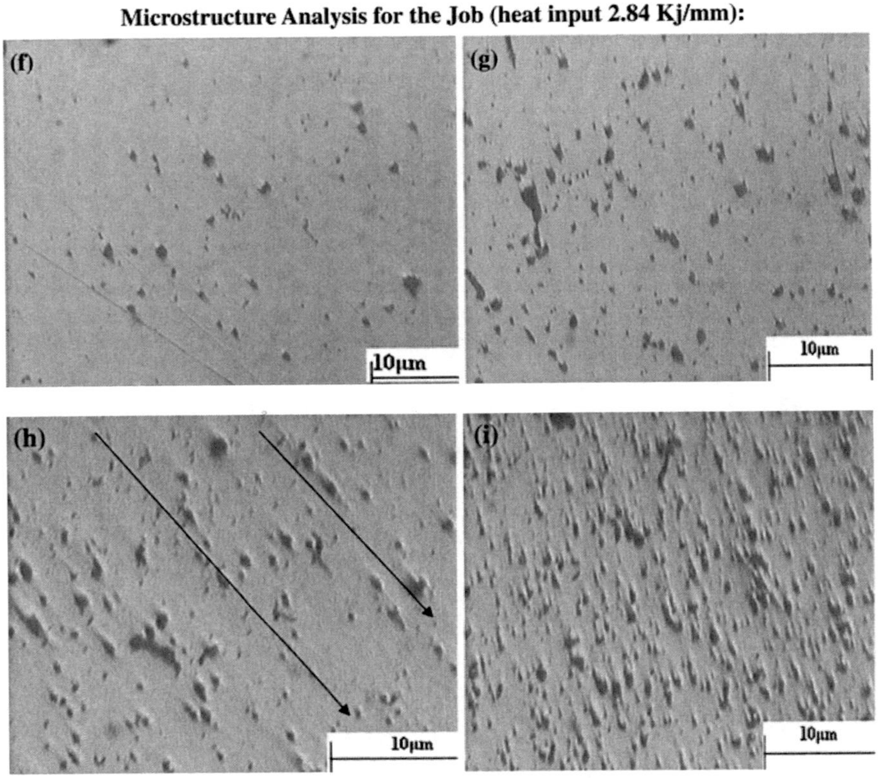

Fig. 1 (continued)

Table 2 Predicted HAZ width(s)

Sl.No.	Job No.	Predicted HAZ width (mm)
1	A1	3.0
2	A2	3.2
3	A3	3.1
4	A4	3.5
5	B1	2.1
6	B2	1.9
7	B3	1.8
8	B4	3.0

Prediction of HAZ width has been made with the help of three dimension transient temperature distribution equation. It is also new technique which is not previously applied.

Existence of prominent grain growth provides the confirmatory evidence of the HAZ softening phenomenon. In the welded portion, grain refinement occurs in most of the region due to the heating and cooling cycle of SAW method. Predominant direction of the grain growth is clearly observed from the photograph of the microstructure. This grain formation is distinctly revealed in the magnification (50,100) for the heat input of 2.84 kj/mm. Hall–Petch equation states the strength of the metal is to vary reciprocally with size of subgrain. The similar phenomenon is also revealed in case of hardness. In the context of this equation one can say that the hardness of the grain growth portion will also manifest lower values related to higher grain sizes. In the grain growth portion of the welded region longer grains have been found depicting the chances of dislocation, slip, low yield strength and low hardness values measured in Rockwell scale B. Validations of calculated HAZ width through analysis of microstructure changes and hardness analysis have been made. Very good agreement has also been found for this case.

4 Digital Image Processing

Digital image processing (DIP) refers to processing digital images by means of a digital computer. DIP methods stems from two principal application areas: improvement of pictorial information for human interpretation and processing of image data for storage, transmission and representation for autonomous machine perception. An image may be defined as a two-dimensional function, f(x, y), where x and y are spatial (plane) coordinates, and the amplitude of 'f' at any pair of coordinates (x, y) is called the intensity or grey level of the image at that point. When x, y and the amplitude values of 'f' are all finite, discrete quantities, we call the image a digital image. A digital image is composed of a finite number of elements, each of which has a particular location and value. These elements are referred to as picture elements, image elements, pels and Pixels. Intensity image is a data matrix whose values represent intensity within some range at a particular location. In binary image each pixel assumes only two discrete values either 0 (Black) or 1 (white). Morphology is a technique of image processing based on shapes. The value of each pixel in the output image is based on a comparison of the corresponding pixel in the input image with its neighbors. By choosing the size and shape of the neighborhood, one can construct a morphological operation that is sensitive to specific shapes in the input image. These functions perform common image processing tasks, such as contrast enhancement, noise removal, thinning, skeletonization, filling, and segmentation. The input image (see Fig. 2) was further processed to assess the grain growth due to rapid heating and cooling process related to submerged arc welding process.

It is quite evident from the Fig. 2 that the grain structure at the heat affected zone of the welding sample experienced a noticeable grain growth due to the heating and cooling effect of the welding process. The alignment of the grain formation confirms the directional component of the grain growth for the sub merged arc welding process. Few points are prominent in this microstructure, these are caused due to grain growth and these are the softening portion of welded zone.

The intensity of the image (see Fig. 2) is adjusted and transformed to refined image (see Fig. 3) so that the grain structure could be extracted from the image.

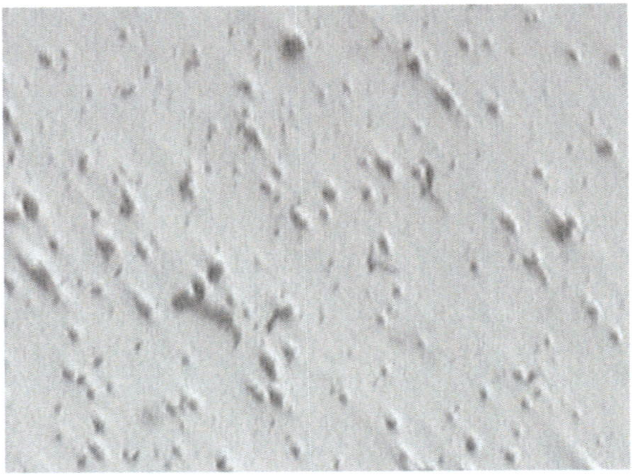

Fig. 2 Microstructure at the welded zone of 50 Magnification

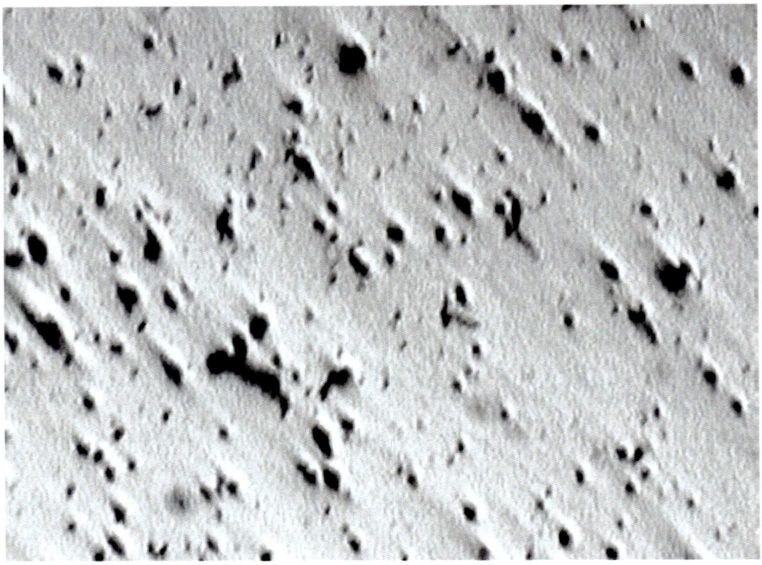

Fig. 3 Transformed Fig. 1 for further extraction of the features of grain

The image (Fig. 3) is complemented so as to convert the dark zones in previous image into bright zones and then it is converted into binary image (Fig. 5) to remove the background.

Fig. 4 Complement image of Fig. 3

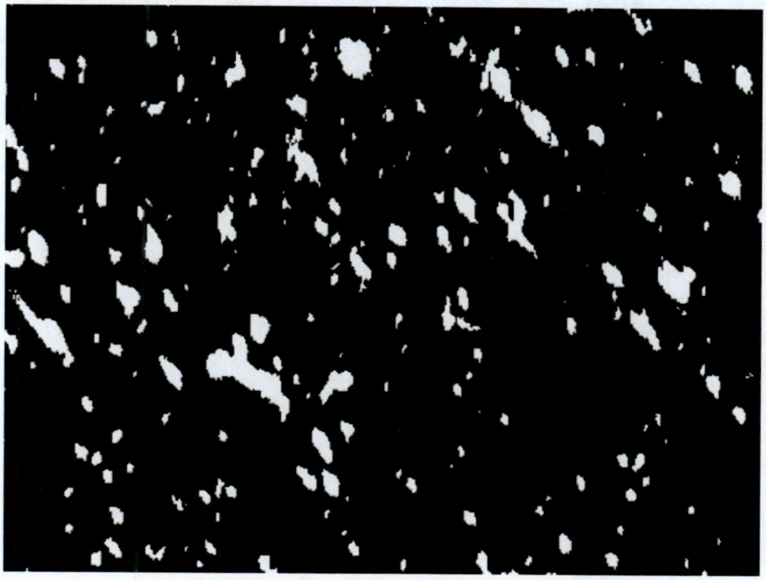

Fig. 5 Binary image of Fig. 4

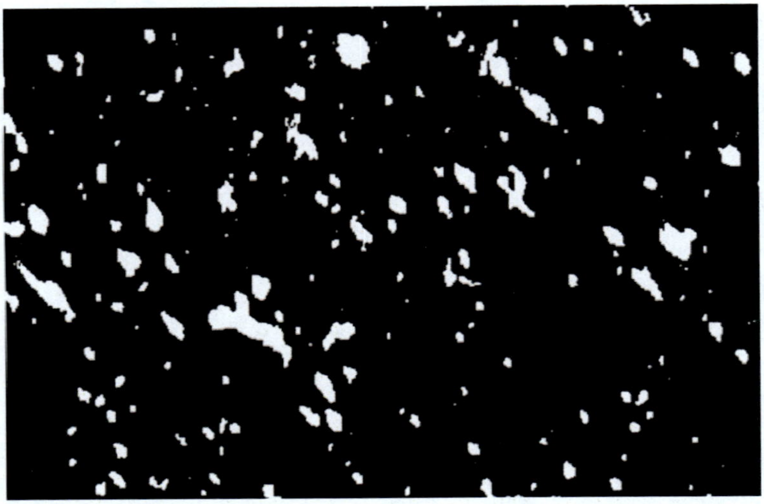

Fig. 6 Image after different morphological operations

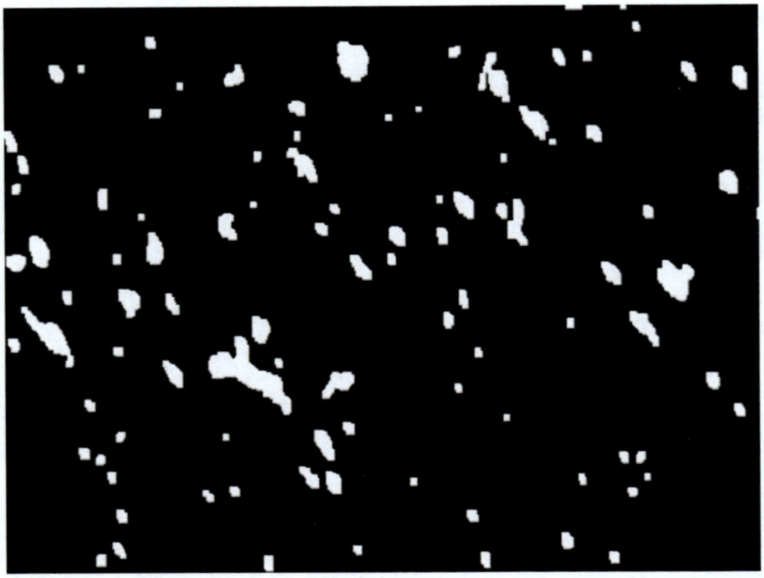

Fig. 7 Image in which stray pixels are removed

The image (Fig. 6) is processed through various operations like cleaning, spurring etc. to improve the quality of the image and to obtain a more accurate result. In the final image (Fig. 7), isolated pixels were removed.

Table 3 Grain distribution as per the Digital Image Processing Tool

Total grains in the image	Area of the grains	Area of the total image	% Area of grains
195	1.0167e + 004	1.3873e + 005	6.8281%

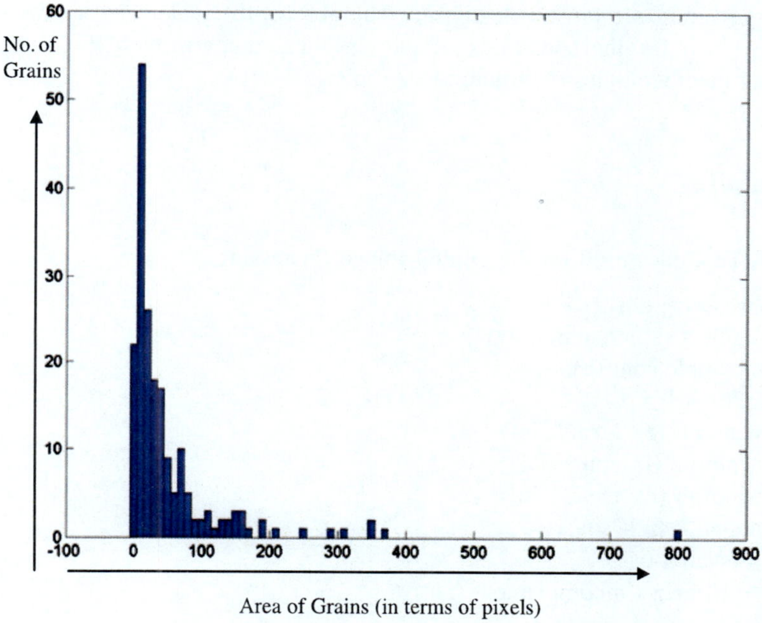

Fig. 8 No. Bar Chart representing distribution of size and area of Grains

5 Analysis

The Heat Affected Zone related to submerged arc welding process is significantly less. The grain growth are even lower than the 10 % (Table 3) of the total area as observed by digital image processing tool and that justifies the low level of heat affected zone of the process. The grains are predominantly of smaller variety and the counts for larger grain are almost negligible as it is reveled in the bar chart (Figs. 7, 8). The absence of larger size grains in the image gives the confirmatory evidence of the narrow HAZ of SAW process. The quality and the soundness of the weld in comparison to the competing welding methodologies of structural steel plates can be convincingly established by the tangible and unambiguous results obtained from digital image processing tools.

6 Conclusions

The Heat Affected Zone related to submerged arc welding process is significantly less. The grain growth are even lower than the 10 % of the total area as observed by digital image processing tool and that justifies the low level of heat affected zone of the process. The grains are predominantly of smaller variety and the counts for larger grain are almost negligible. The absence of larger size grains in the image vouch for the soundness of the weld in comparison to the competing welding methodologies of structural steel plates.

Appendix

MATLAB Codes used for the Digital Image Processing

```
a = imread ('grain1.jpg');
b = imadjust (a, stretchlim(a),[]);
d = imcomplement (b);
e = im2bw (d);
f = bwmorph (e, 'clean', Inf);
h = bwmorph (f, 'spur', Inf);
i = bwmorph (h, 'clean', Inf);
j = imopen(i,strel('square', 2));
area = bwarea (j)
carea = bwarea (imcomplement (j))
area*100/(area + carea)
[L, num] = bwlabel(j, 8);
num
stats = imfeature (L, 'all')
i = 0:10:850
for k = 1:1:num

   p(k) = stats(k). Area;

end
plot(p), figure, hist(p, i), figure
imshow(f), figure, imshow(j)
```

References

1. Pillai, K.R., Ghosh, A., Chattopadhyaya, S., Sarkar, P.K., Mukherjee, K.: Some investigations on the interactions of the process parameters of submerged arc welding, Manufacturing Technology and Research, Jan-June2007, **3**(1–2), 57–67
2. Gunarajan, V., Murugan, N.: Prediction of heat affected zone charecteristics in submerged arc welding of structural steel pipes. Welding Research. 94s–98s (2002)

3. Ghosh, A., Chattopadhyaya, S.: Transient Heat Conduction of Submerged Arc Welded Plates, Proc. Int. Conf. on 'Advanced Topics in manufacturing Engineering and Management', pp. 32–35 University of Czestochowa, Faculty of Management, Czestochowa (2010)
4. Jhon A. Goldak and Mehdi A.: Computational Welding Mechanics, pp. 71–115 Springer (2005)
5. Ghosh, A., Chattopadhyaya, S.: Analytical solution for transient temperature distribution of semi-infinite body subjected to 3-D moving heat source of submerged arc welding process, Proc. Int. Conf. on Mechanical and Electrical Technology, Singapore, September 10–12, 733–737 (2010)
6. Eager, T.W., Tsai, N.S., Temperature fields produced by traveling distributed heat sources. Weld. J. 62(12), 346s–355s (1983)
7. Jeong, S.K., Cho, H.S.: An analytical solution to predict the transient tem-perature distribution in fillet arc welds, Weld. J. 76(6), 223s–232s (1997)
8. Goldak, J., Chakravarti, A., Bibby, M.: A double ellipsoidal finite element model for welding heat source, IIW Doc.No212-603-85
9. Nguyen, N.T., Ohta, A., Matsuoka, K., Suzuki, N., Maeda, Y.: Analytical solution for transient temperature of semi-infinite body subjected to 3-D moving heat sources. Weld. Res. 265s–274s (1999)
10. Fachinotti, V.D., Cardona, A.: Semi-analytical solution of the thermal field induced by a moving double-ellipsoidal welding heat source in a semi-infinite body. Asociacion Argentina de Mecanica Computacional 10–13 Noviembre 2008, 1519–1530
11. Nguyen, N.T., Mai, Y.W., Simpson, S., Ohta, A.: Analytical approximate solution for double ellipsoidal heat source in finite thick plate. Weld. Res. 82s–93s (2004)
12. Ravichandran, G., Raghupathy, V.P., Ganesan, N.: Analysis of temperature distribution during circumferential welding of cylindrical and spherical com-ponents using the finite element method. Comput Struct **59**(2), 225–255 (1996)
13. Chandra, U.: Determination of residual stresses due to girth butt welds in pipes, Trans. ASME J. Press. Ves. Technol. **107**, 178–184 (1985)
14. Akkus, A.: Temperature distribution study in resistance spot welding. J. Sci. Ind. Res. **68**, 199–202 (2009)
15. Veenstra, P.C., Hults, A.: A thermal model of spot welding process. Greve Offset NV, Eindhoven (1969)
16. Bentley, K.P., Greenwod, J.A., Knowlson, P., Baker, R.G.: Temperature distribution in spot welds. Brit Weld J **10**, 613–619 (1963)
17. Kermanpur, A., Shamanian, M., Esfahani, V.: Yeganesh,Three—dimensional thermal simulation and experimental investigation of GTAW circumferentially butt -welded Incoloy 800 pipes. J Mater Process Technol **199**, 110–121 (2009)
18. Maheshwari, A., Arya, H.A., Kumar, S., Singh, C.: Experimental determination of weld pool temperature and to generate temperature profile for GMAW, Proc. 2nd International Conference on 'Production and industrial enginneering', NIT, jalandhar, India, 2010, p. 45
19. Gutierrez, G., Araya, j.G.: Temperature distribution in a finite solid due to a moving laser beam, Proc. IMECE2003-43545, 2003 ASME Int. Mech. Eng. Congr., Washington, November 15–21
20. Bianco, N., Manca, O., Naso, V.: Numerical analysis of Transient Temperature Fields in Solids by a Moving Heat Source, HEFAT 2004, 3rd Int. Conf. On Heat Transfer, Fluid Mechanics and Thermodynamics, paper BN2, 21–24 June 2004, Capetown, South Africa
21. Bianco, N., Manca, O., Nardini, S.: Two dimensional transient analysis of temperature distribution in a solid irradiated by a gaussian lesser source, Proc. ESDA2004-58286 7th Biennial Conf. on Engineering Systems Design and Analysis July 19–22 2004, Manchester, United Kingdom
22. Bianco, N., Manca, O., Nardini, S., Tamburrino, S.: Transient heat conduction in solids irradiated by a moving heat source, excerpt from the Proceedings of COMSOL User Conference 2006 Milano
23. Ohring, S., Lugt, H.J.: Numerical Simulation of a time dependent 3D GMA weld pool due to a Moving Arc. Weld. J. **79**(120), 416s–171s (1999)

24. Mundra, K., et. al.: Weld metal microstructure calculation from fundamentals of transport phenomena in the arc welding of low -alloy steels, Weld. J. **76**, 163s–171s (1997)
25. Biswas, P., Mandal, N.R.: Thermo mechanical finite element analysis and experimental investigation of single -pass single—sided submerged arc welding of cmn steel plates, Proc. IMechE Part B. J. Eng. Manuf. **224**, 627–639
26. Ogwuagwu, Vincent, O.: Simulation of heat transfer in thin plates during arc welding, AU J. T. **10**(1): 52–54 (2006)
27. Postacioglu, N., Kapadia, P., Dowden, J.M.: The thermal stress generated by a moving elliptical weld pool in the welding of thin metal sheets. J. Appl. Phys. **30**, 2304–2312 (1997)
28. Ali, Y.M., Zhang, L.C.: Relativistic heat conduction. Int. J. Heat Mass Transf. **48**, 2397–2406 (2005)
29. Ali, Y.M., Zhang, L.C.: Relativistic heat source. Int. J. Heat Mass Transf. **48**, 2741–2758 (2005)
30. Ghosh, A., Chattopadhyaya, S.: Prediction of weld bead penetration, transient temperature distribution and HAZ width of submerged arc welded structural steel plates, defect and diffusion forum, (Impact Factor- 0.27**), **316–317**, 135–152 (2011)
31. Murugan, N., and Gunaraj, V.: Prediction and control of weld bead geometry and shape relationships in submerged arc welding of Pipes. J. Mater. Process. Technol. **168**(3), 94s–98s (2005)
32. Menaka, M., Vasudevan, M., Venkatraman, B., Raj, B.: Estimating bead width and depth of penetration during welding by infrared thermal imaging. Non-Destr Test Cond Monit **47**(9), 564–568 (2005)
33. Pathak, A.K., Dutta, G.L.: Three dimensional finite element analysis to predict the different zones of microstructure in submerged arc welding; Proc. Institution of Mechanical Engineers, part B, J. Eng. Manuf. **218**(3), 234–239 (2004)
34. Kanjilal, P., Majumdar, S.K., Pal, T.K.: Prediction of submerged arc weld-metal composition from flux ingredients with the help of statistical design of mix-ture experiment. Scandinavian J. Metall. **33**(3), 146–159 (2004)
35. Kanjilal, P., Pal, T.K., Majumdar, S.K.: Combined effect of flux and welding parameters on chemical composition and mechanical properties of submerged arc weld metal. J. Mater. Process. Technol. **171**(2), 125 (2006)
36. Livieri, P., Lazzarin, P.: Fatigue strength of steel and aluminium welded joints based on generalised stress intensity factors and local strain energy values. Int. J. Fract. **133**(3), 247–276 (2005)
37. Bechet, G., Kahraman, N.: Wear behaviour of Bulldozer Rollers welded using a arch submerged arc welding process. Mater. Des. **24**(7), 136 (2003)
38. Trindade, V.B., Mello, R.S.T., Payao, J.C., Paranhos, R.P.R.: Influence of Zirconium on microstructure and toughness of low-alloy steel weld metals. J. Mater. Eng. Perform. **15**(3), 284–286 (2006)
39. Tuek, J., Umek, I., Bajcer, B.: Weld-cost saving accomplished by replacing single-wire submerged arc welding with triple-wire welding. Sci. Technol. Weld. Join. **10**(1) (2005)
40. Kim, S., Jeong, Y.J., Lee, C.W., Yarlagadda, P.K.D.V.: Prediction of welding parameters for pipeline welding using an intelligent system. Int. J. Adv. Manuf. Technol. **22**(9, 10), 713–719 (2003)
41. Zhao, P.C., Wu, C.S., Zhang, Y.M.: Numerical simulation of the dynamic characteristics of weld pool geometry with step-changes of welding parameters. Source: Modell. Simul. Mater. Sci. Eng. **12**(5), 765–780 (2004)
42. Bang, K.S., Kim, W.Y.: Estimation and prediction of HAZ softening in thermo-dynamically controlled and accelerated—cooled steel. Weld. J. AWS. 174s–179s (2002)
43. Goldak, J., Chakravarti, A., Bibby, M.: A double ellipsoid finite element model for welding heat sources. IIW Doc.No. 212-603-85 (1985)
44. Wang, Y., Tsai, H.L.: Impingement of filler droplets and weld pool dynamics during gas metal welding process. Int. J. Heat Mass Transf. **44**(11), 2067–2080 (2001)
45. Ghosh, A., Chattopadhyaya, S.: Prediction of weld bead penetration, transient temperature distribution and HAZ width of submerged arc welded structural steel plates. Defect Diffus. Forum **316–317**, 135–152 (2011)

Modeling the Pore Level Fluid Flow in Porous Media Using the Immersed Boundary Method

Isabel Malico and Paulo J. S. A. Ferreira de Sousa

Abstract This chapter demonstrates the potential of the immersed boundary method for the direct numerical simulation of the flow through porous media. A 2D compact finite differences method was employed to solve the unsteady incompressible Navier–Stokes equations with fourth-order Runge–Kutta temporal discretization and fourth-order compact schemes for spatial discretization. The solutions were obtained in a Cartesian grid, with all the associated advantages. The porous media is made of equal size square cylinders in a staggered arrangement and is bounded by solid walls. The transverse and longitudinal distances between cylinders are equal to two cylinder diameters and at the inlet a fully developed velocity profile is specified. The Reynolds number based on the cylinder diameter and maximum inlet velocity ranges from 40 to 80. The different flow regimes are identified and characterised, along with the prediction of the Reynolds number at which transition from steady to unsteady flow takes place. Additionally, the average drag and lift coefficients are presented as a function of the Reynolds number.

Keywords Porous media · Fluid flow · Numerical simulation · Immersed boundary method

I. Malico (✉) · P. J. S. A. Ferreira de Sousa
Physics Department, University of Évora, R. Romão Ramalho 59,
7000-671 Évora, Portugal
e-mail: imbm@uevora.pt

P. J. S. A. Ferreira de Sousa
e-mail: ferreiradesousa@gmail.com

I. Malico · P. J. S. A. Ferreira de Sousa
Department of Mechanical Engineering, IDMEC/IST, Technical University
of Lisbon, Av. Rovisco Pais, 1049-001 Lisbon, Portugal

J. M. P. Q. Delgado et al. (eds.), *Numerical Analysis of Heat and Mass Transfer in Porous Media*, Advanced Structured Materials 27,
DOI: 10.1007/978-3-642-30532-0_9,

1 Introduction

When considering porous media, most studies apply the volume-average technique to the fluid flow equations and replace the complex geometry of the surfaces that bound the phases by a fictitious model, in which each phase occupies the entire domain [52, 60]. In this case, macroscopic equations are developed and the different phases are treated as overlapping continuous phases that interact with each other. In doing so, extra terms that account for this interaction appear in the transport equations, resulting in complex integro-differential equations. The simplest way of dealing with these equations is to represent the effects of the interfaces by closure models that can be obtained from either experiments or simulations; therefore transferring the problem to obtaining the closure models. The major advantage of this macroscopic approach is that it is relatively cheap as far as computational resources are concerned. However, despite its theoretical and practical importance, only averaged values are obtained and, therefore, the level of detail of the solution is limited. Examples of the use of volume-average models in the study of porous media flows can be found, for example, in [2, 8, 45, 46, 51, 57].

An alternative to the application of the volume-average technique on the fluid flow equations is to solve the pore level flow. In this approach, the solid regions that form the pores are considered, and fluid flow is simulated in the interstices of the matrix. As a result, very detailed solutions are obtained; of course, at the cost of increased computational time. The potential of pore level simulations, also called microscopic simulations, is considerable. Recently, they have been used to predict global (macroscopic) effects such as dispersion or flow resistance through a porous matrix. Since purely theoretical studies of porous media flows are very limited, traditionally, this type of macroscopic quantities was obtained from experiments. One of the problems that arises from using this procedure is that experimental data relies on flow conditions and/or porous media that are often not controlled or well defined in terms of characteristics of the solid matrix, mobility of the particles and other factors.

The prediction of pore level fluid flow through porous media has received considerable attention in recent years. The porous media and typical conditions found in research on porous media flows are broad and cover a large number of Reynolds numbers and porosities [27]; the published works reflect this diversity. The next paragraphs review some of the pore level studies carried out so far.

There exists a large number of studies of Stokes flows in porous media ($Re \ll 1$), of which many focus on pore level simulations. As an illustration of these works, recent examples of microscopic simulations for very small Reynolds numbers are presented next. Manz et al. [36] compared the results of lattice-Boltzmann simulations of the flow field through a three-dimensional random packing of spheres with NMR measurements in the creeping flow regime. They showed a good overall agreement between experiments and simulations. Nabovati et al. [43] presented three-dimensional simulations of Darcy flow through random fibrous media. They used the lattice-Boltzmann method for calculating the permeability of the media for a wide range of porosities and studying the effect of the curvature and aspect ratio of the fibers on this parameter. Additionally, the permeability tensor was determined, yielding a scalar value (as expected, since the medium studied is isotropic). Smolarkiewicz and Winter [53] simulated Darcy flows through samples of three-dimensional simulated pore spaces. They compared an immersed boundary method implementation to a finite-element boundary fitted unstructured mesh model and concluded that the use of an immersed boundary method is both simpler and computational undemanding. To the authors' knowledge this was the only study where the immersed boundary method was applied to realistic random porous media. Maier and Bernard [39] discussed the significance of using interpolation boundary techniques to achieve second-order accurate lattice-Boltzmann methods for the simulation of Stokes flow around solid boundaries and through arrays of spheres. They concluded that interpolations techniques have a minor effect if the grid resolution is low and that, to achieve statistical accuracy in pore level simulations, the grid resolution will be close to the threshold required for resolving secondary flows (where interpolation techniques have only a minor effect). Dersken and Larsen [10] used the lattice-Boltzmann method to study the shear flow over layers of spheres attached to a flat wall. They calculated, for different assemblies and for a Reynolds number much smaller than one, the drag and lift forces exerted on the spheres and realized that the sphere-to-sphere variation of these forces is significant due to the interaction between spheres. Holmes et al. [26] preformed smooth particle hydrodynamics (SPH) simulations of Darcy flow through ordered arrays of cylinders and spheres and proved that the predicted friction coefficients and permeabilities agree well with available benchmarks. Despite this kind of methods being computationally costly, the authors argue that smooth particle hydrodynamics is very powerful when applied to problems involving multi-phase flows and porous media.

All of the above works focus on low Reynolds flows through porous media. When the Reynolds number is small, i.e., the flow is dominated by viscous forces, the Darcy law states that the relationship between the pressure gradient and the volume-averaged fluid velocity is linear. However, as the Reynolds number increases, this relationship becomes nonlinear and inertial effects start to contribute more and more to the pressure gradient. Usually, a quadratic term is used as proposed by Dupuit [11], apud [29], and Forchheimer [19], apud [29]. However, some authors (e.g., Mei and Auriault [40], Hill et al. [24], Chai et al. [9]) state that, at Reynolds numbers $O(1)$, a cubic transitional regime should be considered. The range of Reynolds numbers of this weak inertial regime is small and Fand et al. [14] suggested that this transitional regime can be described in simple mathematical terms if the transition is assumed to occur at a particular artificial Reynolds number (the error being less than 2.8 % for the random packing of spheres experimentally investigated). After this regime, the strong inertial regime (Forchheimer flow) takes place until a turbulent-like regime begins. Note that, at steady state, the average drag force exerted on the solid matrix is proportional to the average pressure gradient; therefore, talking about the dependency of the average drag force or of the average pressure gradient on the Reynolds number is the same.

As already mentioned, in the strong inertial regime, above a certain Reynolds number and for moderate Reynolds numbers $O(10^2)$, the drag force exerted on the porous solid matrix by the fluid phase is proportional to the square of the average fluid velocity (or, in other words, the dimensionless drag force has a linear dependence with the Reynolds number). Examples of widely used relations are the Ergun equation [12] or its modifications (e.g., MacDonald et al.'s [37]). The Ergun equation was derived for packed beds made of several types of particles and does not depend on the orientation, shape and surface of the elements that constitute the porous media and the MacDonald et al.'s, of these parameters, only depends on the surface roughness. Beavers and Sparrow [4], apud Lee and Yang, [31] stated that the Ergun model should depend on the microstructure of the porous medium and could be inappropriate for other classes of porous media. Evidences of that were found on subsequent studies (e.g., [5] and [25]). In fact, flow patterns in the pores are highly-medium specific, therefore, there should be no general model for the drag force as a function of the bulk properties of a porous medium.

As far as pore level simulations are concerned, the number of studies on non-Darcy flow is smaller that on Darcy flow. Hill et al. [24] studied, through theory and lattice-Boltzmann simulations, and for small Reynolds numbers, the effects of fluid inertia in ordered and random arrays of spheres. They showed that, for several porosities, the first inertia contribution to the drag force is proportional to the cube of the average fluid velocity. In the same year, Hill et al. [25] presented a study, based on lattice-Boltzmann simulations and for Reynolds numbers up to $O(10^2)$, on the effects of fluid inertia in ordered and random arrays of spheres. They investigated, for porosities near the close-packed limit up to the dilute limit, the dependence on the Reynolds number of the drag force on the spheres and concluded that at the range of Reynolds numbers studied, the non-dimensional drag force has a more complex dependence on the Reynolds number and porosity than the popular linear scaling of the Ergun correlation [12] (particularly for high porosities, since good agreement was found at the closed-packed limit). Following the previous study, Hill and Koch [23] discuss the transition to unsteady flow in porous media through lattice-Boltzmann simulations. Knowing where the transition to unsteady flow occurs in porous media is important to understand, for instance, dispersion and heat and mass transfer characteristics in porous media. The fact that unsteady velocities exist promotes fluid mixing, that otherwise would be trapped in closed streamlines, and modify the thickness of boundary layers. Hill and Koch [23] focus on the bi-dimensional flow through a square array of cylinders with the porosity of 0.8 and follow two different approaches: considering a periodic unit cell or simulating 26 cylinders with bounding walls. The latter has the objective of comparing the lattice-Boltzmann simulations with experiments (which always involve solid walls). Hill and Koch [23] concluded that the presence of the walls has a significant effect on the spatial and temporal development of the flow and that the presence of the walls results in much smaller critical Reynolds numbers. The Hill and Koch [23] study is an example of an investigation that focuses on unsteady results; however, most computational studies on porous media consider steady state, or when they account for the time dependency, only time-averaged results are analyzed. Beetstra et al. [5] used lattice

Boltzmann simulations to derive the drag force for fluid flow through random arrays of mono- and bi-dispersed spheres. The Reynolds numbers studied were up to 1000. Similarly to Hill et al. [25] for a more limited range of Reynolds numbers, they found a more complex functional form than the linear scaling with the Reynolds number widely used. Teruel and Rizwan-uddin present simulations of flow in porous media in a broad range of regimes (from Darcy to turbulence) [54]. They consider a unit cell representative of square rods in a staggered arrangement and varied the porosity from 5 to 100 %. Using a finite-volume method, they characterized the permeability and pressure drop through the media as a function of the Reynolds number and porosity. Nevertheless, they did not focus on the simulation of the laminar, time dependent, oscillatory flow behavior. As in the previous study, Alshare et al. [1] also studied through unit cell simulations, bi-dimensional cylinder arrays, but in a square arrangement. The porosity was varied from 0.75 to 0.875, the Reynolds number from 1 to 1000 and the flow direction relative to the computational domain from 0 to 90°. They used a finite-volume method to obtain permeabilities, inertial coefficients, Nusselt numbers and thermal dispersion coefficients. Their calculations assumed steady flow. This is a common assumption for this kind of unit cell simulation in porous media (see also, for example, Nakayama et al. [44] or Braga and de Lemos [6]). At the end of Alshare et al.'s [1] paper, unit cell calculations of the unsteady flow are presented, in order to determine the effects of the unsteadiness on their results. They concluded that the pressure gradient steady and unsteady calculations are within 10 %, whereas the dispersion results within 17 %. Chai et al. [9] simulated fluid flow through a random array of cylinders, both for the Darcy and the non-Darcy regimes ($Re \leq 30$). The lattice Boltzmann was used to derive a general correlation for the pressure drop in high porosity porous media.

As an example of the application of the immersed boundary method for moderate Reynolds number fluid flow in porous media we can cite the work of Ghazaryan et al. [21]. These authors simulated the flow through a unit cell representative of square rods in a staggered arrangement. A second-order accurate finite volume method was used to discretize the Navier–Stokes equations. The simulated Reynolds numbers were 1 and 100 and the authors were interested in the consequences of the flow structures on filtration by the porous matrix. In the following year, Lopez Penha et al. [35] computed the apparent permeability of a model porous media composed of a staggered arrangement of square rods. A representative elementary volume was used along with a finite-volume method with an immersed boundary method implementation. They investigated the effects of the Reynolds number ($1 < Re < 600$; based on the distance between two consecutive rods), porosity (0.25, 0.5 and 0.75) and flow direction on the apparent permeability.

Analyzing the numerical methods used in the available literature, Lattice-Boltzmann simulations (using unit cells or not) are the most popular for the prediction of the pore level fluid flow through porous media. Conventional computational fluid dynamics methods, usually with unit cell simulations are also common. On the other hand, immersed boundary methods are not yet easy to find for porous media applications.

For the simulation of the pore level fluid flow in porous media, a description of the solid surfaces is needed. The first studies addressed the flow inside reconstructed porous media built with idealized obstacles (for example, periodic arrays of spheres in Zick and Homsy [64], random arrays of square cylinders in Nabovati and Sousa [42] or staggered arrangements of square rods in Teruel and Rizwan-uddin, [54]). This approach is still common and the complexity of the matrices depends on the nature of the problem analyzed and on the computational resources available. Recently, Navier–Stokes DNS simulations already consider real pore geometries. The real structure of the porous media can be obtained by applied experimental and imaging techniques. For instance, Yamamoto et al. [61] used the three-dimensional computer tomography technique to obtain the porous structure of a Ni–Cr porous metal and to model soot combustion inside this structure. Similarly, Fourie et al. [20] used X-ray computed tomography to create a 3D model of sand. However, this kind of methods may be costly, impose limits on the resolution and only use a limited number of pores space samples. In addition, it is difficult to represent geometrically complex boundaries and to apply to them efficient grid methods (the immersed boundary method avoids the latter; this issue will be addressed in Sect. 3). In the mid way between the idealized porous geometries and the real ones lay the generated random pore spaces. This has been done for instance by Smolarkiewicz and Winter [53].

Another issue that cannot be disregarded when dealing with pore level simulations is the accuracy of the numerical method used. Porous media flows are characterized by the existence of solid boundaries inside the computational domain and the accurate resolution of the near-surface flows is essential for the study of the flow dynamics around these solid boundaries. Besides sufficient accuracy, adequately large samples of the porous media must be used in the pore level simulations. These two requirements compete and a compromise has to be reached. When the porous media is obtained by 3D digital images, the grid resolution is predefined; however, if the porous media is artificially generated, this parameter can be chosen. In this framework, if high-order methods are used for pore level simulations, the accuracy of the solutions is improved and the grid resolution requirements can be reduced.

The most recent pore level studies employ computational experiments to determine the influence of several parameters of the porous structure on the flow through it. The results presented in this chapter are inspired by these works, now using a different combination of the numerical model and porous media geometry: an ordered assembly of bi-dimensional geometrically simple obstacles and an immersed boundary method with a compact high-order finite-differences model [16, 17]. These results are a first step towards the implementation of more complex geometries and intend to be a demonstration of the application of the proposed numerical method to the pore level simulation of flow through porous media. The advantage of using such a numerical approach will be discussed in this chapter.

In the next section, the problem to be solved is presented along with the simulation details. Then, in Sect. 3, the numerical method used is described in detail, while Sect. 4 presents the verification and validation of the immersed boundary method implemented. In Sect. 5 the simulations of flow through a staggered array of square cylinders are reported and finally, the conclusions of this chapter are drawn in Sect. 6.

Fig. 1 Scheme of the porous array of square cylinders simulated

2 Problem Description

The unsteady incompressible form of the continuity and Navier–Stokes equations for a Newtonian fluid were considered

$$\nabla \cdot \mathbf{u} = 0 \tag{1}$$

$$\frac{\partial \mathbf{u}}{\partial t} + (\mathbf{u} \cdot \nabla)\,\mathbf{u} = \nu\,\nabla^2\mathbf{u} - \nabla p + \mathbf{f} \tag{2}$$

where $\mathbf{u}$ is the fluid velocity, p the pressure divided by the density, ν the kinematic viscosity of the fluid and $\mathbf{f}$ is a body-force field of the immersed boundary method that allows for the imposition of the porous matrix.

This set of equations was solved to simulate the flow through an idealized porous media bounded by solid walls. Figure 1 shows the flow configuration. Six bi-dimensional cylinders with square cross-sections and equal sizes, d, compose one out of eight rows displayed in a staggered arrangement. The distance between two consecutive cylinders in a row, s, is the same for all cylinders, so as the spacing between rows, s_r ; both are set to 2 times the cylinders diameter. The cylinders are confined inside a plane channel; therefore Dirichlet boundary conditions are imposed at the walls. A fully developed laminar flow channel flow was prescribed at the inlet and open boundary conditions at the outlet of the channel. Appropriate dimensions to reduce the influence of inflow and outflow boundary conditions were considered. An inflow length of 10 diameters and a total channel length of around 72 diameters were chosen. The number of grid points representing one cylinder dimension was chosen as 10, resulting in a grid with 722×196 nodes.

A series of simulations is conducted for the flow inside the porous array represented in Fig. 1 for different Reynolds numbers ranging from 40 to 80. The characteristic length scale and velocity were taken as d and the maximum velocity of the inlet profile, respectively. The initial flow is stagnant and all calculations are carried past a steady state.

3 The Immersed Boundary Method

The immersed boundary method was originated in the area of computational biomechanics [47]. It is a numerical technique for solving flow problems with immersed bodies using a Cartesian grid Navier–Stokes solver. In detail, the computational grid used does not need to conform to all the boundaries of the domain and the numerical algorithm is modified in the vicinity of the immersed bodies in order to enforce the desired boundary conditions. Beyond reducing the difficulties of generating meshes for complex boundaries and embedded bodies, the immersed boundary method can potentially increase the computational efficiency, as the computational cost per grid node is generally much lower than that of general purpose unstructured (body-fitted) grid solvers.

In the last few years, many immersed boundary methods have been presented. However, even though the capability to calculate complex geometries has grown considerably, the issue of accuracy still stands. In porous media applications, the accurate calculation of both the region near the media and the wake is of upmost importance for understanding specific flow phenomena. One of the main motivations to use high-order finite-difference schemes is to have high accuracy and low numerical dissipation in the region far from the immersed boundary. In several problems in porous media, the wake behind the immersed porous media is of utmost importance (e.g., in shelterbelts [59], windbreaks [63] or urban canopies [33]). Historically, the most common numerical tool used in the study of immersed boundary problems is Peskin's method [48], which in its original version is first-order accurate. Limitations of the original method are pointed out in Lai and Peskin [30], as well as presentation of an improved algorithm. This improved version is formally second-order accurate and does not possess true second order accuracy. This question was addressed in [22], where a new formally second order accurate version of the immersed boundary method was presented and actual second order numerical convergence rates for a prototypical fluid–structure interaction were demonstrated, at least for low and moderate Reynolds numbers and in situations where the true solution appears to be sufficiently regular. More recently, Kim and Peskin [28] presented a penalty immersed boundary method for an elastic boundary with mass which is second-order accurate for a particular test problem. The somewhat lack of accuracy of this and other immersed boundary methods means that the results obtained using this method should be carefully verified in order to prove grid independence.

The class of problems considered in this chapter pertains to stationary boundaries that are rigid. Therefore, the forcing term in the Navier–Stokes equations is used to approximately enforce velocity boundary conditions on the body. One possibility to implement an immersed boundary in a Cartesian Navier–Stokes flow solver is the introduction of a body-force field $\mathbf{f}$ in order to have a desired velocity distribution $\mathbf{V}$ over a boundary. Mohd-Yusof [41] derived a formulation of the forcing that does not affect the stability of the discrete-time equations. In [41], this direct forcing was combined with B-splines to compute the laminar flow over a

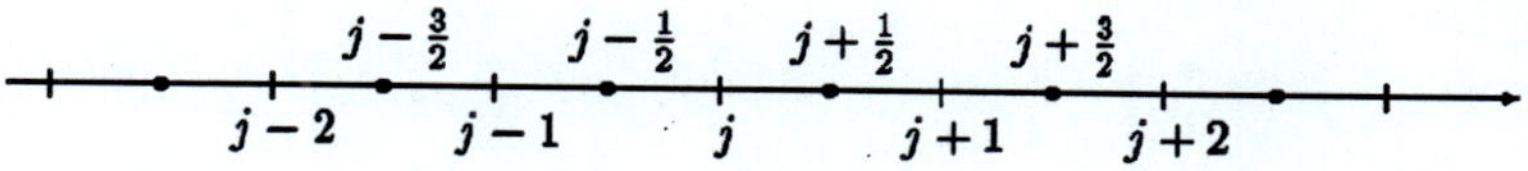

Fig. 2 1D grid for collocated and staggered schemes

three-dimensional ribbed channel. Verzicco et al. [58] used the direct force approach to perform a large-eddy simulation of the flow in a motored axisymmetric piston-cylinder assembly. Fadlun et al. [13] used the same type of forcing to develop a second-order accurate method for simulating unsteady three-dimensional incompressible flows in complex geometries. A similar approach was used in [3] to perform large-eddy simulations around complex boundaries on fixed Cartesian grids. Other direct forcing methods with different ways of treating the immersed boundary conditions include [34], [56] and [62]. An implementation of the forcing of Mohd-Yusof [41] was used in the present method.

In the implementation of the immersed boundary method described in this chapter, the momentum equations are spatially discretized on a Cartesian staggered mesh by finite differences and all derivatives are evaluated with implicit 4th-order accurate compact finite difference schemes [32]. The fourth order accurate Runge–Kutta scheme (RK-4) was used for temporal discretization. The numerical method used belongs to the projection methods class and is presented in the Ferreira de Sousa's papers [16, 17].

For the sake of simplicity, let us consider a uniformly spaced mesh where the nodes are indexed by j, as shown in Fig. 2. The independent variable at the nodes is $x_j = h(j - 1)$ for $1 \leq j \leq N$, where $h = x_{j+1} - x_j$ and the function values at the nodes $f_j = f(x_j)$ are given.

3.1 Approximation of First Derivative

The finite difference approximation f_j' to the first derivative at the node j depends on the function values at nodes near j. These schemes are generalizations of the Padé scheme. These generalizations are derived by writing approximations of the form

$$\beta f_{j-2}' + \alpha f_{j-1}' + f_j' + \alpha f_{j+1}' + \beta f_{j+2}' = a\frac{f_{j+1} - f_{j-1}}{2h} + b\frac{f_{j+2} - f_{j-2}}{4h} + c\frac{f_{j+3} - f_{j-3}}{6h} \quad (3)$$

The relations between the coefficients α, β, a, b and c are derived by matching the Taylor series coefficients of various orders. For the 4th order tri-diagonal scheme (Classical Padé scheme):

$$\alpha = \frac{1}{4}, \beta = 0,\ a = \frac{3}{2},\ b = 0,\ c = 0 \tag{4}$$

3.2 Approximation of Second Derivative

The derivation of compact approximations for the second derivative is similar to the first derivative. Again we start with a relation of the following form

$$\beta f''_{j-2} + \alpha f''_{j-1} + f''_j + \alpha f''_{j+1} + \beta f''_{j+2} = a\frac{f_{j+1} - 2f_j + f_{j-1}}{h^2} + b\frac{f_{j+2} - f_j + f_{j-2}}{4h^2} + c\frac{f_{j+3} - 2f_j + f_{j-3}}{9h^2} \tag{5}$$

where f''_j represents the finite difference approximation to the second derivative at node j. Once again, the relations between the coefficients α, β, a, b and c are derived by matching the Taylor series coefficients of various orders. For the 4th order tri-diagonal scheme:

$$\alpha = \frac{1}{10}, \beta = 0,\ a = \frac{6}{5},\ b = 0,\ c = 0 \tag{6}$$

3.3 Approximation of First Derivative on Cell-centered Mesh

Formulas for calculating the first derivative on a cell-centered mesh are necessary in staggered grids. Referring again to Fig. 2 and starting from an approximation of the form

$$\beta f'_{j-2} + \alpha f'_{j-1} + f'_j + \alpha f'_{j+1} + \beta f'_{j+2} = a\frac{f_{j+1/2} - f_{j-1/2}}{h} + b\frac{f_{j+3/2} - f_{j-3/2}}{3h} + c\frac{f_{j+5/2} - f_{j-5/2}}{5h} \tag{7}$$

The relations between the coefficients α, β, a, b and c are derived yet again by matching the Taylor series coefficients of various orders. For the 4th order tri-diagonal scheme:

$$\alpha = \frac{1}{22}, \beta = 0,\ a = \frac{12}{11},\ b = 0,\ c = 0 \tag{8}$$

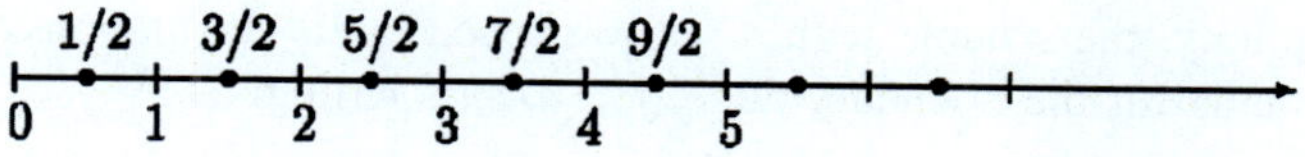

Fig. 3 1D variable arrangement at left boundary

3.4 Boundary Schemes

Most problems of physical interest involve domains with non-periodic boundaries. In order to handle these, boundary schemes need to be implemented for differentiation and interpolation at the boundary nodes. These schemes involve one-sided differencing and need to possess an order of accuracy as high as possible, see for example Mahesh [38]. Boundary schemes satisfying these requirements are presented in this section.

In a staggered grid, two sets of boundary schemes are required. Let us consider the 1D grid shown in Fig. 3 which shows the nodes and the numbering scheme at the left boundary.

3.4.1 Boundary Schemes for the First Derivative

A scheme for the first derivative at the boundary node ($j = 0$) may be written as

$$f_0' + \hat{\alpha} f_1' = \frac{1}{h}\left(af_{1/2} + bf_{3/2} + cf_{5/2} + df_{7/2}\right) \tag{9}$$

In this implementation, we used a third-order scheme given by:

$$\hat{\alpha} = 23,\ a = -25,\ b = 26,\ c = -1 \tag{10}$$

The first derivative at the boundary ($j = 1/2$) can be written as

$$f_{1/2}' + \hat{\alpha} f_{3/2}' = \frac{1}{h}(af_0 + bf_1 + cf_2 + df_3) \tag{11}$$

Again, the third-order scheme given by:

$$\hat{\alpha} = -1,\ a = -1,\ b = 2,\ c = -1 \tag{12}$$

3.4.2 Boundary Schemes for Interpolation

A scheme for interpolating a function to the boundary ($j = 0$) can be written as

$$f_0 + \hat{\alpha} f_1 = af_{1/2} + bf_{3/2} + cf_{5/2} \tag{13}$$

Matching coefficients of the Taylor expansion gives a third-order scheme with

$$a = \frac{1}{8}(3\hat{\alpha} + 15),\ b = \frac{1}{4}(3\hat{\alpha} - 5),\ c = \frac{1}{8}(3 - \hat{\alpha}) \tag{14}$$

For simplicity, the scheme with $\hat{\alpha} = 0$ was used in all calculations. The interpolation scheme for the boundary node ($j = 1/2$) is written as

$$f_{1/2} + \hat{\alpha} f_{3/2} = af_0 + bf_1 + cf_2 + df_3 \tag{15}$$

This allows for a fourth-order scheme with

$$a = \frac{1}{16}(5 - \hat{\alpha}),\ b = \frac{1}{16}(9\hat{\alpha} + 15),\ c = \frac{1}{16}(9\hat{\alpha} - 5),\ d = \frac{1}{16}(1 - \hat{\alpha}) \tag{16}$$

where $\hat{\alpha} = 0$ once again is chosen for simplicity.

3.5 Immersed Boundary

In this immersed boundary problem implementation, we define a body-force field **f** such that a desired velocity distribution **V** can be assigned over a boundary **S** [41]. This way we can solve for **u** explicitly from the body–force **f** added to the Navier–Stokes equations. Theoretically there are no restrictions for the shape and motion of **S** or the velocity distribution **V**. One of the main advantages of this methodology is that **f** can be prescribed on a Cartesian mesh, preserving the accuracy and efficiency of the solution procedure on simple grids. If equation 2 is discretized in time, we have

$$\frac{\mathbf{u}^{t+1} - \mathbf{u}^{t}}{\Delta t} = \mathrm{RHS}^{t+1/2} + \mathbf{f}^{t+1/2} \tag{17}$$

where $\mathrm{RHS}^{t+1/2}$contains the pressure gradient as well as the convective and viscous terms. In order to know which value of $\mathbf{f}^{t+1/2}$will return $\mathbf{u}^{t+1} = \mathbf{V}^{t+1}$ on the immersed boundary, the answer is given from the above equation,

$$\mathbf{f}^{t+1/2} = -\mathrm{RHS}^{t+1/2} + \frac{\mathbf{V}^{t+1} - \mathbf{u}^{t}}{\Delta t} \tag{18}$$

The forcing is direct because the value of the velocity is imposed directly on the boundary without any dynamical process. At every time step, the velocity boundary conditions hold irrespective of the frequencies in the flow.

3.5.1 Immersed Obstacle Implementation

It was already referred at the beginning of this section that the immersed boundary methods have the advantage of allowing the use of simple Cartesian grids. There is a price to pay for this, though. The use of Cartesian grids generically results in a staircase representation of the solid boundaries that compose the porous matrix and hence in a distortion of the original porous media morphology. The deviation from the original structure depends on the grid resolution: the higher the grid resolution, the lower the distortion. Other body-fitted methods do not carry this kind of geometrical deviations.

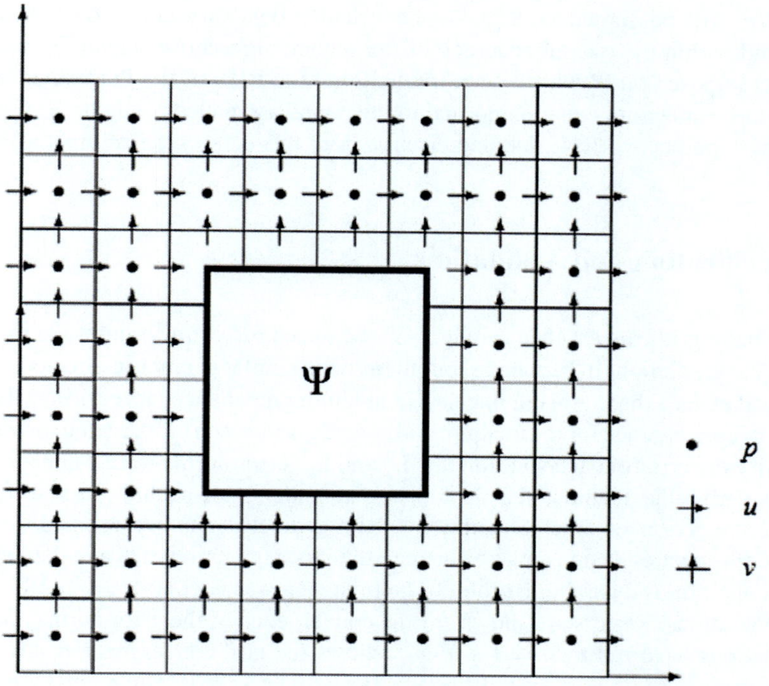

Fig. 4 Immersed boundary method staggered grid arrangement

What was just said loses meaning when addressing geometrical data obtained from digital imaging techniques, as is often the case of porous media applications. In this situation, the solid matrices images are also represented in a matrix of pixels and therefore, if the same grid resolution is used, no deviation from the original image occurs. Another circumstance when no distortion arises is when the solid boundaries are aligned with the Cartesian grid. This is the case of the immersed obstacle implementation presented in this chapter, where the expressions given for the forcing are accurate since the position of the unknowns on the grid coincides with that of the immersed boundary. This requires the boundary to lie on coordinate lines or surfaces. In the present case, the boundary is coincident with the position where the cell-centered velocity is defined (Fig. 4). Since the forcing is used to enforce the velocity on the boundary, the x-direction force is calculated on the u-grid and the y-direction force is calculated on the v-grid.

The forcing procedure used is the following. For each stage of the RK-4 time discretization:

1. Points that lie on coordinate lines corresponding to the shape of the immersed boundary are defined as forcing points.
2. From the Dirichlet boundary condition, the forcing velocity $\mathbf{V}^{t+1}$ is calculated directly.
3. Using Eq. (18), we calculate the body force $\mathbf{f}$.

As it will be shown in Sect. 4, the velocity boundary conditions on Ψ are satisfied within the overall accuracy of the scheme. Pressure boundary conditions are not imposed on Ψ but they are implicit into the RHS of the Poisson equation, since the momentum equation normal to the boundary reduces to $\mathrm{d}p/\mathrm{d}n = 0$ on the boundary points [3] (for a detailed discussion of this detail see Fadlun et al. [13]).

4 Verification and Validation

Systematic grid convergence testing, [49], was used for verification of the method. Since exact solutions of benchmark problems with similar geometries are not known, we need at least three grids (a fine grid, a medium grid and a coarse grid) so that the order of convergence for the medium and coarse grids, p, can be extracted. In order to quantify the error of each solution, the L_2 and L_∞ error norms were calculated.

To verify the method, the flow past a cylinder with square-cross section is calculated. For a detailed description of the geometry see Breuer et al. [7]; the cylinder is centred inside a plane channel, the blockage ratio is 1/8 and the velocity inlet condition is a parabolic profile. The time-step was kept at $\Delta t = 5\times10^{-5}$ s for all simulations. Results are shown for the convergence of the error norms after the simulations have run for $t = 1$ s. Fig. 5 shows the grid convergence results for a set of grids with 2522×641, 1262×321 and 632×161, which corresponds to $h = 2\times10^{-3}$, 1×10^{-3} and 5×10^{-4}, respectively.

As anticipated, the results for the L_∞ norm are slightly better than first order, which in turn impacts the results for the global norm L_2. Even if the formal order of accuracy for the finite difference schemes used is 4th order, that order of accuracy is not recovered in the global due to the first-order convergence of the maximum error. Therefore, the method achieves slightly better than third-order accuracy for the L_2 norm.

As an epilogue to this discussion, we would like to note that we are not trying to fully resolve the issue of convergence of immersed boundary-like methods. Outside the immersed boundary region, the convergence of the L_2 norm approaches 4th order, providing an evident accuracy advantage to the method. We would like to stress that the verified order of accuracy for the present method is significantly better than the second-order accurate methods (that necessarily have first-order convergence for the L_∞ norm as well [55]).

Before studying the fluid flow through the staggered array of square cylinders presented in Sect. 2, the present immersed boundary implementation was validated against the results from other authors. In particular, comparisons with data available in the literature for flows through a square-cylinder and through rows of square-cylinders were made in Ferreira de Sousa and Malico [15] and Ferreira de Sousa et al. [18]. The agreement between the results obtained with the present method and the literature results was good. Note that the geometries chosen in this validation studies can be regarded as the building blocks of the geometry analyzed in this chapter.

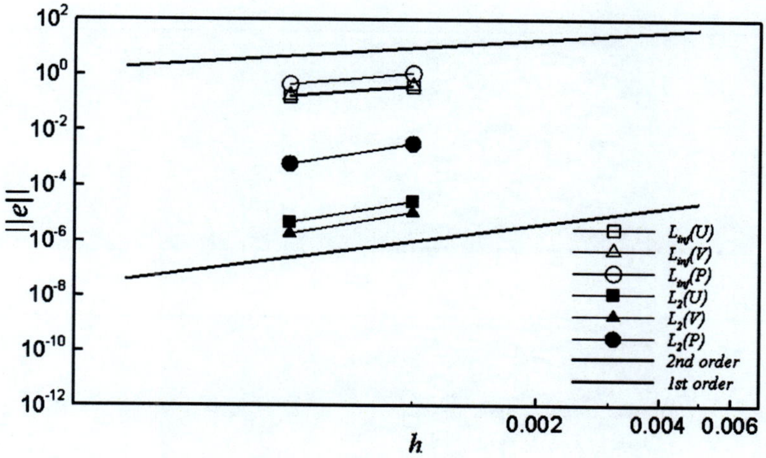

Fig. 5 Grid convergence results

5 Results

This section presents immersed boundary method results for the problem described in Sect. 2. It should be stressed that the current chapter was not meant to be a detailed study of the physics of the flow through a staggered array of cylinders, but rather to present the current immersed boundary method implementation and to demonstrate its potential for the prediction of flow through porous media.

5.1 *Flow Regimes*

Depending on the Reynolds number, different flow regimes can be distinguished for the flow through porous media in general, and through an array of square cylinders in particular. At very low Reynolds numbers ($\mathrm{Re} < 1$), no separation takes place at the surface of the cylinders that constitute the porous matrix. In this regime, viscous forces dominate the flow. As the Reynolds number increases, inertial effects start to be important and the flow starts to separate at the trailing edges of the cylinders. In this regime a steady recirculation region consisting of two symmetrical vortices is observed behind each cylinder. When a critical Reynolds number, Re_c, is exceeded, periodic vortex shedding from the cylinders that compose the downstream row can be detected. When the Reynolds number is further increased, the flow separates at the leading edges of the cylinders. The flow regime is still laminar and bidimensional. With a further Reynolds number increase, 3D structures start to appear in the flow and transition to turbulence takes place.

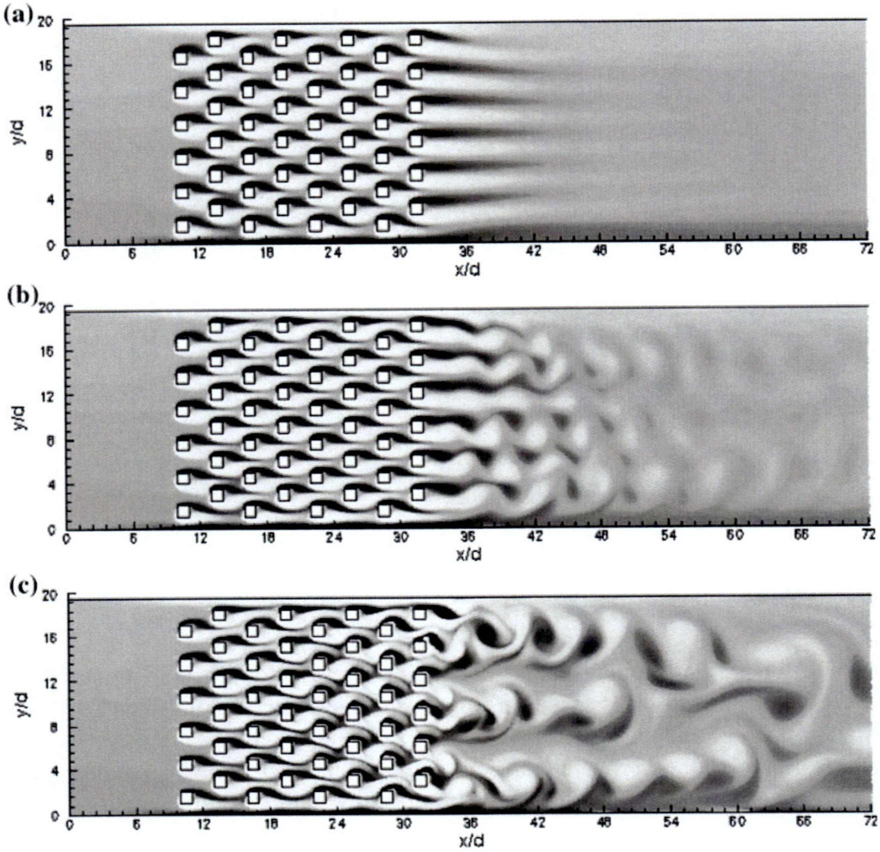

Fig. 6 Instantaneous vorticity contours for **a** Re = 40, **b** Re = 60 and **c** Re = 80

Figure 6 shows the instantaneous vorticity contours at different Reynolds numbers. In the case of Re = 40, the vortices that form behind the cylinders are symmetrical and no vortex shedding occurs. When the Reynolds number is 60, the vortices observed in the last row of cylinders oscillate and shed. As for the wakes of the upstream cylinders, they are reduced and stabilized by the pressure of the downstream obstacles and no periodic oscillations of the shear layers occur. Vortex shedding is also seen at Reynolds 80, but now with clear evidences of vortex merging. Immediately after shedding, two or three jets behind the matrix merge, forming stagnant zones in between. The coalesced groups merge again further downstream.

Figure 7 shows the streamlines in the vicinity of one of the cylinders (the third counting from the bottom wall in the last row) for two different Reynolds numbers (60 and 80). For the lower Reynolds number, the separation point is fixed at the trailing edge and the flow is attached to the side walls. However, for a Reynolds number equal to 80, flow separation already starts at the leading edge of the

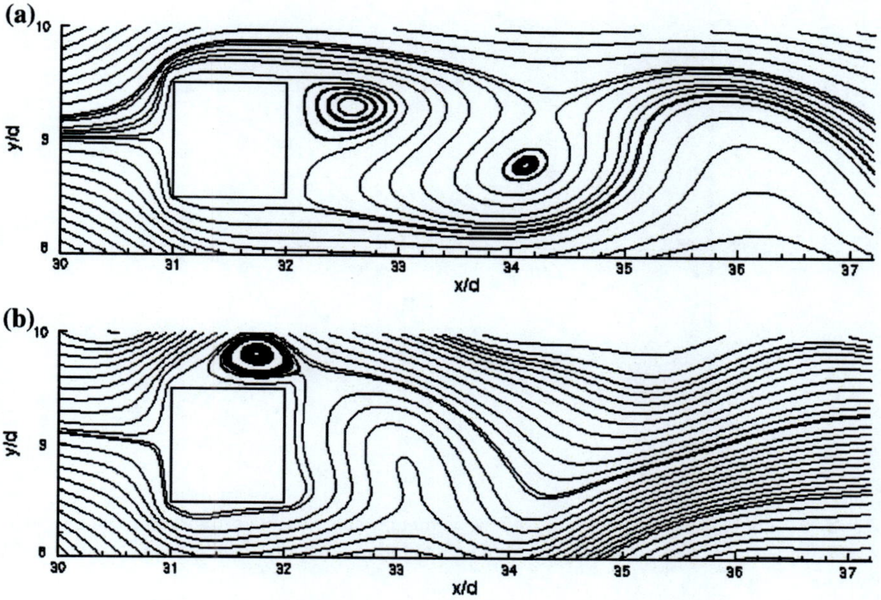

Fig. 7 Streamlines around the third cylinder of the last row for different Reynolds numbers **a** Re = 60, **b** Re = 80

cylinder. This is an important change in the flow structure with a strong influence in the frequency of the vortex shedding. Breuer et al. [7] discuss this issue for the flow past a square cylinder.

5.2 Critical Reynolds Number

The critical Reynolds number, i.e., the Reynolds number at which transition from steady to unsteady flow occurs, has important practical implications, for example, in dispersion and heat and mass transfer. Its determination was based on bifurcation analysis, a procedure also followed, for instance, by Sewatkar et al. [50]. The time series of the transverse velocity at a distance of one diameter behind the last row of cylinders and nine diameters from the bottom wall was monitored for the Reynolds numbers simulated. Figure 8 shows, as a function of the Reynolds number, the maximum and minimum transverse velocities at this location adimensionalized by the maximum inlet velocity, $u_{\max}$.

The interception of the fitted curve with the abscissa gives the critical Reynolds numbers, which is around 52. Above this Reynolds number the flow is unsteady, below steady.

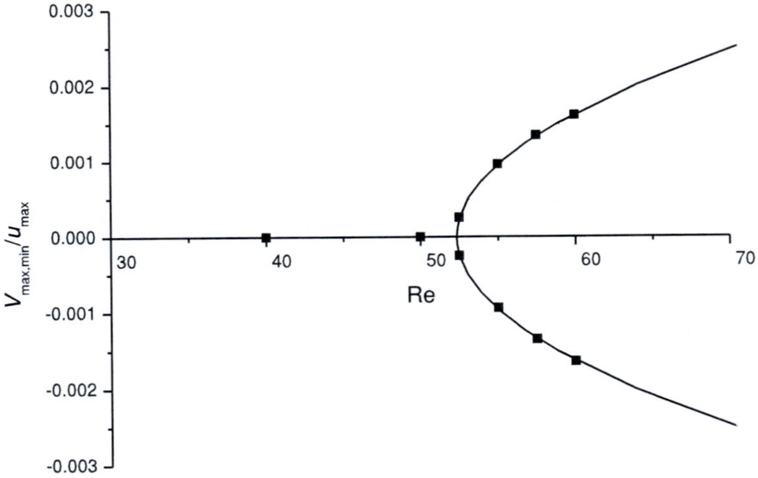

Fig. 8 Reynolds number dependency of the minimum and maximum transverse velocity at $y/d = 9$ and $x/d = 1$ from the porous medium outlet plane adimensionalized by the maximum inlet velocity

5.3 Drag Coefficient

The knowledge of the drag force in a porous matrix is important in engineering applications, since it is related to the pressure drop that occurs in the system where it is integrated. This, in turn, relates to the energy spent, and, therefore, choosing the best ratio between pressure drop and equipment efficiency is of the most importance in equipment/systems design. In this context, a deep understanding of the pressure drop in a porous media is of the most significance. Its calculation is not an easy task due to its complex dependence on the geometric structure of the porous matrices and the complicated flow patterns that are typical of porous media flows.

Figures 9 and 10 show, respectively, the mean time-averaged drag and lift coefficients as a function of the Reynolds numbers (here mean implies average over all the cylinders of the matrix). To quantify how much the forces experienced by the individual cylinders deviate from the average force, the standard deviation of the time-average drag and lift coefficients is also shown.

For the Reynolds numbers range simulated, the mean time-averaged drag coefficient decreases with and increase in the Reynolds number, whereas the mean time-average lift coefficient is almost constant.

The drag and lift forces are strong functions of the relative placement of the cylinders in the matrix and the variability of these forces is considered. This has been shown also, for example, by Derksen and Larsen [10] for a randomly generated, equally sized spheres assemblies supported by a flat wall. In agreement with their results, the variability of the lift force presents large values (even larger than the mean time-average lift coefficient).

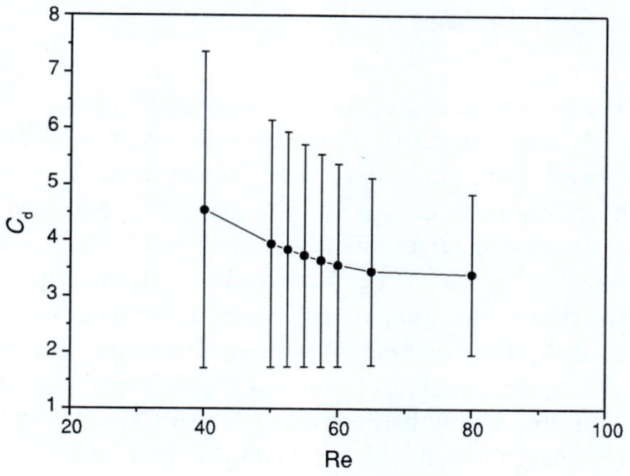

Fig. 9 Variation of the mean time-average drag coefficient and standard deviation with Reynolds number

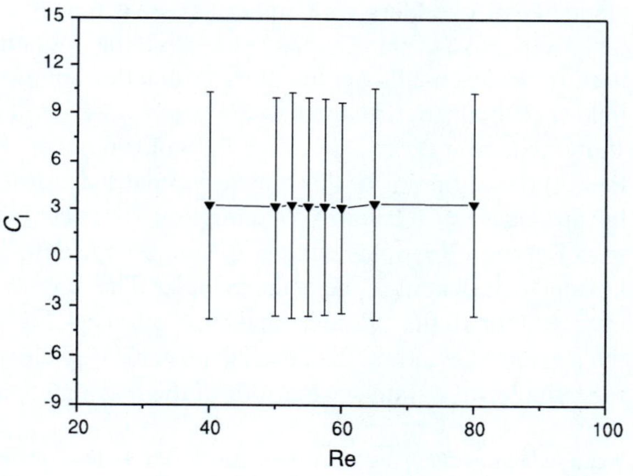

Fig. 10 Variation of the time-average lift coefficient and standard deviation with Reynolds number

For high s/d and s_r/d, the cylinders do not interact and all the obstacles would experience the same drag and lift as a single square cylinder in its position. For denser arrays, the obstacles start to interact with each other and forces in the vertical direction are induced (The velocity around an obstacle contains vertical components induced by other cylinders that contribute to the viscous drag in this direction). The array considered in this study falls in this category and another reflex of the cylinders interaction is the wake structure that is formed downstream the solid matrix.

One of the potentials of the pore level simulations done with the present method is the ability to reproduce dynamic phenomena that depend on accurate resolution of near-surface flows. For the Reynolds numbers considered, separation of the flow at the sharp corners of the square cylinders can be found as seen in the Sect. 5.1. It is evident that the fluid flow possesses significant regions of separated flow and vortical structures. These vortical structures have a large influence on the pressure drop and being able to capture them is essential.

6 Conclusions

The prediction of global (macroscopic) effects such as dispersion or flow resistance from a porous matrix requires the knowledge of the velocity field in an integral control volume that contains the porous media. With the application of the immersed boundary method to a specific geometry, this flow field is known in detail and obtained with no use of closure models. The fluid flow transport equations are valid inside the pores and can be solved with the appropriate set of boundary conditions. Because of the complex geometry of porous media, microscopic simulations have typically been carried out with idealized arrays of geometrically simple obstacles, although examples of simulations with more complex geometries can be found.

In this study, the Ferreira de Sousa's [16, 17] implementation of the immersed boundary method with a 2D compact high-order finite-differences model has been used to carry out direct numerical simulations of the flow through an ordered assembly of cylinders with square cross-section in a staggered arrangement. The major objective of this chapter is to unveil the potential of this method, not so often used in porous media applications. A detailed solution of the time-dependent flow field was obtained, which opens the door to a series of analysis of the problem under study. Different flow regimes for Reynolds numbers between 40 and 80 are identified. At the beginning of this range, symmetrical vortices form behind each obstacle, but around $\mathrm{Re} = 52$, vortex shedding starts to occur in the cylinders that form the last row. Between Reynolds number 60 and 80, the detachment point changes from the trailing to the leading edge of the cylinder. The time-average drag and lift coefficients averaged for all the cylinders that compose the solid matrix is presented, along with the standard deviation. Because the obstacles are close and interact with each other, the cylinder-to-cylinder variations of the calculated drag and lift forces are big.

Acknowledgments This work was developed within the framework of project PTDC/EME-MFE/100178/2008, which is financially supported by the COMPETE program (FCOMP-01-0124-FEDER-010183) and by FCT—Fundação para a Ciência e a Tecnologia. Additionally, Paulo Ferreira de Sousa would like to acknowledge the FCT grant SFRH/BPD/48160/2008.

References

1. Alshare, A.A., Strykowski, P.J., Simon, T.W.: Modeling of unsteady and steady fluid flow, heat transfer and dispersion in porous media using unit cell scale. Int. J. Heat Mass Transf. **53**, 2294–2310 (2010)
2. Antohe, B.V., Lage, J.L.: A general two-equation macroscopic turbulence model for incompressible flow in porous media. Int. J. Heat Mass Transf. **40**, 3013–3024 (1997)
3. Balaras, E.: Modeling complex boundaries using an external force field on fixed cartesian grids in large-eddy simulations. Comput. Fluids **33**, 375–404 (2004)
4. Beavers, G.S., Sparrow, E.M.: Non-Darcy flow through fibrous porous media. J. Appl. Mech. **36**, 711–714 (1969)
5. Beetstra, R., van der Hoef, M.A., Kuipers, J.A.M.: Drag force of intermediate Reynolds number flow past mono and bidisperse arrays of spheres. AIChE J. **53**, 489–501 (2007)

6. Braga, E.J., de Lemos, M.J.S.: Laminar natural convection in cavities filled with circular and square rods. Int. Commun. Heat Mass Transf. **32**, 1289–1297 (2005)
7. Breuer, M., Bernsdorf, J., Zeiser, T., Durst, F.: Accurate computations of the laminar flow past a square cylinder based on two different methods: lattice-Boltzmann and finite-volume. Int. J. Heat Fluid Flow **21**, 186–196 (2000)
8. Breugem, W.P., Boersma, B.J.: Uittenbogaard, R.E.: The laminar boundary layer over a permeable wall. Transp. Porous Media **59**, 267–300 (2005)
9. Chai, Z., Shi, B., Lu, J., Guo, Z.: Non-Darcy flow in disordered porous media: A lattice Boltzmann study. Comput. Fluids **39**, 2069–2077 (2010)
10. Derksen, J.J., Larsen, R.A.: Drag and lift forces on random assemblies of wall-attached spheres in low-Reynolds-number shear flow. J. Fluid Mech. **673**, 548–573 (2011)
11. Dupuit, J.: Etudes Theoriques et Pratiques sur le Movement des Eaux. Dunod, Paris (1863)
12. Ergun, S.: Fluid flow through packed columns. Chem. Eng. Prog. **48**, 89–94 (1952)
13. Fadlun, E.A., Verzicco, R., Orlandi, P., Mohd-Yusof, J.: Combined immersed boundary finite-difference methods for three-dimensional complex flow simulations. J. Comp. Phys. **161**, 35–60 (2000)
14. Fand, R.M., Kim, B.Y.K., Lam, A.C.C., Phan, R.T.: Resistance to the flow of fluids through simple and complex porous media whose matrices are composed of randomly packed spheres. J. Fluids Eng. **109**, 268–274 (1987)
15. Ferreira de Sousa, P.J.S.A., Malico, I.: A 2D compact finite difference immersed boundary method for flow in porous media. In: Proceeding of the 5th European Conference on Computational Fluid Dynamics-Eccomas CFD (2010)
16. Ferreira de Sousa, P.J.S.A., Pereira, J.C.F.: Fourth- and tenth-order compact finite difference solutions of perturbed circular vortex flows. Int. J. Numer. Meth. Fluids **49**, 603–618 (2005)
17. Ferreira de Sousa, P.J.S.A., Pereira, J.C.F., Allen, J.J.: Two-dimensional compact finite difference immersed boundary method. Int. J. Numer. Meth. Fluids **65**, 609–624 (2011)
18. Ferreira de Sousa, P.J.S.A., Malico, I., Fernandes, G.: Numerical Simulation of 2D Flow through a Packed Bed of Square Cylinders. Defect Diffus. Forum **236–328**, 725–730 (2012)
19. Forchheimer, P.: Wasserbewegung durch Boden. Z. Ver Deutsch. Ing. **45**, 1782–1788 (1901)
20. Fourie, W., Said, R., Young, P., Barnes, D.L.: The simulation of pore scale fluid flow with real world geometries obtained from X-ray computed tomography. COMSOL Conference Boston, USA (2007)
21. Ghazaryan, L., Lopez Penha, D.J., Geurts, B.J., Stolz, S., Winkelmann, C.: Simulation of impaction filtration by a porous filter. In: 3rd International Conference Porous Media applications in science, engineering, and industry, Montecatini, Italy (2010)
22. Griffith, B.E., Peskin, C.S.: On the order of accuracy of the immersed boundary method: Higher order convergence rates for sufficiently smooth problems. J. Comput. Phys. **208**, 75–105 (2005)
23. Hill, R.J., Koch, D.L.: Moderate-Reynolds-number flow in a wall-bounded porous media. J. Fluid Mech. **453**, 315–344 (2002)
24. Hill, R.J., Koch, D.L., Ladd, A.J.C.: The first effects of fluid inertia on flows in ordered and random arrays of spheres. J. Fluid Mech. **448**, 213–241 (2001)
25. Hill, R.J., Koch, D.L., Ladd, A.J.C.: Moderate-Reynolds-number flows in ordered and random arrays of spheres. J. Fluid Mech. **448**, 243–278 (2001)
26. Holmes, D.W., Williams, J.R., Tilke, P.: Smooth particle hydrodynamics simulations of low Reynolds number flows through porous media. Int. J. Numer. Anal. Methods Geomech. **35**, 419–437 (2011)
27. Kaviany, M.: Principles of Heat Transfer in Porous Media, 2nd edn. Springer, New York (1995)
28. Kim, Y., Peskin, C.S.: Penalty immersed boundary method for an elastic boundary with mass. Phys. Fluids **19**, 053103 (2007)
29. Lage, J.L., Antohe, B.V., Nield, D.A.: Two types of non-linear pressure-drop versus flow-rate relation observed for saturated porous media. J. Fluids Eng. **119**, 700–706 (1997)
30. Lai, M.-C., Peskin C.S.: An immersed boundary method with formal second order accuracy and reduced numerical viscosity. J. Comput. Phys. **160**, 705–719 (2000)

31. Lee, S.L., Yang, J.H.: Modeling of Darcy-Forchheimer drag for fluid across a bank of circular cylinders. Int. J. Heat Mass Transf. **40**, 3149–3155 (1997)
32. Lele, S.K.: Compact finite difference schemes with spectral-like resolution. J. Comp. Phys. **103**, 16–42 (1992)
33. Lien, F.S., Yee, E., Cheng, Y.: Simulation of mean flow and turbulence over a 2D building array using high-resolution CFD and a distributed drag force approach. J. Wind Eng. Ind. Aerodyn. **92**, 117–158 (2004)
34. Lima e Silva, A.L.F., Silveira-Neto, A., Damasceno, J.J.R.: Numerical simulation of two-dimensional flows over a circular cylinder using the immersed boundary method. J. Comp. Phys. **189**, 351-370 (2003)
35. Lopez Penha, D.J., Geurts, B.J., Stolz, S., Nordlund, M.: Computing the apparent permeability of an array of staggered square rods using volume-penalization. Comput. Fluids **51**, 157–173 (2011)
36. Manz, B., Gladden, L.F., Warren, P.B.: Flow and dispersion in porous media: Lattice-Boltzmann and NMR studies. AIChE J. **45**, 1845–1854 (1999)
37. Macdonald, I.F., El-Sayed, M.S., Mow, K., Dullien, F.A.L.: Flow through porous media—the Ergun equation revisited. Ind. Eng. Chem. Fundam. **18**, 199–208 (1979)
38. Mahesh, K.: A family of high order finite difference schemes with good spectral resolution. J. Comp. Phys. **145**, 332–358 (1998)
39. Maier, R.S., Bernard, R.S.: Lattice-Boltzmann accuracy in pore-scale flow simulation. J. Fluid Mech. **222**, 647–663 (2010)
40. Mei, C.C., Auriault, J.-L.: The effect of weak inertial on flow through a porous media. Ind. Eng. Chem. Fundam. **18**, 199–208 (1991)
41. Mohd-Yusof, J.: Combined Immersed Boundaries/B–splines Methods for Simulations of Flows in Complex Geometries, CTR Annual Research Briefs, NASA Ames/Stanford University, Standford, 317-327 (1997)
42. Nabovati, A., Sousa, A.C.M.: Fluid flow simulation in random porous media at pore level suing the Lattice Boltzmann method. J. Eng. Sci. Technol. **2**, 226–237 (2007)
43. Nabovati, A., Llewellin, E.W., Sousa, A.C.M.: A general model for the permeability of fibrous media based on flow simulations using the lattice Boltzmann method. Compos.: Part A **40**, 860–869 (2009)
44. Nakayama, A., Kuwahara, F., Hayashi, T.: Numerical modelling for three-dimensional heat and fluid flow through a bank of cylinders in yaw. J. Fluid Mech. **498**, 139–159 (2004)
45. Pereira, J.C.F., Malico, I., Hayashi, T.C., Raposo, J.M.F.: Experimental and numerical characterization of the transverse dispersion at the exit of a short ceramic foam inside a pipe. Int. J. Heat Mass Transf. **48**, 1–14 (2005)
46. Pereira, J.C.F, Malico, I., Hayashi, T.C., Raposo, J.M.F.: Study of high Reynolds number pipe flows with porous inserts. J. Porous Media **15**, 549–563 (2012)
47. Peskin, C.S.: Flow patterns around heart valves: a numerical method. J. Comp. Phys. **10**, 252–271 (1972)
48. Peskin, C.S.: The immersed boundary method. Acta Numerica **11**, 479–517 (2002)
49. Roache, P.: Verification of codes and calculations. AIAA J. **36**, 696–702 (1998)
50. Sewatkar, C.M., Sharma, A., Agrawal, A.: On the effect of reynolds number for flow around a row of square cylinders. Phys. Fluids **21**, 083606 (2009)
51. Silva, R.A., de Lemos, M.J.S.: Turbulent flow in a channel occupied by a porous layer considering the stress jump at the interface. Int. J. Heat Mass Transf. **26**, 5113–5121 (2003)
52. Slattery, J.C.: 2-phase flow through porous media. AIChE J. **16**, 345–352 (1970)
53. Smolarkiewicz, P.K., Winter, C.L.: Pores resolving simulation of Darcy flows. J. Comp. Phys. **229**, 3121–3133 (2010)
54. Teruel, F.E., Rizwan-uddin: Characterization of a porous medium employing numerical tools: Permeability and pressure-drop from Darcy to turbulence. Int. J. Heat Mass Transf. **52**, 5878–5888 (2009)
55. Tornberg, A.-K., Engquist, B.: Numerical approximations of singular source terms in differential equations. J. Comp. Phys. **200**, 462–488 (2004)

56. Tseng, Y.H., Ferziger, J.H.: A ghost-cell immersed boundary method for flow in complex geometry. J. Comp. Phys. **192**, 593–623 (2003)
57. Vafai, K., Tien, C.L.: Boundary and inertia effects on flow and heat transfer in porous media. Int. J. Heat Mass Transf. **24**, 195–203 (1981)
58. Verzicco, R., Mohd-Yusof, J., Orlandi, P., Haworth, D.: Large eddy simulation in complex geometric configurations using boundary body forces. AIAA J. **38**, 427–433 (2000)
59. Wang, H., Takle, E.: A numerical simulation of boundary-layer flows near shelterbelts. Bound.-Layer Meteorol. **75**, 141–173 (1995)
60. Whitaker, S.: Diffusion and dispersion in porous media. AIChE J. **13**, 420–427 (1967)
61. Yamamoto, K., Takada, N., Misawa, M.: Combustion simulation with Lattice Boltzmann method in a three-dimensional porous structure. Proc. Combust. Inst. **30**, 1509–1515 (2005)
62. Zhang, N., Zheng, Z.C.: An improved direct-forcing immersed-boundary method for finite difference applications. J. Comp. Phys. **221**, 250–268 (2007)
63. Zhuang, Y., Wilson, J.D.: Coherent motions in windbreak flows. Bound.-Layer Meteorol. **70** 151–169 (1994)
64. Zick, A.A., Homsy, G.M.: Stokes flow through periodic arrays of spheres. J. Fluid Mech. **115**, 13–26 (1982)

Computer Simulation of the Origination Porosity

G. I. Zmievskaya, A. L. Bondareva, V. V. Savchenko and T. V. Levchenko

Abstract The origination of porosity can be imagined as a nonequilibrium heterogeneous phase transition of the first kind in a solid body. The example of the silicon amorphization using processing of substrate by inert gases is presented. The influence of porosity on changes of the elastic stresses in material due to the arising of bubbles consisting of gas and vacancies in the crystal lattice of bilayer Mo/Si has been shown. Models are analyzed by means computer experiment. The structures of defects in the shape of layers and cracks had been formed due to indirect elastic interaction between clusters and their interaction with boundaries of Mo/Si layers.

Keywords Porosity, crystal materials · Computer simulation

G. I. Zmievskaya (✉) · A. L. Bondareva
M.V. Keldysh Institute of Applied Mathematics,
Russian Academy of Sciences, 4 Miusskaya sq,
Moscow 125047, Russia
e-mail: zmig@mail.ru

A. L. Bondareva
e-mail: bal301775@yandex.ru

V. V. Savchenko
Hosei University, 3-7-2 Kajino-chu Koganei-shi,
Tokyo 184-8584, Japan

T. V. Levchenko
Russian Researches Center GeoSystem,
8,Varshavskoe shosse, Moscow 113105, Russia

J. M. P. Q. Delgado et al. (eds.), *Numerical Analysis of Heat and Mass Transfer in Porous Media*, Advanced Structured Materials 27,
DOI: 10.1007/978-3-642-30532-0_10,

1 Introduction

Applications of the mono- and polycrystal materials based on silicon carbide have a great attention [1, 2] due to its perspective semi-conductor properties, high strength, corrosion stability as well optical characteristics et all. As it is known, silicon carbide crystallizes in polytypic forms (about of 200) which contents are defined by conditions of preparation (temperature, pressure and impurities). The paper deals with computer simulation of properties of porosity formation into Si monocrystal which precedes to SiC solid-state epitaxy.

The relevance of the topic in complexity technological processes, which are being implemented the processing of the surfaces using plasma, or in instruments or devices, when plasma it borders with surfaces, as well as the phenomenon of changing the structure over-activities and representing a complex physic-chemical processes, studying of which is in development. At the present time, together with the laboratory and technical experiments have developed methods of computing experiment in the annex to the objectives of materials science and interaction of plasma with the surface. Such strongly non-equilibrium media, as high-temperature plasma, demanded the creation of multidimensional kinetic electromagnetic codes, good practices in solving the fundamental problem of plasma physics [3] and formulated the idea of dusty plasma simulation and other plasma-like media, the movement of the particles which should be modeled in view of the many degrees of freedom and of long-range self-consistent interactions potentials.

Investigation of the properties of materials whose properties change in results blistering (defects formation, such as gas bubbles in the thin layer under the surface of the solid body of the irradiated ions of poorly soluble gas) is presented as numerical models of non-equilibrium physical process in plasma-like media (the crystal lattice defects). The consideration of this process as a phase transition in a solid body, which is made on the basis of the fundamental laws governing this process. The model of fluctuation stage caused by the nucleation of defects and their Brownian motion are described by the equations of the kinetic theory of gas, plasma and plasma-like media, as well as models of Markov random processes. A detailed study of the blistering in a laboratory experiment is limited in connection with small spatial scales (considered region $\sim$hundreds of the lattice parameters, depth from the surface of the ~ 100 nm, the duration of experiment is microsecond), and with a large value of laboratory experiments. The results of numerical experiment can serve as a basis for the development of new blistering theories and planning of laboratory research.

The structure of the nano coatings, with the properties of diamond-like material—carbide silicon SiC and conditions of their formation can be modeled with the help of the software complex, LRnLA/Nano (locally-recursive non-local asynchronous software complex for numerical simulation nano-technology problems named Nano), which is based on the object-oriented model of plasma [4] and plasma-like media model [5]. Important role in the calculations fine-crystalline thin films play phase transitions (condensing, cryistallizalization, melting) which mathematical model requires the solution of a system of quasilinear integro-differential equations mathematical physics in partial derivatives of the second order using stochastic analog method of physical–chemical processes [6].

Other the prospects are related with the creation of the optical elements and devices on the basis of porous of silicon. Porous silicon is a system of nanocrystals separated by the voids. Depending on the size of the voids of the porous silicon, materials can be divided into macroporous, meso-porous and microporous. When the size of the pores may vary from a few to hundreds of nanometers—this structure is typical of low dimensional medium.

A characteristic feature of the porous silicon is a large sum area of its inner surfaces which can be determined for macroporous (large sum area of porous inner surface is from 10 up to 100 m^2/cm^3), mezoporous (from 100 up to 300 m^2/cm^3) and microporous (from 300 up to 800 m^2/cm^3). Porous silicon assists in technological metallic mirrors which used to transmit an optical signal to a detector of diagnostic devices in TNR [7], X-ray lithography [8–10], and other application fields.

Recently the method of SiC film deposition on substrate consisting in processing of Si single-crystal by CO gas (under temperature $T = 1{,}100–1{,}400$ °C and pressure 10–300 Pa during $t = 5–60$ min.) developed [2], where is considered the hypothesis of silicon surface layers chemical transformation: Vacancies into Si interact with CO and layer of SiC experienced the stress relaxation. Authors of [2] supposed that the stresses arise because of a discrepancy between SiC/Si lattices but the role of stress into Si lattice during SiC layer formation was not evaluated in [2]. At the fluctuation stage of epitaxy the stress distribution into Si is able to be caused by nucleation of defects into lattice and it is able to be studied by another way. We study mechanisms of SiC producing process, which provides the solid-state epitaxial silicon carbide films growing on silicon substrate with thickness of 20–100 nm, because here is the possibility to calculate the development of amorphization of Si monocrystal using the computer simulation model of defects clustering into layered sample [11, 12].

Hollow spheres structures are varieties of porous structures which have high stiffness, the ability to absorb high amounts of energy at a relatively low stress levels, potential for noise control, vibration damping and thermal insulation. The effective thermal conductivity of perforated sphere structures in several kinds of arrangements is numerically evaluated for different hole diameters in [13] with using unit cells study and Finite Element Method.

The numerical analysis of the transient heat transfer in composite structures containing phase change materials using Lattice Monte Carlo Method is presented in [14]. In this paper pure paraffin and a paraffin-copper composite structure subjected to heating and cooling processes with phase change of the phase change materials were exampled. This work proved that the composite structure allowed for faster energy transfer, at the same time the pure PCM exhibits a higher maximum thermal energy storage capacity.

The macro- to micro-strain and stress conversion for porous ceramics are analytically modeled in [15] based on stress balance in elastically isotropic media and finite element modeling. In [15] was been shown that the macro-stress to average micro-stress ratio depends linearly on porosity and does not depend on pore morphology factor, while the ratio of macro-strain to average microstrain does.

The difference our model from [2] is follows: Initial stage of epitaxy related with blistering (vacancies gases bubbles formation) into brittle lattice of substrate which is preceded to epitaxy of SiC into lattice.

This our assumption is based on the problem of nucleation investigation or the phase transition at nonequilibrium stage by means computer simulation [5, 11, 12, 16–23], this is introduced by stochastic analogue method [5, 6, 11, 12, 16–25]. New approach is based on the strict results of the probability analysis of equations of mathematical physics, the kinetic theory of plasma and rarefied gases, the theory and practice of numerical experiment in non-linear plasma simulation. The progress in computer engineering allows us to see the overall picture of initial state of substrate thin layer after irradiation with inert gas flux of KeV energy: Pores as well as cracks structures are distributed into lattice in form of layers and/or defects clusters chains. Kinetic model of blistering into layered (Mo/Si) mirror for lithography [12, 22] has been modified for study blistering into silicon substrate planed for SiC epitaxy so thin metal layer of Mo (used in computer simulation model) plays here auxiliary role.

2 Kinetic and Stochastic Equations for Blisters Nuclei into Bilayer Sample

Amorphization of layered sample is represented as model of the Brownian motion of defects with a variable mass. Fluctuations unstable clusters of xenon gas (formation of cluster sizes) are taken into account. Blistering of metal and silicon layers is accompanied by the indirect elastic interaction of defects through the perturbation of acoustic phonons of the crystal lattice and Friedel oscillations of the electrons in the metal lattice. In such an approach, the degree of porosity and a change in the strain within the layers estimated using the kinetic distribution function/DF/ depending on sizes and coordinates in layers, it is possible to estimate roughness at an interface between layers in bilayer model (Mo/Si), which appears because a blisters come to the layers boundary. Craters of cone shape appear after Si lattice monolayer destruction which it is raised by blisters accumulation at the top of sample during its motion into thin layer of substrate in case of gas penetration into Si lattice. Here we give the example of a calculation of the defect formation which accounting the stochastic diffusion of defects into thin layered media.

The migration of blisters from a layer to a layer happens considering the effect of boundaries conditions which affects on the concentration of defects in layer materials. Numerical analysis of helium blistering [16] in nickel has previously detected a non-linear dependence between the average defect size and the sample temperature; porosity layers are perpendicular to the incident flux; the dependence of blistering characteristic on the radiation dose, the temperature, and potential of the sample surface adjacent to the ion flux; and a change in the strain with bulk defect accumulation.

2.1 Model of Blistering into Brittle Thin Layer of Si

The phenomenon of high-temperature blistering is considered as the first-order phase transition; the bubbles in near surface region and clusters are considered as a Brownian particles/BPs/with spherical as well cone shapes with variable masses. The self-organization into phase spaces of model have formed under follows factors: constant flux of ions on the surface, the long-range potentials of indirect interaction defects each with others, with surface as well as with dislocations. The problem of kinetic description of the phase transition at its non equilibrium stage is introduced by stochastic simulation method [6, 11, 12]. The kinetic equations for Brownian particles size clustering and migration into lattice are described by partial integral–differential equations and they have laborious methods for solving. Velocity of bubble/cluster growth (or degradation of size) and its migration have different characteristic time scales which equal for blistering ($\tau_g \sim 10^{-9}$ s for size evolution and $\tau_r \sim 10^{-8}$ s for migration). Thus, kinetic equations on a discrete time grid are solved by technique of splitting in terms of physical processes, and every stage is represented by its stochastic analogue [5, 6, 11, 12, 16–25]. Interaction between bubbles/clusters is indirect, through lattice phonons and electron density oscillations. Kinetic Kolmogorov equation looks like following:

$$\frac{\partial f(g,\vec{r},t)}{\partial t} = \hat{L}f(g,\vec{r},t) + S_\alpha - Q \tag{1}$$

where g is size of blister, g is measured in incompressible unit volumes of vacancy/gas particle; r-coordinates in lattice, x, y, z are dimensionless, measured in parameters of lattice; $f(g, \mathbf{r}, t)$ is DF, that is the detecting probability of blister with size g to be in the point of substrate lattice space with coordinates $\mathbf{r}$ at time moment t. Distribution function $f(g, \mathbf{r}, t)$ is dimensionless and normalized by 1 for each time t: $\int_G \int f(g,\vec{r},t)dgdxdydz = 1$.

In (1) $S_\alpha(f_\alpha)$ is the source of particles forming a nucleus; $S(f_\alpha)$is the source of "monomers" (vacancy or gas particles) with f_α-Maxwellian DF; Q is drain of blisters.

Taking into account difference between characteristic times and availability of a dedicated direction we can write $L = L_g + L_r$, L_g is operator for Kolmogorov–Feller equation for blister size evolution, L_r is operator corresponding to Einstein-Smoluchowski equation which describes motion of brownian particle (or blister). L_g and L_r are linear operators with non-linear functional-coefficients.

After splitting under physical processes the kinetic equations look like as following: the Kolmogorov–Feller equation for blister size evolution at the point with the coordinate $\mathbf{r}$ and the Einstein–Smoluchowski equation for Brownian particle with the mass M_g.

The kinetics equation (Kolmogorov–Feller) for blister size evolution at the point with the coordinate **r**:

$$\frac{\partial f_r(g,t)}{\partial t} = \frac{\partial\left[D_g(g,t)\frac{\partial f_r(g,t)}{\partial g}\right]}{\partial g} + \frac{1}{kT}\frac{\partial\left[D_g(g,t)f_r(g,t)\frac{\partial\{\Delta\Phi(g,\vec{r},t)\}}{\partial g}\right]}{\partial g} + S_\alpha,$$
$$f_r(g,0) = f_{0g}, \left.\frac{df_r(g,t)}{dg}\right|_{g=2} = 0, \left. f_r(g,t)\right|_{g<2} = 0. \tag{2}$$

where $f_r(g, t)$ is the DFs of clusters depends sizes at the point **r** of lattice; T temperature during non equilibrium stage is supposed constant, D_g is the diffusion coefficient in the phase space of nucleus sizes; $\Delta\Phi(g, \mathbf{r}, t)$ is the thermodynamic potential of the nucleation, $\Delta\Phi$ is measured in kT, temperature during fluctuation stage is supposed constant.

So, blister is vacancy-gas bubble into solids body are created by gas ions irradiation, that physical meaning has only $g > 2$, g can not be less 2, $g = 1$ is monomer (atom or vacancy but it is not blisters); the Einstein–Smoluchowski equation for BPs with the mass M_g obtained from Eq. (2)

$$\frac{\partial f_g(\vec{r},t)}{\partial t} = \frac{\partial\left[D_r(\vec{r},t)\frac{\partial f_g(\vec{r},t)}{\partial r}\right]}{\partial \vec{r}} - \frac{\partial\left[\frac{\vec{F}(\vec{r},t)}{M_g\gamma}f_g(\vec{r},t)\right]}{\partial \vec{r}} - Q, \tag{3}$$

$$f_g(\vec{r},t)|_{t=0} = f_{0r},$$
$$f_g(\vec{r},t)|_{x=x_{\mathrm{left}}} = f_g(\vec{r},t)|_{x=x_{\mathrm{right}}}, f_g(\vec{r},t)|_{y=y_{\mathrm{left}}} = f_g(\vec{r},t)|_{y=y_{\mathrm{right}}},$$

where $f_g(\mathbf{r}, t)$ is the distribution function of a BP with mass M_g over the substrate lattice coordinates.

Blister mass M_g is measured in vacancy mass. In common case blister contains atoms of gas and vacancies if vacancies into blister are more than gas atoms into it. Such, $g = g_{\mathrm{atom}} + g_{\mathrm{vac}}$, here g_{vac} is number of untied vacancies in blister and g_{atom} is number of atoms in blister. Blister mass can be calculated as $M_g = g_{\mathrm{atom}}\, m_{\mathrm{at}} + 0.8\, m_{\mathrm{sub}}\, g_{\mathrm{vac}}$, where m_{at} is mass of penetrated gas atom, m_{sub} is mass of substrate lattice atom, here taken into account that vacancy volume is approximately 0.8 of volume substrate atom. Here is the diamond-like face-centered cubic lattice and body-centered lattice. In common case total blisters mass can be written as $M_{bl} = \frac{4\pi}{3}\rho\int\frac{\partial f}{\partial g}g^3 dg$, where ρ is density of gas/vacancies in blister.

Condition of particle preservation is supposed by using of source of particles forming a nucleus S_α and monomers drains and blister drains Q. Monomers and blister drains locate on substratum boundaries including interlayers boundaries, grains boundaries, dislocation and others lattice defects; γ is friction coefficient.

The Fokker–Planck–Kolmogorov equation in the general form is:

$$\begin{cases} \dfrac{\partial f(X,t)}{\partial t} = -\sum\limits_{i=1}^{N} \dfrac{\partial}{\partial X_i} a_i^{\alpha}(X,t) f(X,t) + \dfrac{1}{2} \sum\limits_{i=1}^{N} \sum\limits_{j=1}^{N} \dfrac{\partial^2}{\partial X_i \partial X_j} b_{ij}^{\alpha}(X,t) f(X,t) \\ f(X,t)|_{t=0} = f_0(X) \\ \oint \left[a_i^{\alpha}(X,t) f(X,t) + \dfrac{1}{2} \sum\limits_{j=1}^{N} \dfrac{\partial}{\partial X_j} b_{ij}^{\alpha} f(X,t) \right] d\Gamma = 0, \quad i = 1, \ldots m \end{cases} \tag{4}$$

where Γ is the boundary of all possible values of the Markovian random process, $a_i^{\alpha}(X,t)$ is the drift vector and $b_{ij}^{\alpha}(X,t)$ is the diffusion matrix, which are functionals of MP distributions, should be limited and sufficiently smooth. The functionals $a_i^{\alpha}(X,t)$ and $b_{ij}^{\alpha}(X,t)$ entering the equation have the meaning of meansquare limits in time $\Delta t \to 0$ of the conditional expectation of increment of MP coordinates $\{X(t),\ t \geq 0\}$: $\Delta X(t) = X(t + \Delta t) - X(t)$ (this is $a_i^{\alpha}(X,t)$) and conditional dispersion—$b_{ij}^{\alpha}(X,t)$.

The relation between coefficients of Fokker–Planck–Kolmogorov equation and coefficients of the SDE is follows:

$$a_i = \underset{\Delta t \to 0}{l.i.m.} \frac{1}{\Delta t} \mathrm{E}[X(t + \Delta t) - X(t) | X(t) = x] \tag{5}$$

$$b_{ij} = \underset{\Delta t \to 0}{l.i.m.} \frac{1}{\Delta t} \mathrm{E}\left[[X(t + \Delta t) - X(t)]^2 | X(t) = x \right]$$

The stochastic analogue of kinetic equation (1) in the Ito–Stratonovich form can be written as

$$X = X_0 + \underbrace{\int_{t_0}^{t} H(X,\tau) d\tau}_{drift} + \underbrace{\int_{t_0}^{t} \sigma(X,\tau) dW(\tau)}_{diffusion} + X_{coll} \tag{6}$$

$$X|_{t=0} = X_0 = X(t = 0),\ t \in [t_0, T_{\text{finish}}],$$

where X is the stochastic dynamical variable, accounting in (4) diffusion and jump-like part. $dW(t)$ is increment of Wiener stochastic process, which is used in model of nuclei clustering as well in BM description. X_{coll} is jump-like stochastic process contribution into superposition of processes in (4). The consideration of chemical reactions can be calculated using point-vise.

The functionals $H(\tau,\ X(\tau))$ and $\sigma(\tau,\ X(\tau))$ should be limited and smooth. $H(\tau,\ X(\tau))$ and $\sigma(\tau,\ X(\tau))$ are nonanticipating functionals of the MP, $W(t)$ is the stochastic Wiener process. The existence of the unique strong solution of (6) is really proved. The solution can be simplified by replacing the Wiener process with white noise only if the Ito SDE takes the Stratonovich's form [26–28], i.e., the use of white noise in algorithms simplifies stochastic integration, since the limiting transition to the model described by the SDE is correct only if SDE takes the Stratonovich's form. In view of this fact, in the one-dimensional case, the H and σ coefficients take the form

$$H(E(X),t) = a_i(E(X),t) - \frac{1}{4}\frac{\partial b_{ij}(E(X),t)}{\partial X} \tag{7}$$

$$\sigma(E(X),t) = \frac{1}{q}\sqrt{b_{ij}(E(X),t)}$$

where q is white noise intensity.

Since, for Kolmogorov–Feller equation analogue (4) according the strong relation between coefficients of the both: kinetic and stochastic equation model drift H_g and diffusion σ_g can be written:

$$H_g = -\frac{1}{kT}D_g\frac{\partial\Delta\Phi}{\partial g} - \frac{1}{2}\frac{\partial D_g}{\partial g}, \quad \sigma = \frac{1}{q}\sqrt{2D_g} \tag{8}$$

Initial condition is $g(t_0) = g_0 \in [g_{0\min}; g_{0\max}]$. $g_{0\min}$ and $g_{0\max}$ are the minimum and maximum initial values of the bubble size g_0, which are chosen using the conditions $\Delta\Phi(g_{0\min}) = \Delta\Phi(g_{0\max}) = \Delta\Phi(g_{cr})$-kT, g_{cr} is the critical nucleus size determined by the expression $\left.\frac{\partial\Delta\Phi}{\partial g}\right|_{g=g_{cr}} = 0$.

For Einstein–Smoluchowski equation can be written:

$$H_r = -\frac{1}{\gamma M_g}D_r\frac{\partial U(r)}{\partial r} - \frac{1}{2}\frac{\partial D_r}{\partial r}, \quad \sigma = \frac{1}{q}\sqrt{2D_r}\,. \tag{9}$$

The variables D_g and $\Delta\Phi$ are functional-coefficients, i.e., its depends from mathematical expectation of blister size decided on previous time step (time step for blister size change) using distribution function of blisters.

$F_x = -\frac{\partial U(x,y,z)}{\partial x}$ where $U(x, y, z)$ is the superposition of potentials of the indirect elastic interaction of BPs among themselves, with layer boundaries, and so on.

$$U(r) = U_{ij}(r) + U_{\text{surf}}(r) + U_{\text{ph}}(r) + U_{\text{pore}}(r); \tag{10}$$

Here $U_{ij}(\mathbf{r})$ corresponds to an indirect interaction of blisters with each other; $U_{\text{surf}}(\mathbf{r})$ corresponds to the blister interaction with all surfaces of the layered structure (the interaction mechanism is the same), including the irradiated surface; $U_{\text{ph}}(\mathbf{r})$ corresponds to the interaction with lattice defects of the dislocation type and grain boundaries; and $U_{\text{pore}}(\mathbf{r})$ is the potential of the blister–pore interaction for porous materials. It is worth noting that the interaction with pores also is meaningful only inside one porous Si layer. The condition $U|_{Mo} - U|_{Si} = \Delta U < U|_{Mo}, U|_{Si}$ is satisfied between Mo and Si layers. The potential $U(x, y, z)$ is long-range and sign-variable; interaction occurs through the perturbation of the lattice acoustic phonons and Friedel oscillations of the lattice electron density. For

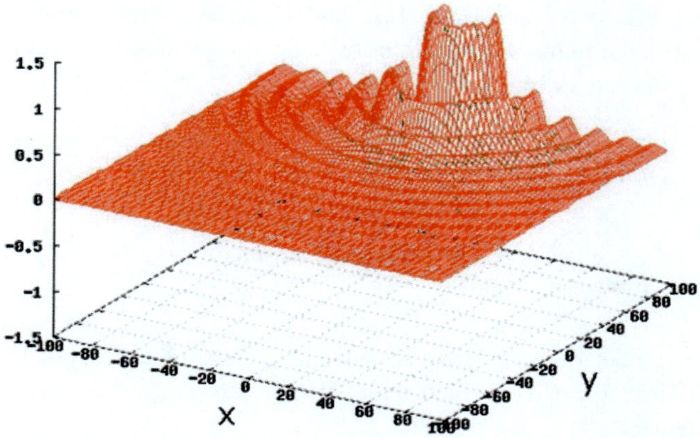

Fig. 1 2D projection of equal potential (only blister–blister interaction) values as surface U(x,y,z = const) in Si layer is presented. Distance from Mo/Si surface is 20 lattice parameters of Si. Blister-surfaces interaction and blister-lattice defects interaction are not taken into account for this picture

example, the blister–blister interaction potential similarly to the potentials deduced in [29–32], is presented:

$$U_{ij}(x,y,z) = \sum_{i \neq j}^{N} M_{gi} \left[\underbrace{\frac{b_r \left[\frac{3}{5} - \frac{(x_i - x_j)^4 + (y_i - y_j)^4 + (z_i - z_j)^4}{(\vec{r}_i - \vec{r}_j)^4} \right]}{|\vec{r}_i - \vec{r}_j|^3}}_{\substack{\text{Interaction through the perturbation of} \\ \text{lattice acoustic phonons}}} + \underbrace{\frac{a_r \cos(c_r |\vec{r}_i - \vec{r}_j|)}{|\vec{r}_i - \vec{r}_j|^3}}_{\substack{\text{Interaction through} \\ \text{Friedel oscillations} \\ \text{of the lattice electron density}}} \right] \tag{11}$$

where *i,j* are numbers of interacting blisters if we have numerate them from 1 to *N*, M_{gi} is the BP mass and b_r, a_r, and c_r are lattice model parameters of the respective layer. The blister–blister interaction is much weaker than the blister–boundary interaction. 2D projection of equal potential (blister–blister) values as surface U(x,y,z = const) in Si layer for distance from Mo/Si surface 20 lattice parameters of Si is presented on Fig. 1.

The blister–blister interaction is much weaker than the blister–boundary interaction. So, potential (11) is taken into account only inside one layer. It is worth noting that the interaction with pores also is meaningful only inside one porous Si layer, i.e., at the upper and lower boundaries of the layer

$$U_{ij}^{+}\Big|_{l} = U_{ij}^{-}\Big|_{(l+1)} = 0 \tag{12}$$

The interaction potentials $U_{\rm surf}$, $U_{\rm ph}$, and $U_{\rm pore}$ are written similarly to (11). The potential is continuous in the interior; i.e., at each interface between the ith and jth layers the condition

$$U^{+}_{surf}\Big|_{l} = U^{-}_{surf}\Big|_{(l+1)} \tag{13}$$

is satisfied.

$$U_{surf} = \sum_{l=1}^{N_l} m_{surf_l} \frac{-0.4 b_{surf_l} + a_{surf_l} \cos(c_{surf_l}(z_l - z_i))}{(z_l - z_i)^3} \tag{14}$$

where N_l is number of boundaries in system, z_l is boundary depth from surface under irradiation, $z = 0$ is surface under irradiation, m_{surf}, b_{surf}, a_{surf} and c_r are model parameters.

At the upper and lower boundaries of the sample which are consisting of several layers, we consider only the first-order boundary condition $\frac{1}{V_l}\int g f(g,x,y,z)\, dg dx dy dz = 0$, where V_l is the layer volume. Total flux is equal sum fluxes between surface Mo/Si (reflected flux, crossed flux and stayed on interface surface between layers).

The thermodynamic potential of the nucleus formation (Gibbs energy into (3)) can be presented as:

$$\Delta\Phi(g, r, t) = -(a_\Phi - c)g + b g^{2/3} + \Delta\Phi_r - \Delta\Phi_{\rm break} + \Delta\Phi_d, \tag{15}$$

Where a_Φ is the difference of chemical potentials of the phases (gas in the form of injected atoms and a blister); the coefficient $b = \eta \cdot \sigma_{\rm bl}$; $\sigma_{\rm bll}$ is the surface tension at the bubble–lattice interface; η is form factor, and c is the elastic reaction of the lattice to the nucleus formation. $c = 2\mu^{\alpha} \cdot \frac{3K}{3K+4\mu^{\alpha}} \cdot \frac{(V_{sub}-V_{gas})^2}{3V_{gas}}$ where μ^{α}—modulus of rigidity of substrate material, K—modulus of condensability of substrate material, $V_{\rm sub}$—volume on atom of substrate material, $V_{\rm gas}$—atom volume of irradiating material. Conventional elastic site bonds in the lattice are assumed to break simultaneously with a growing blister, it is taken into account in $\Delta\Phi_{\rm break}$. If $\frac{\Delta\Phi}{N_b \Delta_{break}} = k$, where $\Delta_{\rm break}$ is the energy of one bond rupture (1.76 eV for Si, and ~2 eV for Mo), and $N_{\rm b}$ is the number of bonds at one lattice site (14 for Si and Mo), k is integer number, k ≥ 1 that $V_{blister} = V_{blister} + 14 \cdot V_{vac}$, where $V_{blister}$ is blister volume, V_{vac} is vacancy volume.

$\Delta\Phi_{\rm d}$ is part of Gibbs energy corresponding taking place in tensions field (discrepancy between materials lattices Mo and Si, tensions as result of blistering). In common case it can be written:

$$\begin{gathered} \Delta\Phi_d = -K U_{ll} + \mu\left(U_{ik} - \frac{1}{3}\delta_{ik} U_{ll}\right)^2 + \frac{K}{2} U_{ll}^2, \\ K = \frac{E}{3(1-2\delta)}, \quad \mu = \frac{E}{2(1+\delta)}, \quad -1 < \delta < 0.5, \end{gathered} \tag{16}$$

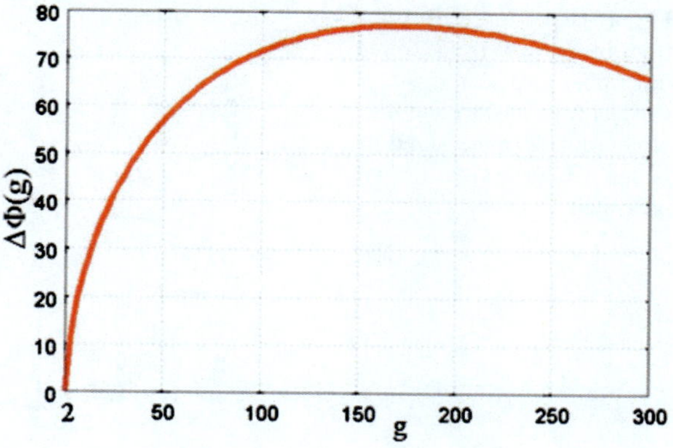

Fig. 2 Common form of part of Gibbs energy without coordinate part $(-(a_\Phi - c)g + bg^{2/3})$ is presented

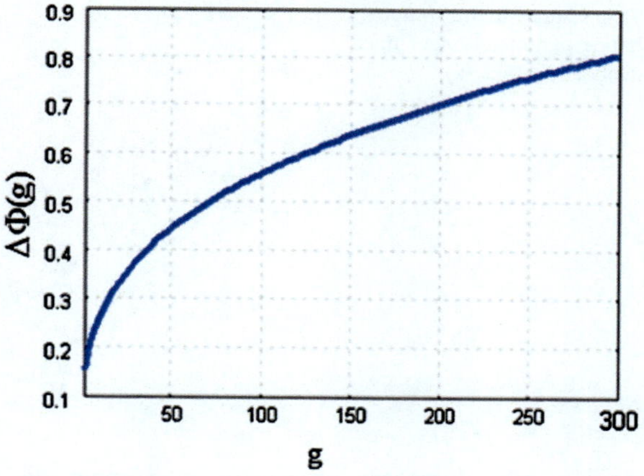

Fig. 3 Module of ratio of "volume" part of Gibbs energy $((a_\Phi - c)g)$ to "surface" part of Gibbs energy $bg^{2/3}$ is presented

where δ is Poisson's ratio, K is compression modulus, μ is modulus of rigidity, E is Young's Modulus, U_{ik} is deformation tensor.

Common form of part of Gibbs energy without coordinate part and lattice break looks like following (Fig. 2).

Module of ratio of part of Gibbs energy corresponding difference of chemical potentials of the phases (gas in the form of injected atoms and a blister) to part of Gibbs energy corresponding surface tension at the bubble–lattice interface is presented on Fig. 3.

The difference between lattice points and interstices is presented in $\Delta\Phi_r$:

$$\Delta\Phi_r = \frac{\Psi_r}{\sqrt{g}}\left[k_x \cos(2\pi(x - \phi_x)) + k_y \cos(2\pi(y - \phi_y)) + k_z \cos(2\pi(z - \phi_z))\right] \quad (17)$$

where k_x, k_y, k_z, φ_x, φ_y, φ_z, Ψ_r are lattice parameters. Dependence of $\Delta\Phi_r$ from blister size (g measured in number of gas atoms/vacancies in blister) and one of coordinate (z measured in lattice parameters) is presented on Fig. 4.

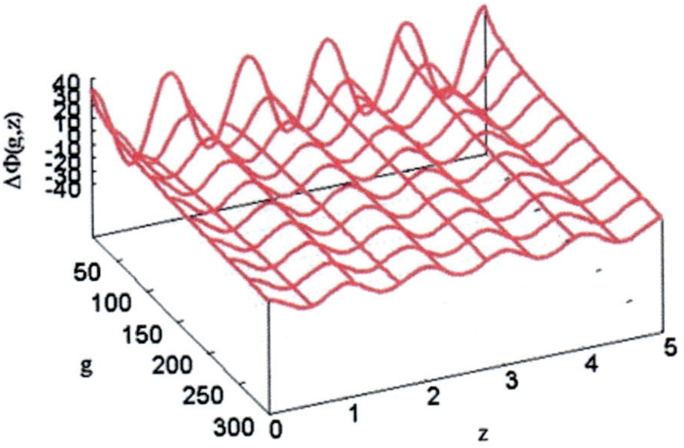

Fig. 4 Dependence of $\Delta\Phi_r$ from blister size (g measured in number of gas atoms/vacancies in blister) and one of coordinate (z measured in lattice parameters) is presented

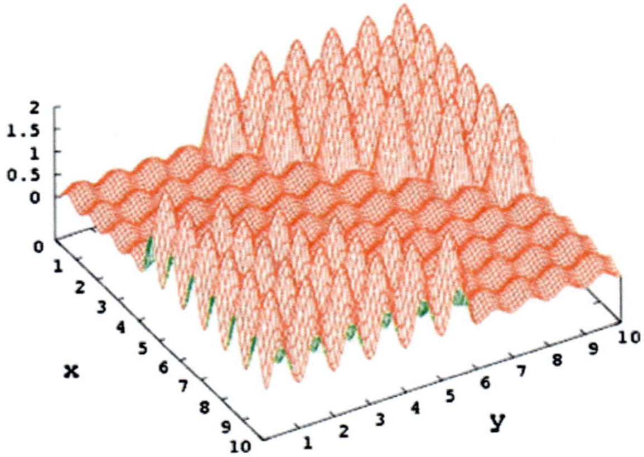

Fig. 5 Dependence of $\Delta\Phi_r$ from two coordinates (x and y) measured in substrate lattice parameters near linear dislocation (y = x)

The $\Delta\Phi_r$ decreases at several times near dislocation or grain boundaries (see Fig. 5).

Diffusion coefficient in space of substrate lattice for one of coordinate can be written as

$$D_x = \begin{cases} \frac{D_{r0}e^{-E_m/kT}}{\gamma M_g}\left(1 + \alpha_x(x - x_{ins})^2\right) & if \quad |x - x_b| > |x_b - R_b| + \Delta_b \\ A_r \frac{D_{r0}e^{-E_m/kT}}{\gamma M_g}\left(1 + \alpha_x(x - x_{ins})^2\right) & if \quad |x - x_b| \leq |x_b - R_b| + \Delta_b \end{cases} \tag{18}$$

where x is coordinate of mass center of blister; x_{ins} is coordinate of nearest interstice, D_{r0} is diffusion coefficient of Xe atom into substrate, $D_{r0} = 1.8{\cdot}10^{-8}$ cm^2/s, $E_m = 0.1$ eV is the energy of migration of Xe atom into substrate, α is model parameter, γ is friction coefficient, x_b is coordinate of the internal boundary layers or coordinate of grain boundaries. k is Boltzmann constant, T is temperature of substrate, $R_b = r_{Xe}\sqrt[3]{g}$ is blister radius, $r_{Xe} \approx 2.18$ Å is Xe atom radius, g is number of Xe atoms/vacancies in blister. $\Delta_b \approx 3$ is model parameter measured in lattice parameters of substrate $a = 5.43$ Å for Si. M_g is blister mass. A_r is model parameter corresponding diffusion increase near dislocation, layers surfaces (for example, Mo/Si) and grains boundaries.

Characteristic time for blister motion is $\tau_r = \frac{r_{Xe}^3}{D_{r0}a} \approx 1.06 \cdot 10^{-8}$ s.

Diffusion coefficient in space of blister sizes is proportional velocity of blister size change. So, $D_g \sim c_{Xe}\beta(g)4\pi r_{Xe}^2 \bar{v}_{Xe}$ can be written, where β is the coefficient of heterogeneous condensation on the surface of blister, c_{Xe} is Xe density in substrate, $\bar{v}_{Xe}$ is middle thermal velocity of Xe atoms into substrate. Diffusion coefficient in space of blister sizes is

$$D_g = 12\pi r_{Xe}^2 \beta(g) c_{Xe} D_0 e^{-Em/kT} a g^{2/3} = D_{g0} g^{2/3} \tag{19}$$

Characteristic time for blister size change is given by $\tau_g = \frac{a}{12\pi r_{Xe}^2 \beta(g) c_{Xe} D_0 e^{-Em/kT}} \approx 10^{-9} s$ and D_{r0} and D_{g0} are measured in $a^2/_g$.

3 Numerical Scheme for Solution of Ito-Stratonovich SDEs System

Numerical method of non-linear SDE solution is based on Taylor series expansion of exact solution of Ito SDE X(t + Δt) [26–28, 33].

For the *i*-th trajectory of the diffusion MP, the values of g_{n+1} and, for example, x_{n+1} the instant $n + 1$ are calculated using these formulas

$$\begin{aligned} g_{n+1}^i &= g_n^i + \left[\hat{I} - \frac{h}{2}\frac{\partial H_n^i}{\partial g}\right]^{-1} \left[hH_n^i + \sqrt{h}\sigma_{ng}^i \xi_{ng} + \frac{h}{2}\frac{\partial \sigma_{ng}^i}{\partial g}\sigma_{ng}^i \xi_{ng}^2\right], \\ H_{ng}^i &= -\frac{1}{kT} D_g^i(g_n^i,\ t_n) \frac{\partial \Delta \Phi^i(g_n^i, x_n^i, y_n^i, z_n^i, t_n)}{\partial g_n^i} - \frac{1}{2}\frac{\partial D^i(g_n^i,\ t_n)}{\partial g_n^i}, \\ \sigma_{ng}^i &= \sqrt{2D_n^i(g_n^i,\ t_n)}. \end{aligned} \tag{20}$$

$$\begin{cases} x_{n+1}^i = x_n^i + \left[\hat{I} - \frac{h_x}{2} \cdot \frac{\partial H_{xn}^i}{\partial x_i}\right]^{-1} \left[hH_{xn}^i + \sqrt{h_x}\sigma_{xn}^i \xi_{xn}\right], \\ H_{xn}^i = \frac{1}{M(g_{n+1}^i)\gamma} \cdot \frac{\partial U^i(x_n^i, y_n^i, z_n^i)}{\partial x} - \frac{1}{2}\frac{\partial D_{xn}^i}{\partial x}\xi_{xn}^2, \\ D_{xn}^i = D_{x0}(1 + \alpha(x_n^i - x_{its}^i)^2), \\ \sigma_{xn}^i = \sqrt{2D_{xn}^i}, \end{cases} \tag{21}$$

where I is the unit matrix, g_n^i is the solution to SDE at the grid point, which corresponds to the time t_n and trajectories i, x_{its} is the coordinate of the nearest lattice interstice, h is the constant steps in the algorithm time for calculating the nucleus size evolution (equals characteristic time for blister size change) and h_x is the constant steps in the algorithm time for calculating the nucleus motion in the substrate space (equals characteristic time for blister migration). Set ξ_n is normally distributed independent random number of the succession, with zero mathematical expectation and unit dispersion. $\xi_n = \sqrt{-2\ln\alpha_1}\cos(2\pi\alpha_2)$, where α_1 and α_2 are random numbers uniformly distributed in the interval (0,1).

As we can see, the resulting system of four interconnected SDEs in Ito-Stratonovich form is solved using a modified method having a second-order (or higher) accuracy [34]. Every step of time functional-coefficients of problem are iterated taking account data about DF on previous step of time.

4 Interpretation of the Numerical Experiment

The mathematical model of a random process {g(t), t $\geq$ 0}, $\forall$g$\in${G}, where g is the nucleus size in phase space $\{G\}$, considers the following conditions for the implementation of a Markovian process (MP): a nucleus should contain more than two particles ($g > 2$). For a one-dimensional MP, the blister-size phase space *{G}* contains the MP value $\{g(t)\}$. By definition, the distribution function over sizes g (namely, $f_r(g,\ t)$ specifies the probability of finding the average number of discrete medium elements in a given subregion $\{G\}$ of phase space and can be calculated as

$$f_g(G) = \sum_i p_i\{g_i(t) \in G\} = \sum_i \int_G p_i(g,t)dg \tag{22}$$

Here, p_i are finite-dimensional measures calculated using the Ito SDEs and the subscript i stands for the infinite countable set of elements of a discrete medium. According to the conditions of the numerical experiment, we consider Markovian paths and, for each partial phase volume of phase space $\{G\}$, we replace the infinite countable set with a finite number of elements. Using the paths of the random process $X_i(t),\ i = 1,\ldots,N$, we calculate the mathematical expectation $M(X) = \frac{1}{N}\sum\limits_{i=1}^{N} X_i$, dispersion $Dis(X) = \frac{1}{N-1}\sum\limits_{i=1}^{N}(X_i - M(X_i))^2$, and the higher moments of the random process $X(t)$; and using on the paths of the process $X(t)$, we reconstruct the distribution function $f(X)$ of the MP and calculate the macroscopic physical parameters of the medium by integrating the nonequilibrium distribution function $f(X)$.

Distribution function $f_r(g,\ t)$ is normalized by 1 for each time t: $\int\limits_G f(g,t)dg = 1$.

Total number of blisters is $N_{bl} = \int\limits_G f(g,t)dg$

The mathematical expectation of blister size is $\langle g\rangle(t) = \frac{1}{N}\int\limits_G f(g,t)g\ dg$,

Dispersion of blister size is $Dis(t) = \frac{1}{N_l}\sum\limits_{i=1}^{N}(g_i - \langle g\rangle)^2$

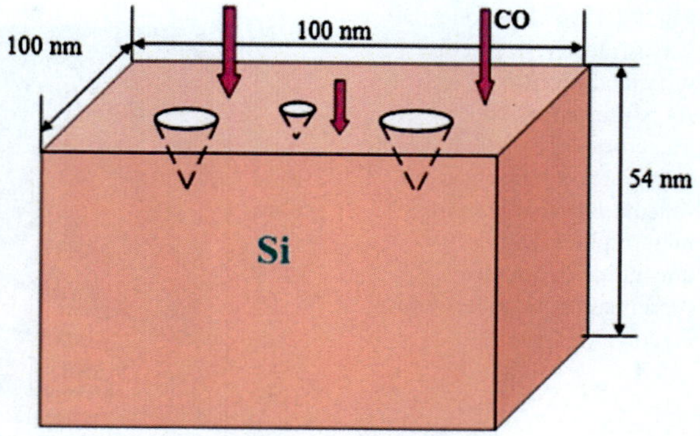

Fig. 6 Schematic presentation of solid-phase epitaxy according [2] is presented at initial time

The layer porosity with z-coordinate z_j can be calculated as:

$$p(z_j, t) = \frac{V_a}{V_l} \frac{\sum_{i=1}^{N} g_i^3 f(g_i, z_j, t)}{\sum_{i=1}^{N} g_{0i}^3 f(g_0)} \tag{23}$$

where V_a is total examined volume, V_l is layer volume, N is number of blister, g_0 is initial blister size in number Xe in blister, $f(g_0)$ is initial distribution function from blister sizes, g_i is size of blister at examined time moment, $f(g, z, t)$ is distribution function at examined time, z is distance between surface under irradiation (Mo surface) and centre of blister, z is measured in lattice parameter, $z = 0$ is surface under irradiation.

5 Stochastic Simulation Results

Results of numerical simulation of cone blister-pores evolution with number of trajectories 10^7, time step for size change is 10^{-9}s, time step for migration is 10^{-8}s, number of time step for size change is 10^5, number of time step for migration is 10^4. Initial base radius of cone blister size is 8 Å, cone height is equal cone base radius. Substrate is Si with sizes 100 nm × 100 nm × 54 nm under irradiation by CO with energy 3 keV, substrate temperature is 900 K, irradiation dose is 10^{16} cm^{-2}. Schematic presentation of initial stage of solid-phase epitaxy according [2] is presented on Fig. 6. Visualization of dependence of evolution of cone blister-pores distribution function from their sizes (in Å) is presented on Fig. 7 as results of numerical experiment of solid-phase epitaxy of SiC on Si according [2]. During fluctuation stage average cone height is 25 Å and its filled depth by SiC is 16.5 Å.

Results of numerical simulation of blistering development in the two-layer substrate consisting of layers Mo (width is 30 lattice parameters $a_{Mo} = 3.147$ A)

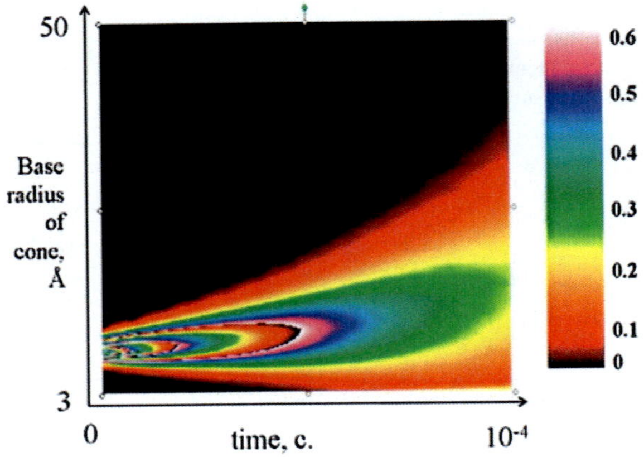

Fig. 7 Time dependent DF versus cluster sized color visualization. Blister sizes (in Å) is presented. 10^7 trajectories are used. Layer Si with pores is presented on scheme on *right side* of picture pores with unconditied form are presented on boundary of Si layer

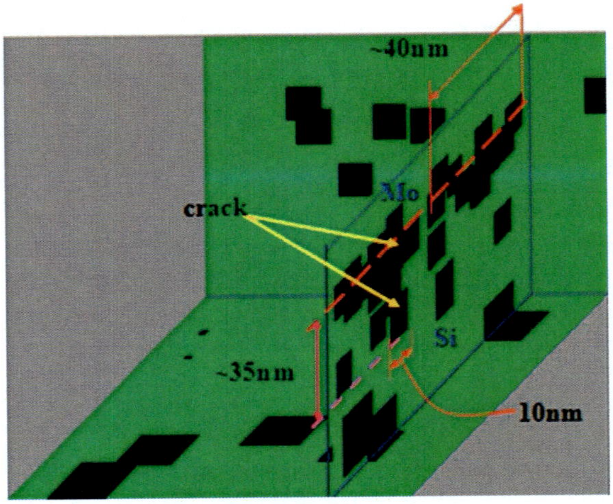

Fig. 8 Illustration of blistering develops with the two-layer substrate consisting of layer Mo (width is 94 A) and layer of crystal Si (width is 543 A) is presented. Black color corresponds porosity, and green color lack of porosity in this point. Temperature is 900 K, Xe + ions energy is 5 keV, irradiation dose is 10^{16} cm^{-2}

and crystal Si (width is 100 lattice parameters $a_{si} = 5.431$ A) are presented on Fig. 8–14. Number of trajectories is 10^7, time step for size change is 10^{-9}s, time step for migration is 10^{-8}s, number of time step for size change is 10^5, number of time step for migration is 10^4. Square of substrate is 100 nm × 100 nm, substrate depth is 70 nm approximately. Temperature is 900 K, Xe^+ ions energy is 5 keV, irradiation dose is 10^{16} cm^{-2}. Initial blister radius [6.4Å; 9.4 Å]. At initial time moment blisters locate in Si on depth from boundary Mo/Si [16 nm; 35 nm] uniformly.

Illustration of porosity into layers and long-size structure defects formation is presented on Fig. 8. Black color corresponds porosity and green color lack of porosity in this point. Formation of porosity layer perpendicular to irradiating stream and formation of cracks along irradiating stream are observed.

Lengths of long-size structure defects, which located along Xe + flow, reach to 30÷35 nm, widths reach to 10 nm. Lengths of long-size structure defects which

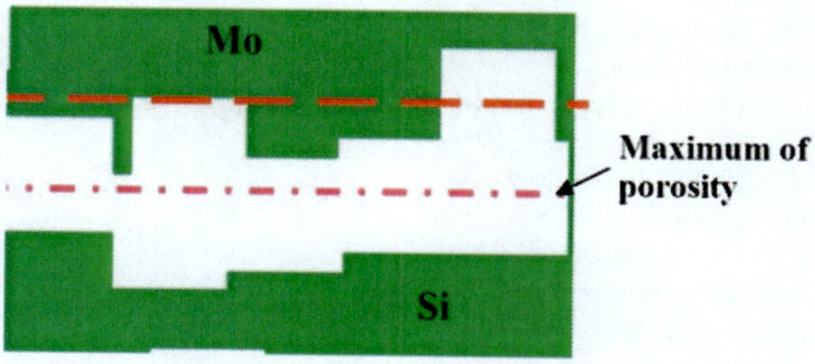

Fig. 9 Illustration of long-size defects structure formation, which perpendicular Xe + flow, in the two-layer substrate consisting of layer Mo (width is 94 A) and layer of crystal Si (width is 543 A) is presented. *White color* corresponds porosity, and *green color* lack of porosity in this point. Temperature is 700 K, Xe + ions energy is 5 keV, irradiation dose is 10^{16} cm^{-2}

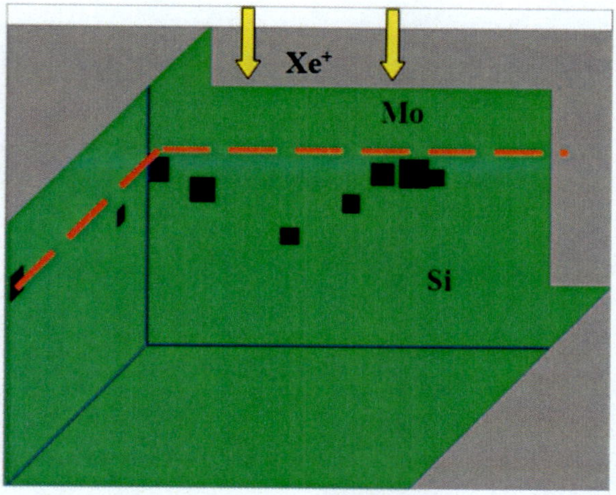

Fig. 10 Illustration of blisters formation in the two-layer substrate consisting of layer Mo (width is 94 A) and layer of crystal Si (width is 543 A) is presented. *Black color* corresponds porosity, and *green color* lack of porosity in this point. Temperature is 1,000 K, Xe^+ ions energy is 5 keV, irradiation dose is 10^{15} cm^{-2}

are perpendicular to Xe^+ flow reach 40 ÷ 50 nm. Dependence of lengths of nucleus of long-size structure defects, which located along Xe^+ flow, from temperature are non-linear and has maximum which temperature is ≈0,53 fusion temperature of Si. Maximum of porosity migrates to Mo/Si surface when temperature increases. When substrate temperature decreases, maximum of porosity migrates locates in Si below Mo/Si surface. Location of maximum of layer porosity which perpendicular to Xe^+ flow is presented on Fig. 9 for following physical parameters: temperature is 700 K, Xe^+ ions energy is 5 keV, irradiation dose is 10^{16} cm^{-2}.

Lengths of long-size structure defects, which locate perpendicular to Xe^+ flow, and which located along Xe^+ flow, increase with increase of irradiation dose.

Figure 10 illustrates porosity in same layer of Mo/Si substrate when the irradiation dose is 10^{15} cm^{-2}, we can see that long-size defects structures don't format. The formation of long-size defects structures is presented on Fig. 6 when irradiation dose is 10^{16} cm^{-2}. As we can see from Fig. 11 the porosity of substrate

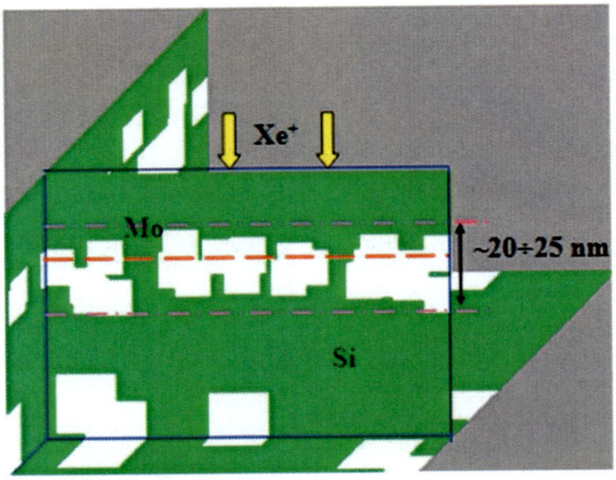

Fig. 11 Illustration of formation of long-size structures of defects in the two-layer substrate consisting of layer Mo (width is 94 A) and layer of crystal Si (width is 543 A) is presented. White color corresponds porosity, and green color lack of porosity in this point. Temperature is 1,000 K, Xe^+ ions energy is 5 keV, irradiation dose is 10^{16} cm^{-2}

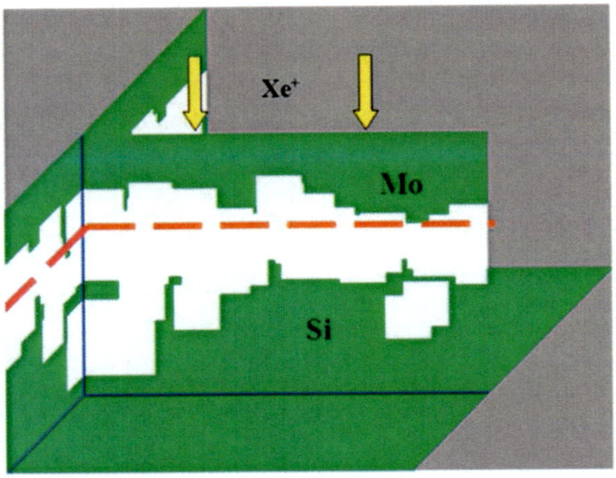

Fig. 12 Illustration of formation of long-size structures of defects in the two-layer substrate consisting of layer Mo (width is 94 A) and layer of crystal Si (width is 543 A) is presented. White color corresponds porosity, and green color lack of porosity in this point. Temperature is 1,000 K, Xe^+ ions energy is 5 keV, irradiation dose is 10^{17} cm^{-2}

and length and branching of long-size defects structures are increase with increase of irradiation dose (10^{17} cm^{-2}). Depth of amorphous substrate reaches $30 \div 35$ nm when temperature is 1,000 K and dose is 10^{16} cm^{-2}, but average amorphization depth is 25 nm. Depth of amorphous substrate reaches 50 nm when temperature is 1,000 K and dose is 10^{17} cm^{-2}.

As can see from Fig. 12–13 formation of nucleus of long-size structure defects which are perpendicular to Xe^+ flow can leads to peeling of Mo layer and formation of long-size structure defects, which located along Xe^+ flow, become permeable. If after peeling surface are under irradiation of gas mix contained carbon then SiC islands several politypes format on Si surface and on long-size structures walls.

Mo islands remain after Mo layer peeling. These islands can be catalyst for nanowhiskars formation. Nanowhiskar is wire with the diameter of the base of the order of a few nanometers.

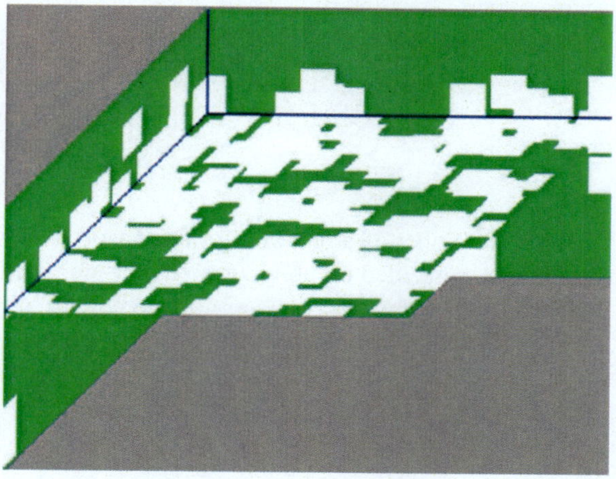

Fig. 13 Illustration of porosity on Mo/Si surface is presented

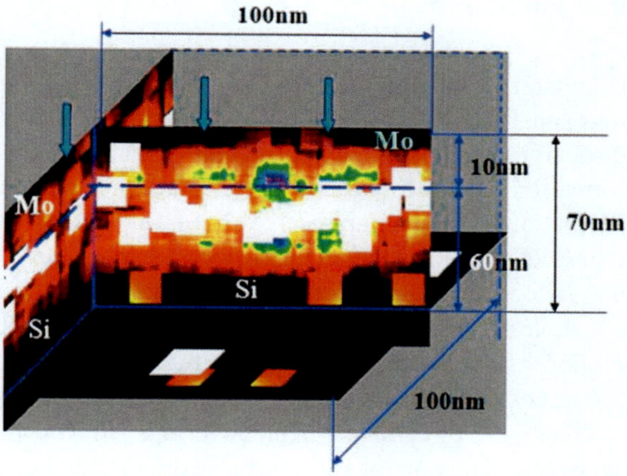

Fig. 14 Stress corresponding with bubbles in cross-section of substrate is presented. Scale of corresponding colors and values is presented on right side of picture. White color corresponds stress bigger that 10^8 Pa. Temperature is 900 K, Xe^+ ions energy is 5 keV and irradiation dose is 10^{16} cm^{-2}

Stress corresponding with bubbles in some sections of substrate is presented on Fig. 4. For example, and stress in j-layer of Si is calculated by

$$\sigma_j = 3.8 \times 10^8 \times \int \left(\begin{array}{r} \sum_i \left(\dfrac{0.33 \ln(<g_i>) - 0.1}{\left((x-x_i)^2 + (y-y_i)^2 + (z-z_i)^2\right)^{3/2}} \langle g \rangle_i^{2/3} \right) + \\ + \sum_i \left(\langle g \rangle_i^{2/3} (0.33 \ln(<g_i>) - 0.1) \right) \end{array} \right) dxdydz\,,\ Pa\ . \tag{24}$$

where $i = 1 \ldots N$, and stress in j-layer of Mo is calculated by:

$$\sigma_j = 2.435 \times 10^9 \times \int \left(\sum_j \left(\frac{0.33\ln(g_i)+0.57}{\left((x-x_i)^2+(y-y_i)^2+(z-z)^2\right)3/2} \langle g \rangle_i^{2/3} \right) + \atop + \sum_i \left(\langle g \rangle_i^{2/3} (0.33\ln(g_i)+0.57) \right) \right) dxdydz, Pa, \quad (25)$$

where N is number of trajectories of stochastic process X(t), $< g_i >$ is expectation of blister size decided on previous size time step using DF of blisters solution of a problem (2–3).

At initial time moment stress near blister reaches $3.7 \cdot 10^9$ Pa, at finish of fluctuation stage stress near blister reaches $9 \cdot 10^{12}$ Pa, so during fluctuation stage stress increase more than 10^3.

6 Conclusions

Computer simulation models of phase transition at nonequilibrium stages find various appendices. As models reflect properties of nucleation in the presence of plasma discharge. Models development can be claimed for precision of mechanisms of formation nano-size structures of surfaces on times period not exceeding 10^{-4} s. Formation of germanium islets [11, 16] and carbide of silicon [5], and also defects into solids (similar to a metal mirror [23], which are consisting of thin layers of metal and silicon) were simulated with application of the kinetic theory of fluctuations which leads to phase transition, considering in model of a nonlinear Brownian motion of nuclei and a method of the stochastic analogue leaning against strict properties of mathematical model and efficiency of its numerical realization.

Receiving of carbide silicon coverings which can be used as in electronics and as the strengthening of solid surfaces draws attention that dearness of its receiving. Leaves this problem among actual and comparison of efficiency of receiving of carbide of silicon by method of solid-phase epitaxial covering [2] with others, and also the porosity forecast in a thin layer of carbide of silicon of heat-resistant coverings and properties of a roughness of such surface at influence on it of chemically active gases. Parity of formation of amorphous and crystal form of nuclei, mechanisms of processes had been discussed.

Solution of systems of stochastic equations allows:

- find distribution function of blisters on sizes and layer depth, i.e., degree of Si amorphization;
- find porous layers and its depths from surface under irradiation. Layers with maximum porosity is located near surface Mo/Si;
- find effect of self-organization i.e., porous structure—nucleus of cracks perpendicular to falling stream and along them. Length of structures perpendicular to falling stream (along boundary Mo/Si layers) reaches 60 nm, its width reaches 10 nm. Length of structures along stream (perpendicular to boundary

Mo/Si) reaches 35 nm, its width reaches 10 nm. Pressure in such structure can reach $10^9 \div 10^{12}$ Pa, it can ensure development of cracking of substrate (for Si Young's modulus is $1.89{\cdot}10^{10}$ Pa, modulus of rigidity is $7.99{\cdot}10^{10}$ Pa, modulus of rupture is $1.25{\cdot}10^8$ Pa);

- define efficiency of carbon implantation into pores walls and filling of pores by SiC.

Acknowledgments This study was supported in part by the Russian Foundation for Basic Research (projects no. 09-01-00798-a and 11-01-00282-a), the Department of Mathematics of the Russian Academy of Sciences (program no. 3), and the Russian Federal Target Program "Scientific and Educational Personnel of the Innovative Russia" (state contract nos. 02.740.11.0615 and 02.740.11.0475).

References

1. Andrievski, R.A.: Synthesis, structure and properties of nanosized silicon carbide. Rev. Adv. Mater. Sci. **22**, 1–20 (2009)
2. Kukushkin, S.A., Osipov, A.V.: Phys. Sol. State **50,** 1238 (2008) (English transl.)
3. Sigov, Y.S.: Computer Simulation: The Link Between Past and Future of Plasma Physics, Selected Works, Composed by Zmievskaya, G.I., Levchenko V.D. (Nauka, Fizmatlit, Moscow, 2001) [in Russian]
4. Sigov, YuS, Levchenko, V.D.: In: Maino, G., Otta-viani, M. (eds.) Dynamics of Trans-port in Plasmas and Charged Beams, p. 55. World Scientific, Singapore (1996)
5. Zmievskaya, G.I., Bondareva, A.L.: Plasma Phys. Rep. **37**(1), 87–95 (2011)
6. Zmievskaya, G.I.: Plasma Phys. Rep. **32**, 337 (1997)
7. Voitsenya, V.S., Belyaeva, A.I., Bardamid,A.F. et al.: Nucl. Mater. **345**(2–3), 101 (2005)
8. Fan, Y._J., Yankulin, L., Antohe, A., et al.:Proc. SPIE Int. Soc. Opt. Eng. **7271**, 72713U-1–9 (2009)
9. Allain, J.P., Hassanein, A., Allain, M.C. et al.: Nucl. Instrum. Methods Phys. Res. B. **242**, 520 (2006)
10. Koshelev, K.N., Banin, V.E., Salashchenko, N.N.: Usp. Fiz. Nauk **177**, 777 (2007)
11. Zmievskaya, A.L., Bondareva, V.D., Levchenko, T.V.: J. Phys. D: Appl. Phys. **40**, 4842–4849 (2007)
12. Bondareva, A.L., Zmievskaya, G.I.: J. Surf. Investig. X-ray, Synchrotron and Neutron Tech. **4**(3), 480–487 (2010)
13. Ochsner, A., Hosseini, S.M., Merkel, M. In: Proceedings of the International MultiConference of Engineers and Computer Scientists 2009 (IMECS 2009), vol. II, March 18–20, 2009, Hong Kong, ISBN: 978-988-17012-7-5
14. Fiedlerl, T., Belova, I.V., Ochsner, A., Murch, G.E. In: Proceedings of the 4th WSEAS International Conference on Applied and Theoretical Mechanics (Mechanics '08), pp. 20–25 (2008). ISSN: 1790-2769, ISBN: 978-960-474-046-8
15. Bruno, G., Efremov, A.M., Levandovskyi, A.N., Clausen, B.: J. Mater. Sci **46**, 161–173 (2011)
16. Bondareva, A.L., Zmievskaya, G.I.: Eur. Phys. J. D: At. Mol. Opt. Plasma Phys. **38**, 143 (2006)
17. Zmievskaya, G., Bondareva, A., Levchenko, V., Goryachev, I., Perepelkina, A. In: IS Plasma 2011 3rd Interantional Symposium on Advanced Plasma Science and its Applications for Nitrides and Nanomaterials, Nagoya Institute of Technology, Nagoya, Japan, P4-078C, p. 214. March 6–9, 2011
18. Zmievskaya, G.I., Bondareva, A.L.: Bulletin of the Russian academy of sciences. Physics **68**(3), 378–382 (2004) (ISSN: 1062-8738)

19. Bondareva, A.L., Levchenko, T.V., Zmievskaya, G.I.: Defect and Diffusion Forum. V. 297–301: Diffusion in Solids and Liquids V: Trans Tech Publications, Switzerland (2010), pp. 502–507
20. Bondareva, A.L., Zmievskaya, G.I. In: Lab. de Phys. des Plasmas, Ecole Polytechnique-CNRS (ed.) 63rd Gaseous Electronics Conference and 7th International Conference on Reactive Plasmas (GEC/ICRP Meeting 2010), Oct 4–8, 2010, Paris (CD), KWP-078
21. Zmievskaya, G.I., Bondareva, A.L. In: Lab. de Phys. des Plasmas, Ecole Polytechnique-CNRS (ed.) Bulletin of the American Physical Society, Oct 2010. 63rd Gaseous Electronics Conference and 7th International Conference on Reactive Plasmas (GEC/ICRP Meeting 2010). Oct 4–8, 2010, Paris, France. vol. 55(7), p. 24 (2010)
22. Bondareva, A.L., Zmievskaya, G.I. In: Lab. de Phys. des Plasmas, Ecole Polytechnique-CNRS (ed.) BULLETIN of the American Physical Society, Oct 2010. 63rd Gaseous Electronics Conference and 7th International Conference on Reactive Plasmas (GEC/ICRP Meeting 2010). Oct 4–8, 2010, Paris, France. vol. 55(7), p. 122 (2010)
23. Zmievskaya, G.I.,Bondareva, A.L. Porosity in solids: computer simulation of blistering at nucleation fluctuation stage. In: Mejo University (ed.) Proceeding of the 3rd International Conference of PLAsma- NanoTechnology and Science, March 11–12 2010, Nagoya, Japan, 2010, P. 48
24. Zmievskaua, G.I., Levshenko, T.V., Soboleva, T.K.:Teplofiz. Aeromekh. **9**, 365 (2002)
25. Zmievskaya, G.I., In'kov, L.V., Levchenko, V.D., Levchenko, T.V.: Poverkhnost' vol. 5, p. 106 (2004)
26. Kloeden, P.E., Platen, E.: Numerical Solution of Stochastic Differential Equations, p. 36. Springer, New York (1992)
27. Kloeden, P.E., Platen, E., Schurz, H.: Numerical Solution of SDE Through Computer Experiments, p. 292. Springer, Berlin (1994)
28. Pugachev, V., Sinitsin, I.: Stochastic Differential Systems: Analysis and Filtering. Wiley, New York (1987)
29. Morozov, A.I., Sigov, A.S.: Zh. Eksp. Teor. Fiz. **95**, 170 (1989)
30. Morosov, A.I., Sigov, A.S.: Solid State Commun. **67**, 841 (1988) [Sov. Phys. JETP **68**, 97 (1989)]
31. Morozov, A.I., Ovchenkov, P.A., Sigov, A.S.: Discrete Plasma Simulation, Collection of Works of Appl. Math. Inst. AN SSSR. In: Sigov, Y.S. (IPM, Moscow, 1990) [in Russian]
32. Berzin, A.A., Morozov, A.I., Sigov, A.S.: Phys. Solid State **38**, 747 (1996)
33. Savchenko, V.: Bull. Inst. Theor. Astron. **XV**6(169), 324–328 (1982) (in Russian)
34. Artem'ev, S.S., Averina, T.A.: Numerical Analysis of Systems of Ordinary and Stochastic Differential Equations (VSP, Utrecht, 1997), p. 176

Unsaturated-Saturated Flow in Porous Media Under Centrifugation

Jozef Kačur, Benny Malengier and Pavol Kišon

Abstract Numerical modeling of one dimensional two-phase flow (water/oil) under centrifugation is presented. A new method is analysed to determine capillary-pressure curves (which is an inverse problem). This method is based on modeling the interface between the zone containing only wetting liquid and the zone containing wetting and non wetting liquids. This interface appears when into a fully saturated sample with non-wetting liquid we inject a wetting liquid. The mathematical model of the interface movement (which results in an ODE) is substantial in the construction of an efficient and correct numerical approximation of the original mathematical model for one dimensional two-face flow. The resulting method is a good candidate to be used in the solution of the inverse problem. Both liquids are assumed to be immiscible and incompressible. Some numerical experiments are presented. As a special case, single phase flow (water/air) of a wetting liquid is considered and some numerical experiments concerning inverse problems are presented.

Keywords Two-phase flow · Inverse problem · Porous media flow · Determination of soil parameters

J. Kačur · P. Kišon
Faculty of Mathematics, Physics and Informatics,
Comenius University Bratislava, Bratislava, Slovakia
e-mail: kacur@fmph.uniba.sk

P. Kišon
e-mail: Kison@fmph.uniba.sk

B. Malengier (✉)
Department of Mathematical Analysis, Research Group NaM2,
Ghent University, Ghent, Belgium
e-mail: bm@cage.UGent.be

J. M. P. Q. Delgado et al. (eds.), *Numerical Analysis of Heat and Mass Transfer in Porous Media*, Advanced Structured Materials 27,
DOI: 10.1007/978-3-642-30532-0_11,

1 Introduction

To predict the flow and solute transport in soils concerns the soil hydraulic properties in terms of capillary-pressure, and the hydraulic permeability versus saturation curve. These relationships can be formulated in terms of soil parameters, of which one of the most used is the Van Genuchten-Mualem ansatz [11]. The soil parameters are then the required input data to solve the governing mathematical model, which is given by Richards' equation, a nonlinear and degenerate parabolic equation. Determination of the soil parameters from experimental data is a so-called inverse problem, which are ill-posed. To solve the inverse problem requires some regularization and an efficient and correct solver of the direct problem (which is the Richards' equation where the soil parameters are known).

Our main goal is to present a new method which can then be used in the inverse problem to determine the soil parameters. The direct problem considered first is a sample, fully saturated with non-wetting liquid (oil). Into this sample we inject (from the front) the wetting liquid, see Fig. 1, and perform some measurements on the sample undergoing two-phase flow (see [1]). Application of centrifugation substantially speeds up the flow through the porous media, so transient effects are more pronounced, and measurements can be finished in a fraction of the time required outside of the centrifuge.

For inverse problems, the general rule is that more experimental data leads to better reconstruction of the soil parameters. However, doing more measurements increases the cost of the experiment. In this work we show that for the flow considered, it suffices to perform only simple measurements, that is, we measure only global characteristics such as the rotational momentum, the gravitational center, and the amount of expelled non-wetting liquid.

Centrifugation was introduced 40 years ago. Most centrifugation scenarios are based on collecting measurement data at equilibrium states obtained at a series of different rotational speeds. To reach equilibrium can take a long time (hours or even days) depending on the error tolerance, since equilibrium is reached in an infinite time. The distribution profile of wetting and non-wetting liquids at the equilibrium state can be obtained very simple and avoids difficult numerical computation of the governing equations. Most of these methods are based on the Hassle and Brunner formula [3], see [6, 7, 9] and for an overview [10]. Of course, also the centrifugation scenario considered here admits measurements from equilibria states, but if one requires fast results, transient measurements as suggested in this Chapter are needed.

To simplify the determination of the soil parameters we consider the saturated hydraulic conductivity K_s as known. This is not a problem as this parameter can be obtained very efficiently by centrifuging separately a fully saturated sample with wetting fluid when in the injection chamber (in front of the sample) there is also wetting fluid. Once K_s is known, the intrinsic saturated permeability of the porous structure follows. A direct formula can be used to determine K_s, see [8], or an inverse method based on a mathematical model, see our article [5]. A full treatment of single phase flow (air–water) can be found in this last reference.

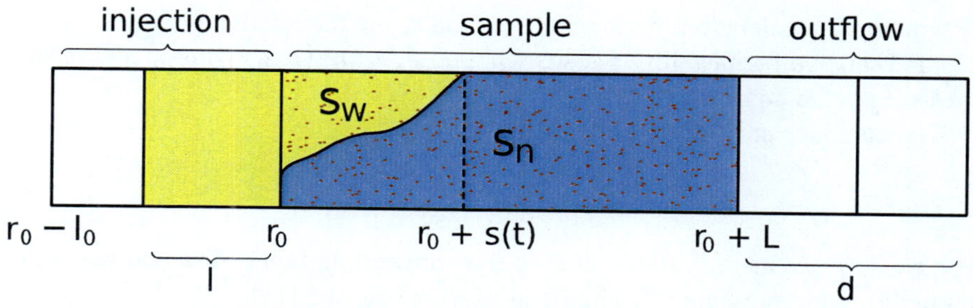

Fig. 1 Centrifugation scenario

We now introduce the mathematical model, after which we present the numerical method used. Numerical experiments are discussed in the last Section.

2 Mathematical Model

2.1 Injection of Wetting liquid into a Sample Saturated with Non-Wetting Liquid

We consider a one dimensional fully saturated sample with a non-wetting liquid with effective saturation $S_n(0 lt; S_n \leq 1)$. The sample in the form of a tube, starts (with top boundary) at the distance $r = r_0$ from the centrifuge axis and ends at the distance $r = r_0 + L$. In the position $r \in (r_0 - \ell_0, r_0)$ of the centrifuge an injection chamber is placed with wetting liquid. By centrifugation the wetting liquid enters the sample and is replacing the non-wetting liquid, which is expelled into the outflow chamber. The wetting liquid penetrates the sample giving rise to a wetting liquid saturation S_w $(0 < S_w \leq 1)$, and pushing out non-wetting liquid at the base. The expelled water is collected in an outflow chamber situated at $(r_0 + L, r_0 + L + d)$, with L, $d > 0$. In the fully saturated sample we have

$$S_w + S_n = 1, \tag{1}$$

and at any time t we have $S_n > 0$ in the sample. Moreover, $S_w > 0 \in (r_0, r_0 + s(t))$ and $S_w = 0 \in (r_0 + s(t), r_0 + L)$ where $s(t) \in (0, L)$ (for $t \in (0, T)$) is an interface. The mathematical model reads as follows, see [1],

$$\begin{aligned} \Phi \partial_t S_w + \nabla q_w = 0, \qquad q_w &= -K_s \tfrac{k_w}{\mu_w} \left(\partial_x h_w - \tfrac{\omega^2}{g} (r_0 + x) \right), \\ \Phi \partial_t S_n + \nabla q_n = 0, \qquad q_n &= -K_s \tfrac{k_n}{\mu_n} \left(\partial_x h_n - \rho \tfrac{\omega^2}{g} (r_0 + x) \right), \end{aligned} \tag{2}$$

where $x \in (0, s(t))$, Φ is the porosity, ω the angular speed of rotation (in radians per second), $K_s = \frac{K_i \rho_w g}{\mu_b}$ and $\rho = \dot{\rho}_n / \rho_w$, where K_i is the intrinsic saturated permeability, ρ_w, ρ_n the density of the wetting and non-wetting phase, μ_b a base viscosity and g the gravitational acceleration. Furthermore $\mu_\bullet$ and $k_\bullet$ are the relative viscosity (versus μ_b) and relative permeability (versus K_i), and $h_\bullet = p_\bullet/(\rho_w g)$ with p_n and p_w the pressures in the liquids.

The capillary pressure p_c is defined as

$$p_c = p_n - p_w \geq 0. \tag{3}$$

For $x \in (s(t), L)$ there is Darcy flow of the non-wetting liquid. The soil hydraulic properties are represented by empirical expressions, see [11],

$$\begin{aligned} S_w \equiv u = & \frac{1}{\left(1 + (\bar{\delta} p_c)^n\right)^m} = \frac{1}{\left(1 + (\alpha h_c)^n\right)^m}, \\ k_w(u) = & u^{1/2}\left[1 - \left(1 - u^{1/m}\right)^m\right]^2, \\ k_n(u) = & (1-u)^{1/2}\left(1 - u^{1/m}\right)^{2m}, \end{aligned} \tag{4}$$

where $m = 1 - 1/n$, $n > 1$ and $\alpha > 0$ are empirical soil parameters, and where we used $h_c = p_c/(\rho_w g)$. If we define $q_t := q_w + q_n$ (total flux), then we deduce

$$\partial_x q_t = 0, \quad x \in (0, L),$$

(i.e., q_t depends only on t) and

$$q_w = \frac{K_s}{\mu_w} \frac{k_w k_n}{\mu k_w + k_n} \left(\partial_x h_c + (1-\rho) \frac{\omega^2}{g} (r_0 + x) \right) + \frac{q_t}{\mu_w} \frac{k_w \mu_n}{\mu k_w + k_n}, \tag{5}$$

where

$$\mu = \frac{\mu_n}{\mu_w}; \quad \rho = \frac{\rho_n}{\rho_w}.$$

In (5), we replace

$$\partial_x h_c = \frac{\partial h_c}{\partial S_w} \partial_x S_w = f_c(S_w) \partial_x S_w,$$

where

$$f_c(S_w) := -\frac{1}{\alpha(n-1)} \frac{1}{\left(1 - S_w^{1/m}\right)^m S_w^{1/m}},$$

and we can solve only one PDE in terms of S_w which reads

$$\Phi \partial_t S_w = -\frac{K_s}{\mu_w} \partial_x \left(D(S_w) \partial_x S_w + \frac{q_t}{K_s} \mu_n \frac{k_w}{\mu k_w + k_n} + \frac{k_w k_n}{\mu k_w + k_n} (1-\rho) \frac{\omega^2}{g} (r_0 + x) \right), \tag{6}$$

where

$$D(u) = \frac{a(u)}{b(u)} f_c(u),$$

with

$$\begin{aligned} a(u) &= (1-u)^{1/2}\left(1-u^{1/m}\right)^{2m} u^{1/2}\left(1-\left(1-u^{1/m}\right)^{m}\right)^{2}, \\ b(u) &= (1-u)^{1/2}\left(1-u^{1/m}\right)^{2m} + \mu u^{1/2}\left(1-\left(1-u^{1/m}\right)^{m}\right)^{2}. \end{aligned} \tag{7}$$

We can verify that $D(u) \to 0$ when $u \to 1$ or $u \to 0$. So the parabolic Eq. (6) is strongly nonlinear and degenerates at $S_w = 0$. Since $S_n > 0$ in $(0, s(t))$ for $t < T$, the second degeneracy, at $u = 1$, is not reached. Thus, (6) is a "porous media type" equation and there arises an interface $s(t)$ so that $S_n = 1$ for $x \in (s(t), L)$. Due to this fact we shall solve Eq. (6) on $(0, s(t))$. The flux q_t is linked with the water level $\ell(t)$ in the injection chamber

$$\dot{\ell}(t) = -q_t. \tag{8}$$

and we assume that ℓ can be measured at some time points. The interface s can be modelled by analyzing the degeneracy of the diffusion term D in (6). We can find easily that

$$\frac{D(u)}{u^{1/2+1/m}} \to \frac{m^2}{\alpha(n-1)}, \quad \textit{for } u \to 0.$$

Then, following the idea in [2], one can develop a time evolution model for the interface s in terms of an ordinary differential equation (ODE).

$$\dot{s}(t) = -\frac{K_s m^2}{\alpha(n-1)\mu_n(1/m+1/2)} \partial_x S_w^{1/m+1/2}\Big|_{x\to s(t)^-}, \quad \text{for } S_w = 0. \tag{9}$$

In the case that q_t is known (obtained by measurements), one can solve the single Eq. (6) in a moving domain $(0, s(t))$ completed with the boundary condition $S_w(s(t), t) = 0$. The boundary condition at $x = 0$ is set in an implicit form $S_w(0, t) = u_0(t)$, where $u_0(t)$ is linked with the global mass balance of water, M_w, in the sample, given by

$$M_w(t) = -\int_0^t q_t = \ell(0) - \ell(t), \tag{10}$$

which implicitly determines $u_0(t)$ (as M_w is obtained by integrating S_w over the sample). Next, apply the transformation $y = \frac{x}{s(t)}$ and rewrite (6) in the fixed domain $y \in (0, 1)$ using $\partial_x = \frac{1}{s(t)} \partial_y$

$$\Phi\partial_t S_w = -\frac{K_s}{s(t)\mu_w}\partial_y\left(\frac{1}{s(t)}D(S_w)\partial_y S_w + \frac{q_t}{K_s}\mu_n\frac{k_w}{\rho k_w + k_n} + \frac{k_w k_n}{\mu k_w + k_n}(1-\rho)\frac{\omega^2}{g}(r_0 + ys(t))\right)$$
$$+\Phi\frac{\dot{s}(t)}{s(t)}y\partial_y S_w, \tag{11}$$

and replace $\dot{s}(t)$ by the right hand side of (9). This equation is then completed with (10) and

$$S_w(0,t) = u_0(t), \quad S_w(s(t),t) = 0. \tag{12}$$

It is suitable to approximate (11) by and ODE system (using the Method of Lines, MOL) and then the complex system (10–12) is reduced to the solution of a DAE system (Differential and Algebraic Equations).

Note that as indicated this requires that $\ell(t)$ can be obtained by measuring the inflow of wetting liquid or the expelled non-wetting liquid up to the time that s reaches the end of the sample, $x = L$. In other words, q_t is obtained, see (8), and can hence be considered known.

2.2 *Injection of Non-Wetting Liquid into a Sample Saturated with Wetting Liquid*

We proceed as in the previous subsection and assume that q_t is known. The governing equation for S_n is obtained from (6) by inserting $S_w = 1 - S_n$, which gives

$$\Phi\partial_t S_n = -\frac{K_s}{\mu_w}\partial_x\left(D(1-S_n)\partial_x S_n - \frac{q_t}{K_s}\mu_n\frac{\bar{k}_w}{\mu\bar{k}_w + \bar{k}_n} - \frac{\bar{k}_w\bar{k}_n}{\mu\bar{k}_w + \bar{k}_n}(1-\rho)\frac{\omega^2}{g}(r_0 + x)\right), \tag{13}$$

where $\bar{k}(S_n) = k(1 - S_n)$. This is again a porous media type equation with degenerating diffusion $D(1 - S_n)$ for $S_n = 0$. Thus, there appears an interface $s_n(t)$ separating the zone in $(0, s_n(t))$ with wetting and non-wetting liquid (with $S_n > 0$), and the wetting liquid zone in $(s_n(t), L)$ where $S_w = 1$. One can reduce S_n governed by (13) to the moving subdomain $(0, s_n(t))$ with the boundary condition $S_n(s(t), t) = 0$ at $x = s_n(t)$, and the boundary condition $S_n(0, t) = u_{n,0}(t)$, where $u_{n,0}(t)$ is linked with the mass balance of the non-wetting liquid (the mass M_o in the sample, similarly as in the previous subsection). It remains to develop the mathematical model for the time evolution of the interface $s_n(t)$. To this end we analyze the degeneracy of $D(1 - S_n)$ for $S_n \to 0$. This allows to verify that $D(w)$, $w = 1 - z$, asymptotically for $z \to 0$ equals

$$D(1-z) \approx -\frac{z^{1/2}(1-(1-z)^{1/m})^m}{\alpha(n-1)} \approx \frac{z^{1/2+m}}{\alpha(n-1)m^m}. \tag{14}$$

Following the idea in [2], the time evolution model for $s_n(t)$ is obtained as

$$\dot{s_n}(t) = -\frac{K_s}{\alpha(n-1)m^m \mu_w (m+1/2)} \partial_x S_n^{m+1/2}\bigg|_{x \to s_n(t)^-}.$$

Hence, the mathematical model for S_n is complete.

2.3 *Single phase Flow: Water–Air*

Single phase flow under centrifugation has been studied by many different authors. In this, only water and air are present, and one assumes the air to have a constant pressure throughout the sample. We present here the connection of this case with the two-phase flow studied in the previous two subsections.

As only water is considered, we retain the Eq. (2) for S_w. Assume further $p_n = 0$, $\mu_n = 0$, $\rho_n = 0$, $k_n = 0$, so (2) now coincides with (6). There are two centrifugation scenarios possible, which are linked with the previous subsections. In the first one there is infiltration of water into a dry sample, and in the second one water is expelled from an originally fully saturated sample by centrifugation with zero water influx at the top (front). This Section hence follows the results in [4].

Both possible scenarios differ only in boundary conditions. The first case is more difficult since there appear two interfaces, s_1 and s_2, separating the fully saturated zone from the partially saturated zone and the dry zones in the sample. They correspond to the degeneracy of the diffusion term D when $S_w \to 0$ and $S_w \to 1$. The mathematical model for the time evolution of s_2 (describing the wetting front) is the same as for $s(t)$ in (9). But the mathematical model for the interface s_1 (describing the fully saturated front) is difficult to model and also difficult to identify in an experiment. It typically creates problems during the numerical modelling.

Equation (6) in the moving domain $(s_1(t), s_2(t))$, equipped with the boundary conditions $S_w(s_1(t), t) = 1$ and $S_w(s_2(t), t) = 0$, becomes a closed system, if s_1 and s_2 can be determined. As mentioned, the mathematical model for s_2 is known, see (9). A first approach to determine s_1 is applying the mass balance argument to obtain

$$\dot{s}_1(t) = q_t - q_{int}(t), \tag{15}$$

where $q_{int}(t)$ is the flux of water entering into the partially saturated zone from the fully saturated zone, and q_t as before the total influx of water. First, q_t can be determined. For this problem we consider the flow in the fully saturated zone $(0, s_1(t))$ which is governed by Darcy's equation, with $\partial_t u = 0$, and consequently with a constant flux along $(0, s_1(t))$. The piesometric pressure at $r = r_0$ (top of the sample) can be computed from the rotational speed and the water level $\ell(t)$ in the injection chamber. The piesometric pressure at $x = s_1$ is zero. Note that the piesometric head is positive in $(0, s_1(t))$ and negative in $(s_1(t), s_2(t))$. Thus, the piesometric head is a quadratic function which can be fully determined by the values on the boundaries $r = r_0$ and $r = r_0 + s_1(t)$. Hence, we obtain

$$q_t = K_s \frac{\omega^2}{2gs_1(t)} \Big(2r_0 s_1(t) + s_1(t)^2 + \ell(t)(2r_0 - \ell(t)) \Big). \tag{16}$$

Next, the flux $q_{int}(t)$ can be computed from the head (instead of saturation) profile of the solution inside $(r_0 + s_1, r_0 + s_2)$. However, numerically it is difficult to obtain a correct approximation of this flux at $r = s_1(t)$.

A second approach, which leads to a more correct approximation of $s_1(t)$, comes from considering a global mass balance condition instead of (15),

$$\ell(t) + \Phi s_1(t) + \Phi \int_{r_0+s_1(t)}^{r_0+s_2(t)} u(x,t)\,dx = \ell(0). \tag{17}$$

In this way a closed system is obtained in $(s_1(t), s_2(t))$. The governing PDE for $S_w = u$ can be transformed to the fixed domain $y \in (0,1)$ using the transformation

$$y = \frac{r - (r_0 + s_1(t))}{s_2(t) - s_1(t)},$$

which gives

$$\begin{aligned} \partial_t u = &- \frac{K_s}{\Phi(s_2 - s_1)^2} \partial_y \left(D(u)\partial_y u + (s_2 - s_1)k(u)\frac{\omega^2}{g}(r_0 + s_1 + y(s_2 - s_1)) \right) \\ &+ \frac{1}{s_2 - s_1}(\dot{s}_1(1-y) + \dot{s}_2 y)\partial_y u. \end{aligned} \tag{18}$$

For the numerical approximation this PDE is reduced to a system of ODE which is completed by the ODE (9) and the algebraic equation (17).

The second scenario of drainage of a sample saturated with water can be derived likewise from Sect. 2.2, with one moving interface between partially saturated zone at the front, and saturated zone at the base.

3 Numerical Method

We only go into details for the numerical method to solve the injection of wetting liquid. Apply a space discretization (in the transformed equation) to then solve the resulting ODE system (MOL method). The space discretization can reflect the sharp front of the solution near the front s, which corresponds to the point $y = 1$. Denote the grid points as

$$0 = y_0 < y_1 < \ldots < y_i < \ldots < y_N = 1,$$

with increasing density towards $y = 1$, and define $\delta_0 = 0$, $\delta_i := y_i - y_{i-1}$, $i = 1, \ldots, N$. This discretization corresponds to moving grid points $x_i(t) = r_0 + s(t)y_i$ in the sample. The used transformation guarantees a good approximation accuracy of space derivatives at the sharp front during the entire computation. For the space discretization we integrate the transformed Eq. (11) over the domain $I_i := (y_{i-1/2}, y_{i+1/2})$, $\forall i = 1, \ldots, N-1$, where $y_{i-1/2} := y_i - \frac{\delta_i}{2}$ and $y_{i+1/2} := y_i + \frac{\delta_{i+1}}{2}$.

Denote $u_i(t) \approx S_w(y_i, t)$, $\forall i = 1, \ldots, N-1$ and approximate $\partial_t S_w(y,t) \approx \dot{u}_i(t)$ in the interval I_i. Then, with y_i we link the ODE

$$\dot{u}_i = -\frac{2}{\Phi s(t)(\delta_i + \delta_{i+1})}(q_{i+1/2} - q_{i-1/2}) + y_i \frac{\dot{s}}{s} \partial_y u\Big|_{y=y_i}, \tag{19}$$

for $i = 1, \ldots, N-1$. Here $-q_{i+1/2}$ is the approximated flux (at the point $y_{i+1/2}$) given by

$$q_{i+1/2} = \frac{K_s}{\mu_w}\left(\frac{1}{s} D(u_{i+1/2})\partial_y u\Big|_{y=y_{i+1/2}} \right.$$
$$\left. + \frac{q_t}{K_s}\mu_n \frac{k_w}{\mu k_w + k_n}\Big|_{y=y_{i+1/2}} .+ \frac{k_w k_n}{\mu k_w + k_n}\Big|_{y=y_{i+1/2}} (1-\rho)\frac{\omega^2}{g}(r_0 + y_{i+1/2}s)\right),$$

with

$$\partial_y u\Big|_{y=y_{i+1/2}} \approx \frac{u_{i+1} - u_i}{\delta_{i+1}}, \quad \partial_y u\Big|_{y=y_{i-1/2}} \approx \frac{u_i - u_{i-1}}{\delta_i}.$$

At the point $y = y_i$ we use the tree point approximation

$$\partial_y u\Big|_{y=y_i} \approx \frac{d}{dz}\mathscr{L}_i(z)\Big|_{z=y_i},$$

where $\mathscr{L}_i(z)|_{z=y_i}$ is the second order Lagrange polynomial crossing the points (y_{i-1}, u_{i-1}), (y_i, u_i), (y_{i+1}, u_{i+1}). The same approximation is used in (9) to approximate the term $\partial_y S_w^{1/2+1/m}$ at the point $y = 1$. In that case the polynomial is crossing the points (y_{N-2}, u_{N-2}), (y_{N-1}, u_{N-1}) and $(1, 0)$. The global mass balance (10) on the other hand is approximated using the trapezoid rule, which gives

$$\ell(t) + \Phi s(t)\left(\frac{\delta_1}{2} + \sum_{1}^{N-1} u_i(t)\frac{\delta_{i+1} + \delta_i}{2}\right) = \ell(0). \tag{20}$$

From this algebraic equation it is possible to obtain the unknown $s(t)$. Finally, we obtain an N dimensional DAE system

$$M(t,w)\dot{w} = f(t,w), \tag{21}$$

$w = [u_1, u_2, \ldots, u_{N-1}, s]$. The N-th row of M consists of zero components (because the relationship between s and the other variables is expressed by an algebraic condition). This system can be solved with a suitable DAE solver, like ode15 s in MATLAB, or IDA from the sundials library.

Numerical approximation of the mathematical model describing displacement of wetting liquid by non-wetting liquid Eq. (13) goes along the same lines, as does the single phase flow. For the last there are $N+2$ unknowns, $w = [u_1, u_2, \ldots, u_{N-1}, s_1, s_2, \ell]$, as here ℓ can be considered part of the unknowns. Again the Nth equation is an algebraic condition, now given by

$$\ell(t) + \Phi s_1(t) + \Phi(s_2(t) - s_1(t))\left(\frac{\delta_1}{2} + \sum_1^{N-1} u_i(t)\frac{\delta_{i+1} + \delta_i}{2}\right) = \ell(0). \tag{22}$$

4 Numerical Experiments

In the numerical experiments we shall use $K_s = \frac{K_i \rho_w g}{\mu_b} = 2.4 \times 10^{-5}$ cm/s, $\mu_w = 1$, $n = 2.81$, $\alpha = 0.0189$, $\ell(0) = 2$ cm, $r_0 = 30$ cm, $\omega = 30$ rad/s, $\mu = \frac{\mu_n}{\mu_w} = 2$, $\rho = \frac{\rho_n}{\rho_w} = 0.8$. In some experiments (in two-phase flow) we shall also use $K_s = 2.4 \times 10^{-4}$ cm/s and $\ell(0) = 3$ cm. For the two-phase flow experiments, the total flux q_t is required known. We generate q_t from discrete values WL of the water level ℓ in the injection chamber. We consider

$$WL = [3.00, 2.68, 2.42, 2.19, 1.99, 1.82, 1.67, 1.52, 1.38,$$
$$1.25, 1.12, 0.99, 0.86, 0.75, 0.65, 0.59].$$

Note that although we consider transient measurements, it is not needed to measure $\ell(t)$ more precise that the values given in WL. These values correspond to the time points

$$t_{span} = [0, 65.33, 130.66, 196.00, 261.33, 326.66, 392.00, 457.33, 522.66$$
$$588.00, 653.33, 718.66, 784.00, 849.33, 914.66, 980.00],$$

which is roughly every minute. Then $\ell(t)$ is obtained from a cubic spline interpolation, from which q_t can be determined.

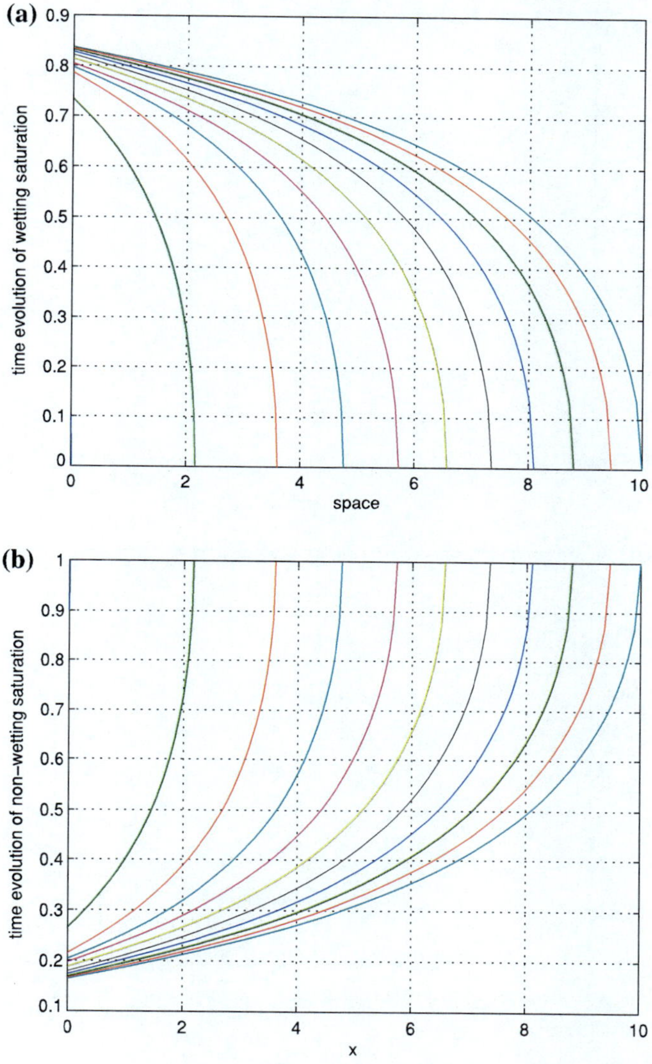

Fig. 2 Time evolution of the liquids for the generic soil parameters, $t \in (0, 980)$. **a** Wetting liquid, **b** non-wetting liquid

4.1 Experiment 1: Displacement of Non-Wetting by Wetting Liquid

The time evolution of wetting and non-wetting liquids for the standard soil parameters are drawn in Fig. 2 up to the time when the interface s reaches the boundary of the sample. The time evolution of the water level $\ell(t)$ and the

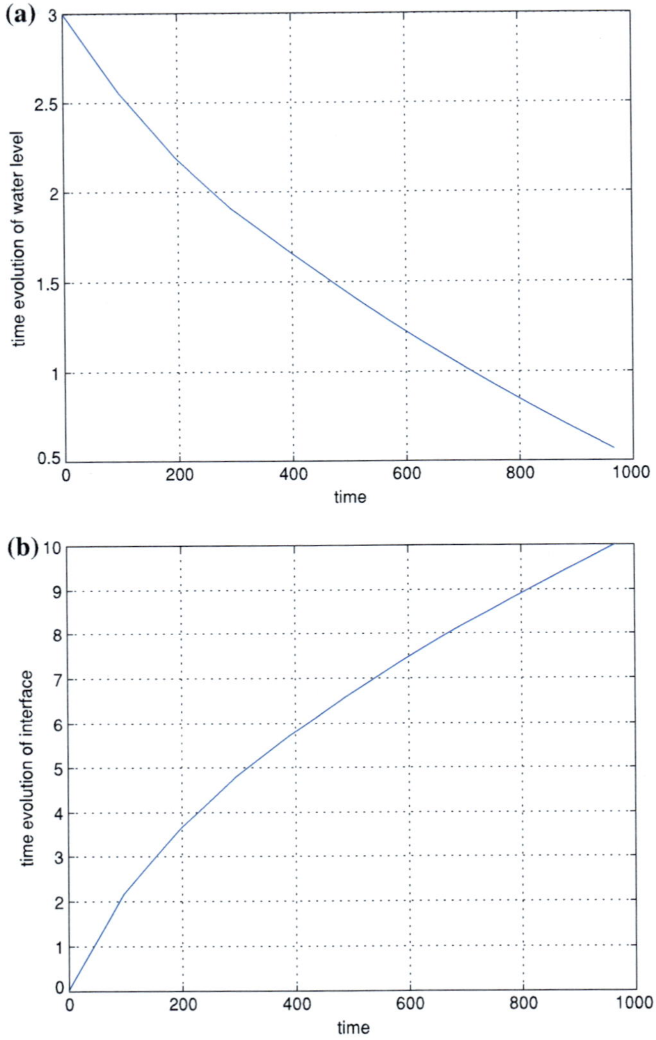

Fig. 3 Time evolution of water level ℓ and interface s for the generic soil paramters, $t \in (0, 980)$. **a** Water level, **b** interface

interface $s(t)$ are drawn in Fig. 3. As can be seen, we indeed obtain that $S_n > 0$ for the entire duration of the sample. The model must be stopped when the interface reaches the end of the sample, as this is not contained in the current model. The water level change obtained corresponds with the values *WL*, as is expected.

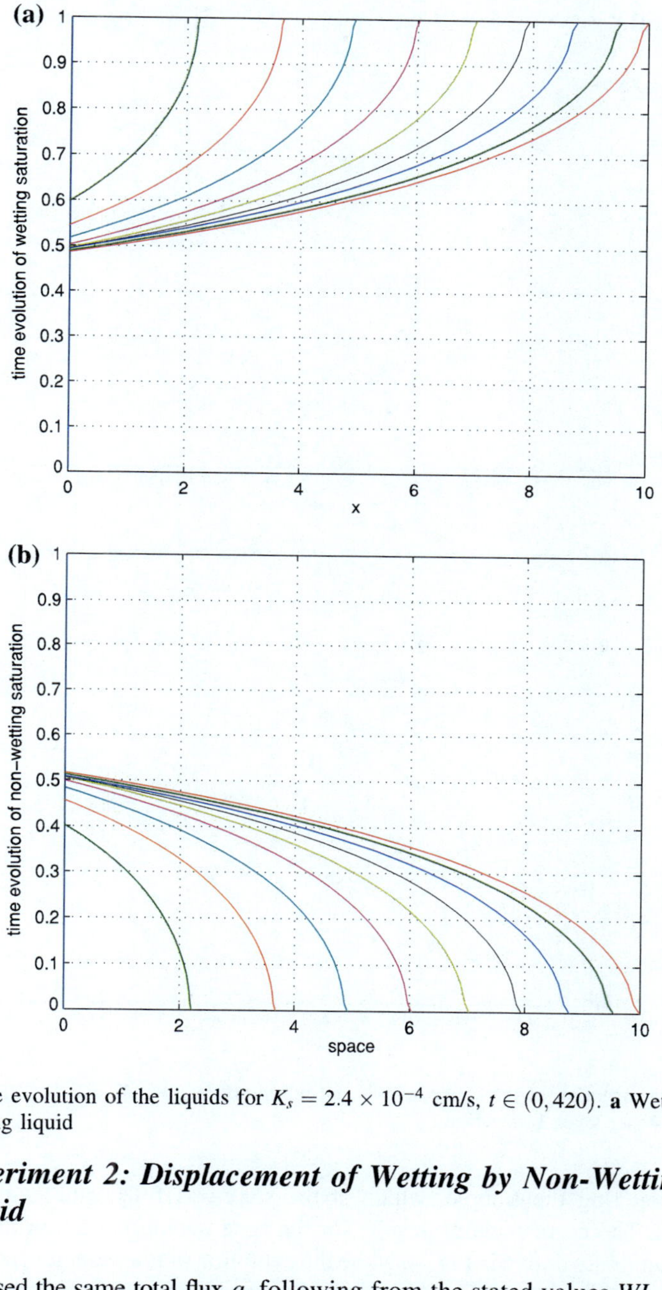

Fig. 4 Time evolution of the liquids for $K_s = 2.4 \times 10^{-4}$ cm/s, $t \in (0, 420)$. **a** Wetting liquid, **b** non-wetting liquid

4.2 Experiment 2: Displacement of Wetting by Non-Wetting Liquid

Since we used the same total flux q_t following from the stated values *WL*, using the generic soil parameters would result in very similar graphs due to the fact that $S_w + S_n = 1$. Therefore it is more interesting to use (for comparison) a higher hydraulic permeability $K_s = 2.4 \times 10^{-4}$ cm/s. The time when the interface s_n reaches the boundary of the sample is hence shorter. In this experiment non-wetting

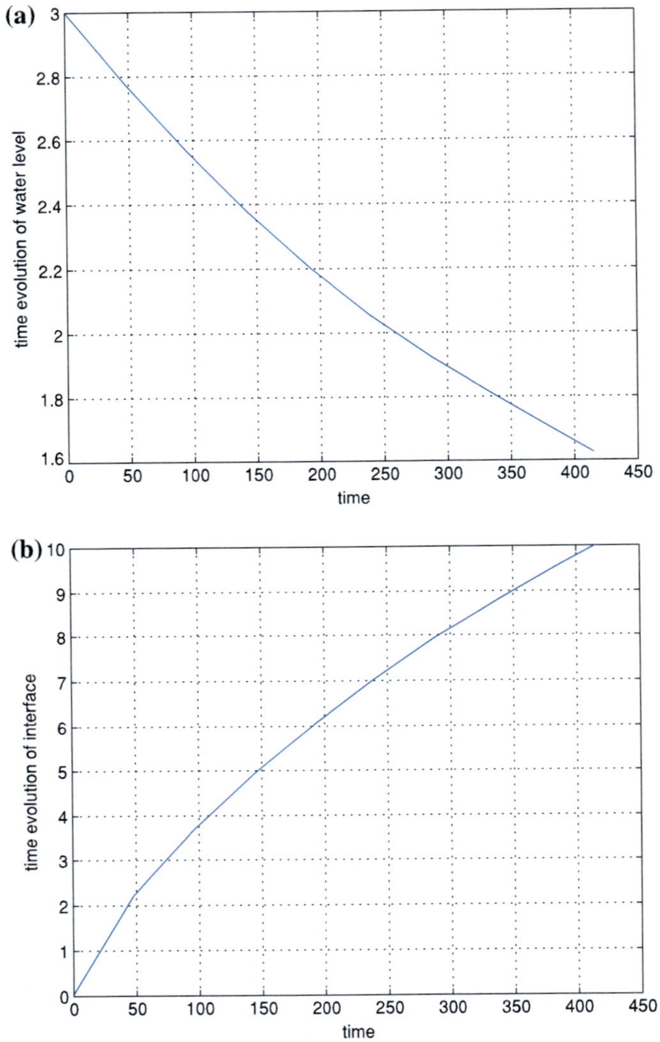

Fig. 5 Time evolution of the water level ℓ and the interface s for $K_s = 2.4 \times 10^{-4}$ cm/s, $t \in (0, 420)$. **a** Water level, **b** interface

liquid is entering the sample, which at the start was fully saturated with wetting liquid. The corresponding results for the time evolution of wetting and non-wetting liquids are drawn in Fig. 4, while the evolution of the water level $\ell(t)$ and the interface $s(t)$ are drawn in Fig. 5. For this case we observe that the wetting liquid remains at a saturation of 0.5 at the front of the sample, while in the previous experiment the non-wetting liquid was displaced up to a saturation of 0.2. This is a consequence of the difference in the liquids (the wetting liquid adheres to the soil particles).

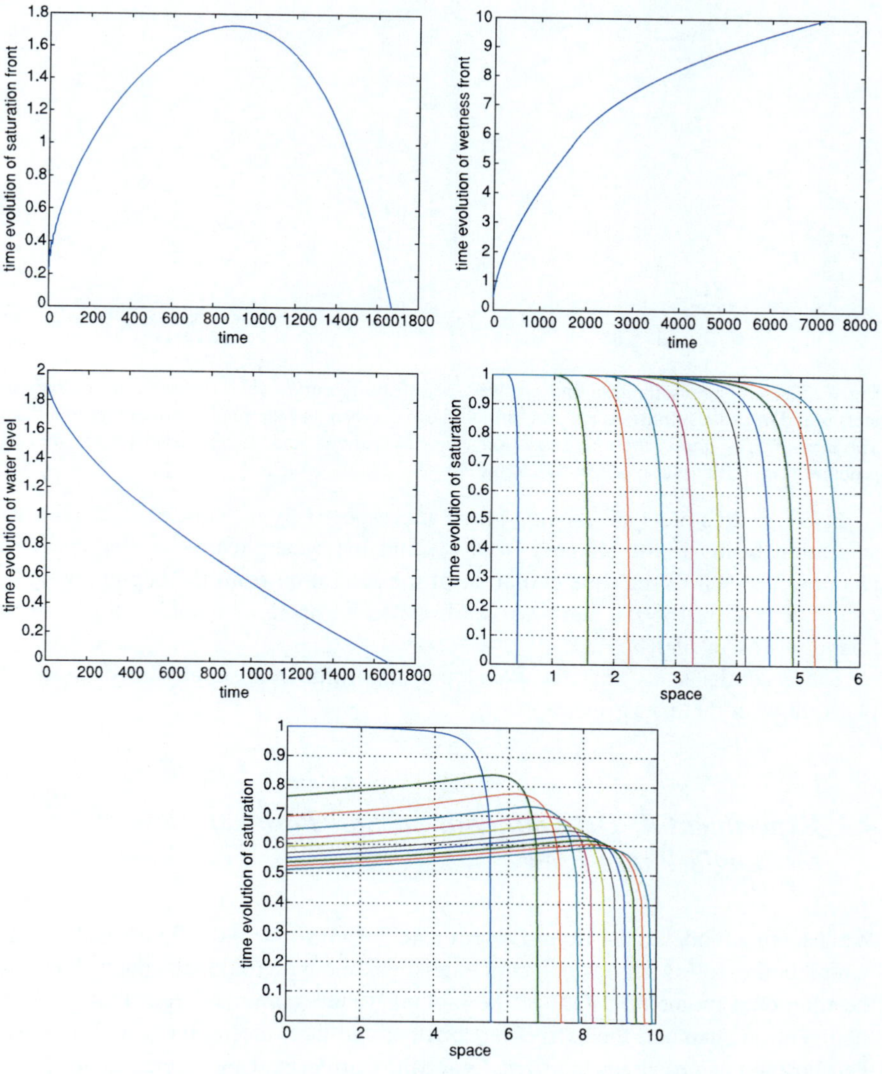

Fig. 6 The evolution of s_1, s_2 and ℓ. The fourth and fifth figure are the saturation over the sample for equidistant time points in $t \in (0, 1650)$ and $t \in (1650, 8650)$

4.3 Experiment 3: Single Phase Flow (Water–Air)

Infiltration into an (originally) dry sample creates after a very short time a fully saturated and a partially saturated zones separated by the interface s_1. The wetting front is modelled by s_2. In this experiment we assume $K_s = 2.4 \times 10^{-5}$ cm/s and $\ell(0) = 2$ cm. At $t = 1650$ s the injection chamber is empty, after which point the top boundary becomes a no-flux boundary. At that moment one also has that the interface s_1 equals zero, and hence disappears.

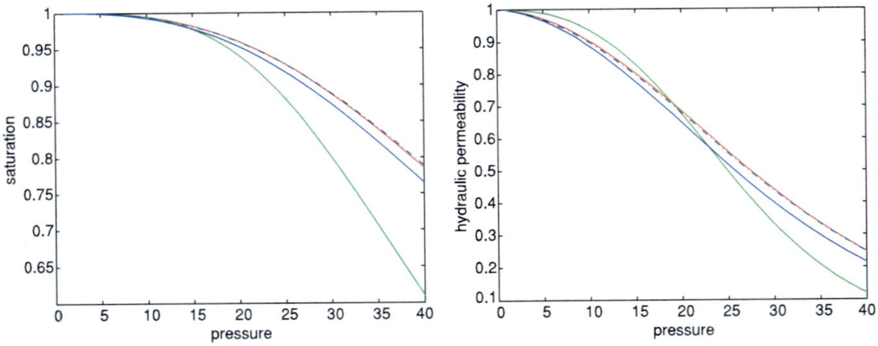

Fig. 7 The pressure–saturation and pressure–permeability curve when determining the parameters α and n; exact solution is *red*, the initial guess is *green* and the final approximation is *blue*. The *dashed blue line* is the approximation when s_2 without noise is also used for the inverse problem. The initial guess is $\alpha = 0.025$ and $n = 3.5$

In Fig. 6, five plots are shown, which are respectively the time evolution of the saturation front s_1, the wetness front s_2, and the water level ℓ in the injection chamber, as well as the time evolution of the saturation from the beginning up to $\ell(t) = 0$, ($t \in (0, 1650)$), and after s_1 disappeared up to s_2 reaching the outflow boundary ($t \in (1650, 8650)$).

These results show that the developed algorithm realistically models single phase flow in the centrifuge.

4.4 Experiment 4: Determination of Soil Parameters for Single Phase Flow

We use the model data from Experiment 3 (with $\alpha = 0.0189$, $n = 2.81$) and compute the following global characteristics: the gravitational centre (G) and the rotational momentum (M) of the moving water (combined contribution from the injection chamber, the extraction chamber and the water in the sample). These we considered measured and then forget the correct soil parameters α, n. Now, these soil parameters must be determined via the solution of the inverse problem.

We use as starting values for the inverse problem $\alpha = 0.025$, $n = 3.5$, and use the Levenberg-Marquardt iteration procedure (minimizing the distance of computed and measured global characteristics) to restore the correct soil parameters. Before doing this, the generated measurement data has been perturbed with 3 % noise. In Fig. 7 the results of the LM iterations are depicted in terms of the starting and recovered pressure–saturation and pressure–permeability curves (which are modelled by means of the soil parameters). The curve corresponding to the initial guess ($\alpha = 0.025$, $n = 3.5$) is shown in *green*, the final approximation using LM iterations is in a *blue full line*.When, in addition to G and M, also measured values of s_2 are added without noise, then the final result is drawn in a *blue dashed line*.

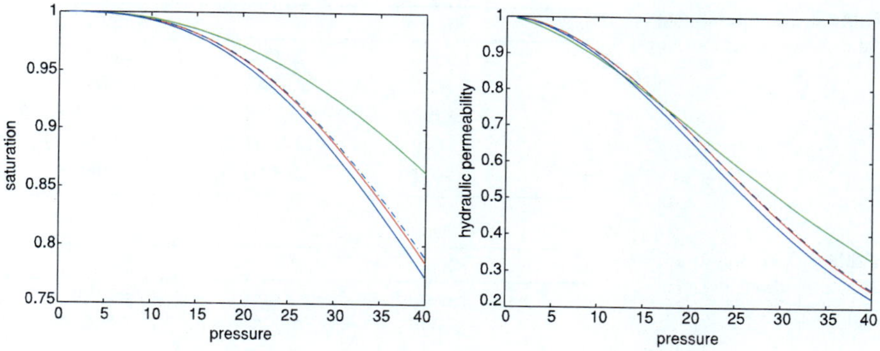

Fig. 8 The pressure–saturation and pressure–permeability curve when determining the parameters α and n; exact solution is *red*, the initial guess is *green* and the final approximation is *blue*. The *dashed blue line* is the approximation when s_2 without noise is also used for the inverse problem. The initial guess is $\alpha = 0.015$ and $n = 2.5$

Table 1 LM -iterations; data : G, M, s with 3 % noise

it	$100 \times \alpha$	n	*RMS*
0	1.6000	2.000	22.933
2	1.5723e	2.3421	13.026
4	1.5452	2.7665	8.900
6	1.5565	2.8657	8.7794

The curve corresponding to the exact solution is given in *red*. Starting from the different initial guess $\alpha = 0.015$, $n = 2.5$, the results are presented in Fig. 8.

The experiment confirms that rotational momentum and gravitational centre measurements are sufficient to accurately approach the correct constitutive laws. This experiment confirms also that the information concerning movement of the wetness front (when they are available) can increase the reliability in determination of the soil parameters.

4.5 Experiment 5: Determination of Soil Parameters in Two-Phase Flow

Determination of the soil parameters α and n for the two-phase flow setting is a more difficult task than the single phase flow in the previous experiment. The main reason is that the global characteristics (gravitational center, rotational momentum, s_2) are not so dynamic as in the previous case. The mass density of non-wetting liquid (oil) does not differ substantially from the one of the wetting liquid (water). Thus, the movement of water/oil in the sample does not contribute much to the evolution of the gravitational centre and rotational momentum. The largest contribution will come from the decrease of liquids in the injection chamber, and the

Table 2 LM -iterations; data : G, M, s without noise

it	$100 \times \alpha$	n	*RMS*
0	1.6000	2.000	13.802
2	1.1675	2.1945	0.48818
4	1.8299	2.7877	3.4833×10^{-3}
6	1.8649	2.7964	1.679×10^{-4}
8	1.8671	2.7956	1.5969×10^{-5}

Table 3 LM -iterations; data : s without noise

it	$100.\alpha$	n	*RMS*
0	1.6000	2.000	4.7348
2	1.1627	2.2632	2.7816×10^{-2}
4	1.7308	2.7962	4.1126×10^{-3}
6	1.8494	2.7711	2.3791×10^{-4}
8	1.8852	2.8068	3.4946×10^{-6}

Table 4 LM -iterations; data : s with 3 % noise

it	$100.\alpha$	n	*RMS*
0	1.6000	2.000	0.15214
2	1.5904	2.2372	0.10870
4	1.5808	2.4806	8.8399×10^{-2}
6	1.5750	2.6200	8.5048×10^{-2}

Table 5 LM -iterations; data : G with 3 % noise

it	$100.\alpha$	n	*RMS*
0	1.6000	2.000	21.741
2	1.5830	2.3374	12.857
4	1.5627	2.7351	8.6991
6	1.5687	2.7711	8.3201
8	1.5714	2.8767	8.1201

Table 6 LM -iterations; data : G, M with 3 % noise

it	$100.\alpha$	n	*RMS*
0	1.6000	2.000	31.521
4	1.4651	2.4815	15.381
6	1.5128	2.5613	8.6991
8	1.5687	2.6017	1.5969×10^{-1}

increase in the extraction chamber. In case it is possible to measure the movement of the interface s (e.g. via X-rays, ...), the determination of the soil parameters can be greatly improved. This will be clearly visible in the Levenberg-Marquardt (LM) iteration procedure, since there are many local (flat) minima. Especially, when the measured data are perturbed with noise.

We present some result of LM iterations, starting with $\alpha = 0.016$ and $n = 2.$ in Table 1. Here, we use $K_s = 2.4 \times 10^{-5}$ cm/s, $\ell(0) = 3$ cm and the same values *WL* as in Experiment 1. The error measure used in the LM method is the Root Mean

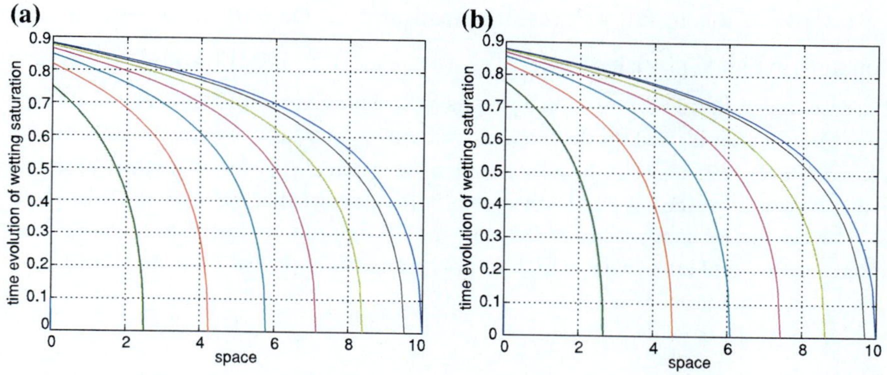

Fig. 9 Time evolution of saturations using q_t and $q_{5,t}$, respectively **a** saturation based on 3 % noise, **b** saturation based on 5 % noise

Square error (RMS) with respect to the measurements. The obtained values for α and n approach the true values $n = 2.81$, $\alpha = 0.0189$, showing that the soil parameters can be obtained if the measurements are sufficiently precise and also s is considered.

Removing the noise, very accurate results are obtained after a few iterations, as seen in Table 2.

In Table 3 we use only s measurements without noise, and in Table 4 s measurements with 3 % noise. This shows that accurately measuring s is more beneficial than using G or M.

Indeed, in Table 5 only the measurements of G with 3 % noise are used, while in Table 6, G and M with 3 % noise are used.

The minimum obtained in these is however part of a large flat region. As a proof of this, the parameters $\alpha = 0.018788$, $n = 2.9713$ which are close to the correct solution, give $RMS = 2.4537$, which the LM method rejected since this RMS is higher as the choice in iteration 8 in Table 6. We can conclude that the soil parameters can be determined from global measurements, provided the errors on the measurements are under control, or the interface s can be accurately tracked.

4.6 Experiment 6: Applicability of Measured q_t

Applicability of the Method for Two-Phase Flow is Based on q_t which needs to be constructed from measurements represented by the discrete values of WL. These values are affected by measurement error, and in the previous experiments we used 3 % noise. For the current experiment, we consider instead 5 % noise to the data WL, and show the algorithm is sufficiently robust. Denote the corresponding new measurements by WL_5 and realize the same computation (as in Experiment 1) by means of the corresponding total flux $q_{5,t}$. In Fig. 9 we compare the corresponding obtained saturations.

To analyse this result, we present an indicator for the error E between the two solutions in Fig. 9, given by $E = \sqrt{\frac{1}{R}\sum_{i,j}(u_{q_t} - u_{q_{5,t}})^2} = 0.0213$, with $R = 7.N$, and the sum over all grid points of the 7 graphs (corresponding with t_{span}). The same type of error indicator is used for the front movement $s(t)$, so for $s_{q_t} - s_{q_{5,t}}$. In that case we obtain $E = 0.142$. These indicators confirm the applicability of the method based on using measured values of ℓ and consequently for q_t. Note that the error made in q_t can still be decreased using some additional smoothing of the data WL to improve their monotonicity and convexity, which in practise is the expected behaviour for $\ell(t)$.

5 Conclusions

An efficient numerical method is proposed for one dimensional two-phase flow. Wetting and non-wetting liquids in a porous media under centrifugation are assumed to be incompressible and immiscible. The proposed method is a good candidate for solving the inverse problem of determining the soil parameters arising in the capillary-pressure curve. The improvements do however require some additional smoothing of the measured data.

For the single phase flow (e.g. water/air), and for the two-phase flow, suitable centrifugation scenarios are considered in which only global characteristics (evolution of gravitational center, rotational momentum and amount of expelled liquid) of the moving liquid are required. The determination of soil parameters for single phase flow is more reliable than in the case of two-phase flow. The applicability of the method has been demonstrated in a series of experiments.

Acknowledgments The authors thank for financial support to the Slovak Research and Development Agency under contract APVV-0184-10 and APVV-0743-10.

References

1. Bear, J., Bachmat., Y.: Introduction to modelling of transport phenomena in porous media. Kluwer Academic Publishers, Dordrecht-Boston-London (1991)
2. Constales, D., Kačur, J.: Determination of soil parameters via the solution of inverse problems in infiltration. Comput. Geosci. **5**, 25–46 (2004)
3. Hassler, G.L., Brunner, E.: Measurement of capillary pressures in small core samples. Trans. AIME **160**, 114–123 (1945)
4. Kačur, J., Budacova H.: Time measurement scenario for inverse problems in infiltration. Submitted in Computers & Fluids (2012)
5. Kačur, J., Malengier, B., Kišon., P.: A numerical model of transient unsaturated flow under centrifugation based on mass balance. Trans. Porous Media **87**, 793–813 (2011)
6. Rajan R.R., Brunner E.: Theoretically correct analytical solution for calculating capillary pressure-saturation from centrifuge experiments. In: SPWLA 27th Annual Logging Symposium Transactions, Texas (1986)

7. Ruth D., Wong S.: Calculation of capillary pressure curves from data obtained by the centrifuge method. In: SCA Conference, Number 8802 (1988)
8. Sharma, J.S., Samarasekera, L.: Effect of centrifuge radius on hycraulic conductivity measured in a falling-head test. Can. Geotech. J. **44**, 96–102 (2007)
9. Subbey, S., Nordtvedt, J.E.: Capillary pressure curves from centrifuge data: a semi-iterative approach. Comput. Geosci. **6**, 207–224 (2002)
10. van den Bert E.H., Perfect E., Tu C., Knappett P.S.K., Leao T.P., Donat R.W.: Vadose Zone J. **8**:531 (2009)
11. van Genuchten M.Th.: A closed-form equation for predicting the hydraulic conductivity of unsaturated soils. Soil Sci Soc. Am. J. **44**:892-898 (1980)

Contaminant Transport in Partially Saturated Porous Media

Erika Trojakova and Jela Babusikova

Abstract We discuss the numerical modeling of unsaturated flow in porous media with contaminant transport, dispersion and adsorption. The mathematical model for unsaturated flow is based on the Richard's nonlinear and degenerate equation. The model of contaminant transport is based on the Fick's law and the mass balance equation. We present the operator splitting method for the numerical solution of this problem. In our numerical approximation we reduce the solution of our problem, successively along a small time interval, into three subproblems: unsaturated flow, transport and dispersion, adsorption. Our numerical solution is based on an implicit time and space discretization. The convergence of our numerical solution to the weak solution of the original problem is discussed. Finally, we demonstrate in our numerical experiments and comparisons with the benchmark solution (in 1D) the effectiveness of our method.

Keywords Contaminant transport · Operator splitting · Saturated-unsaturated flow · Contaminant transport with adsorption

---

E. Trojakova · J. Babusikova (✉)
Faculty of Mathematics, Physics and Informatics, Comenius Univerzity,
Mlynska dolina, 84248 Bratislava, Slovakia
e-mail: Jela.Babusikova@fmph.uniba.sk

E. Trojakova
e-mail: erikatrojakova@gmail.com

J. M. P. Q. Delgado et al. (eds.), *Numerical Analysis of Heat and Mass Transfer in Porous Media*, Advanced Structured Materials 27,
DOI: 10.1007/978-3-642-30532-0_12,

1 Introduction

The protection of clean water in underground aquifers creates an important part in many projects of research teams. Many of scientists and engineers are focused on this field. Many of monographs and articles, appeared last decade, singnificantly contribute to this field. The contaminated surface water infiltrates into the unsaturated underground and there, contaminant, dissolved in the water, obeys the rules of convection, dispersion and adsorption. These are complex geophysical and geochemical processes and there are many mathematical models describing them realistically (see [1]). In fact, these mathematical models include many data (hydrological and geochemical) strongly related to the specific site where the model is applied. These data must be measured and some of them must be determined by solution of inverse problems (via black box) after some additional measurements of some characteristics become available. Many of these model data can be obtained in laboratory conditions with 1D samples (for example some tubes of the corresponding underground). When all needed data are determined the complex models can be applied also to 3D cases. Solution of the inverse problem requires very accurate and effective numerical solution of the direct problem (when data are given). The main goal of our contribution is to propose such a numerical method. The obtained numerical results confirm that our method is a good candidate for solving the above complex problem. Due to this fact we are also ambitious to present a mathematical proof of the convergence of our numerical solution to the exact one.

In our contribution we focus on the numerical modeling of contaminant transport with adsorption in unsaturated-saturated porous media. The corresponding mathematical model is strongly nonlinear, linking convection, diffusion-dispersion and adsorption together. Especially, when the contaminated water infiltrates into the originally dry porous media, there appear saturated, partially saturated and (remaining) dry zones, which are separated by moving interfaces. These interfaces are not known originally and they are implicitly included in the solution. Moreover, the contaminant dissolved in the water can also generate zones where it is present or absent, due to (generally) nonlinear adsorption. It is well-known that the front of saturation in the neighbourhood of a dry zone is very sharp. On the other hand it is very difficult to determine the interface between saturated and partially saturated zones. It is a difficult task to create an efficient numerical scheme to obtain a correct solution in 3D. Our main goal is to demonstrate that the numerical method based on "operator splitting" is suitable for solving our problem in spite of the fact that it is only of the first order. We demonstrate this in our numerical experiments (in 1D) which are very promising. The main reason is that the speed of the flow is very low and the adsorption dynamics is significantly higher than the dynamics of convection-diffusion. This fact is substantially implemented in our "operator splitting" approximation. Due to the practical applicability of our method we also devote attention to the theoretical background of this method by proving its convergence. There we follow the ideas from [6] and [8]. Unlike as in [8], we do not split up convection and diffusion-dispersion, but we keep them together. This enables us to obtain stronger convergence results. The proof of the convergence is sketched only since it is too lengthy. The details can be found in [14].

The mathematical model for contaminant transport in unsaturated-saturated porous media is based on the Richard's-Darcy's equation for flow and the convection-diffusion-adsorption equation is based on the Fick's law and the mass ballance argument for a contaminant. The flow in unsaturated porous media is modeled by the hydraulic conductivity K, where $K(h) = K_s \cdot k(h)$ and $k(h)$ is a function describing the dependency of conductivity on piezometric head h, or on corresponding effective saturation (see Van Genuchten [15]). Here, K_s is the hydraulic conductivity in fully saturated porous media,

$$K_s = \kappa_0 \frac{\rho g}{\mu},$$

where ρ and μ are the density and the dynamical viscosity of water and the coefficient κ_0 depends only on the structure of the porous medium, g is the gravitational acceleration. The Richard's equation determines the changes in saturation θ in the unsaturated zone. In basic form it can be written as

$$\partial_t \theta = \nabla \cdot (K(h)\mathbf{A}(x)\nabla(h+z)) = \operatorname{div}(K(h)\mathbf{A}(x)\nabla(h+z)), \tag{1}$$

where the matrix $\mathbf{A}$ describes the changes in conductivity according to the space structure (in our experiments $\mathbf{A} = \mathbf{I}$) and $K(h)$ is given by

$$K(h) = K_s k(\bar{\theta}), \quad k(\bar{\theta}) = \bar{\theta}^{\frac{1}{2}}(1 - (1 - \bar{\theta}^{\frac{1}{m}})^m)^2, \tag{2}$$

where $\bar{\theta}$ is the effective saturation defined as

$$\bar{\theta} = (\theta - \theta_r)/(\theta_s - \theta_r), \tag{3}$$

and

$$\bar{\theta} = \frac{1}{(1 + (\alpha h)^n)^m}, \tag{4}$$

where $n > 1$, $m = 1 - \frac{1}{n}$ are the soil parameters and $\alpha < 0$ is the van Genuchten-Mualem (empirical) ansatz. Here, θ_s and θ_r are fully saturated and residual water contents, respectively. In the saturated zone we have (Darcy's law) $k(h) \equiv 1$ and $\theta \equiv \theta_s$.

There are also other models (Brooks and Corey, Brutsaert, Vauclin), but we will use the van Genuchten-Mualem model, because the fundamental relations (hydraulic conductivity versus saturation and saturation versus head) reflect the appearence of interfaces separating saturated and unsaturated zones. The contaminant transport is the superposition of convective and diffusive-dispersion parts—see [1]. The water flux is given by $\mathbf{v} = -K(h)\mathbf{A}(x)\nabla(h+z)$ and the corresponding contaminant flux is given by

$$\mathbf{J}_{\text{adv}} = \mathbf{v}w \tag{5}$$

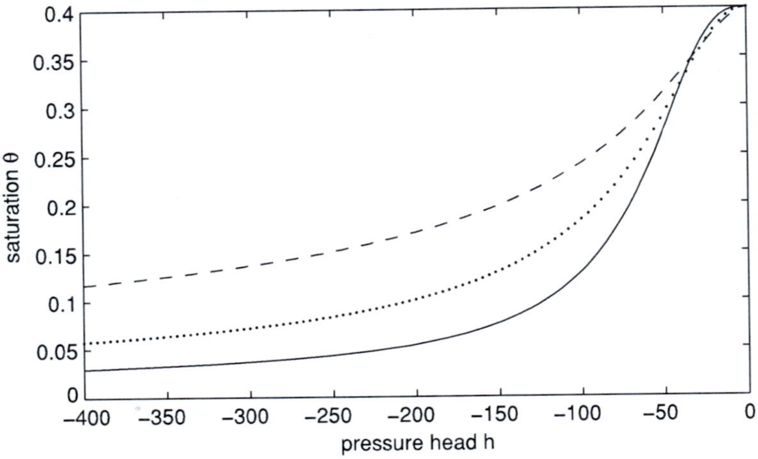

Fig. 1 Retention curve—graph of pressure head h vs. saturation θ for $m_1 = 0.6441$ (*solid*), $m_2 = 0.5327$ (*dotted*), $m_3 = 0.4002$ (*dashed*)

where w is the concentration of the contaminant dissolved in the water. The contaminant flux due to the diffusion-dispersion is given by

$$\mathbf{J}_{\text{dis}} = -\mathbf{D} \cdot \nabla w.$$

Here, $\mathbf{D}\,[L^2T^{-1}]$ is a positive definite symmetric matrix. The components of $\mathbf{D}$ can be expressed as

$$D_{ij} = (D_0 + \alpha_T|\mathbf{v}|)\delta_{ij} + \frac{v_i v_j}{|\mathbf{v}|}(\alpha_L - \alpha_T),$$

where α_L, α_T are longitudinal and transversal dispersivities, respectively δ_{ij} is the Kronecker delta and D_0 is the molecular diffusion coefficient (Figs. 1 and 2).

Using the mass balance equation for the total flux $\mathbf{J} = \mathbf{J}_{\text{adv}} + \mathbf{J}_{\text{dis}}$ we get the equation for the transport of the contaminant with diffusion,

$$\frac{\partial(\theta w)}{\partial t} = -\nabla \cdot \mathbf{J},$$

which leads to the well-known *convection-diffusion* equation

$$\theta\partial_t w + \mathbf{v}\nabla w - \operatorname{div}(\mathbf{D}\nabla w) = -\rho\partial_t S \tag{6}$$

where the conservation of water [see (1)] has been taken into account. The adsorption of the contaminant is govered by ODE

$$\partial_t S = \kappa(\Psi(w) - S), \tag{7}$$

where S represents the adsorbed contaminant by a unit mass of porous media. Here, κ is the sorption rate coefficient and describes the velocity of the adsorption.

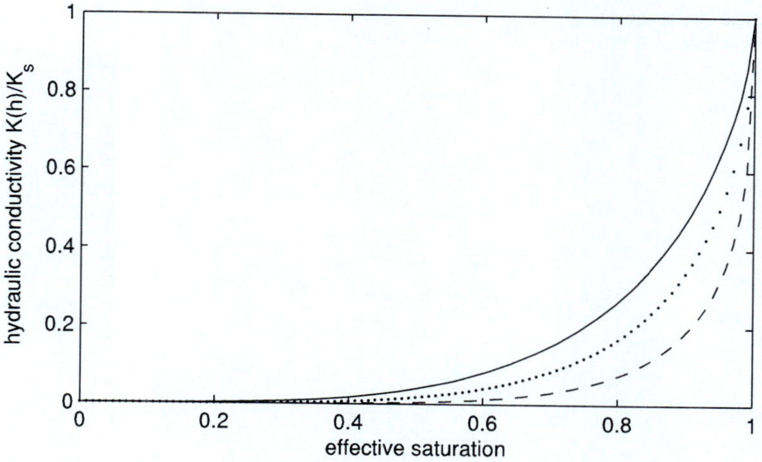

Fig. 2 Permeability curve—graph of conductivity $K(h)/K_s$ vs. effective saturation $\bar{\theta}$ for $m_1 = 0.6441$ (*solid*), $m_2 = 0.5327$ (*dotted*), $m_3 = 0.4002$ (*dashed*)

The mathematical model (7) is a very special one and all the results obtained here can be extended to other models. The most common isotherms are

- linear: $\psi(s) = as$;
- Freundlich: $\psi(s) = as^p$;
- Langmuir: $\psi(s) = \frac{as}{1+bs}$;
- Mixed Freundlich-Langmuir: $\psi(s) = \frac{as^p}{1+bs^p}$;

In (7) the reversible adsorption mode is considered when the contaminant can be fixed to the solid and also dissolved back from the solid to the water. It is possible to consider also the irreversible mode, when the contaminant cannot be dissolved back into the water. Then, the following modification of (7) can be used:

$$\partial_t S = \kappa \max(0, \Psi(w) - S).$$

Another extension can be considered, when some compounds of the solid matrix adsorb the contaminant in an equilibrium mode which correspons to $\kappa \to \infty$. In that case we have $S = S_e = \Psi_e(s)$, where $\Psi_e(s)$ is the sorption isotherm in the equilibrium mode and can have one of the above mentioned forms. In that case the R.H.S. term $\partial_t S_e = \partial_t \Psi_e(w)$ can be added to the term $\partial_t w$ on the L.H.S. of (6). In general, the contaminant can be adsorbed both in the equiulibrium and the non-equilibrium modes. In that case (due to superposition) we obtain (7) , where we replace $\partial_t w$ by $\partial_t(w + \Psi_e(w))$. Then $\Psi(s)$ in (7) represents the sorption isotherm in the nonequilibrium mode with the corresponding S—see [3, 5, 7, 11].

Applying Kirchhoff's transformation $u := \beta(h) = \int_0^h K(z)dz, \quad b(u) := \theta(\beta^{-1}(u))$ the system (1), (6) and (7) takes the form

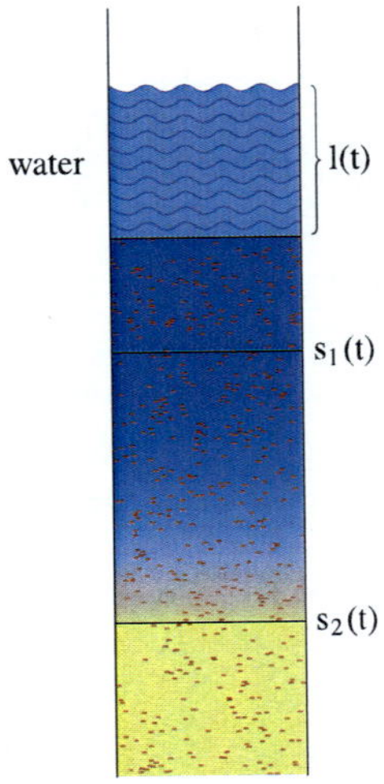

Fig. 3 The sketch of unsaturated/saturated flow in a sample

$$\partial_t b(u) + \mathrm{div}_z F(x,u) - \mathrm{div}\mathbf{A}(x)\nabla u = 0, \tag{8}$$

$$b(u)\partial_t w + v(u,\nabla u)\cdot\nabla w - \mathrm{div}(D(u,\nabla u)\nabla w) = -\rho\partial_t S, \tag{9}$$

$$\partial_t S = \kappa(\Psi(w) - S), \tag{10}$$

where the unknown $b(u) = \theta$ is the moisture content and the remaining functions are defined (see [1]) as $F(x,b(u)) = -\mathbf{A}(x)eK_s k(b(u))$, e being a unit vector in direction z, $v(u,\nabla u) = -K_s k(b(u))A(x)\nabla(h+z)$ (flow flux) and consequently $\mathbf{D} = D(u,\nabla u)$. For our experiments we use the Freundlich's isotherm $\Psi(s) = as^p$.

The presented model is completed by initial and boundary conditions. In our theoretical part we consider our problem in terms of (8), (9), (10) which is more suitable for the convergence proof. In the numerical experiments we approximate our system in the form (1), (6) and (7).

Unsaturated-saturated flow in the sample is drawn in the Fig. 3 .

2 Operator Splitting

Although all these processes are realized simultaneously, we split our problem to three subproblems, each of them realized successively along a small time interval τ. These subproblems are easier to handle and the numerical approximation can be computed more precisely, but the question of the convergence of such an approximation to the weak solution of the original problem is very important and it is a difficult task to answer it. Moreover, we analyze the error of such a numerical approximation with respect to the splitting length τ by comparison with some benchmark solutions of our complex system. We split the system in time intervals (t_k, t_{k+1}), $\tau = t_{k+1} - t_k$, in the following three subproblems. First we solve the unsaturated flow

$$\partial_t b(u) + \operatorname{div}(F(x,u) - \mathbf{A}(x)\nabla u) = 0 \tag{11}$$

and as a result we obtain updated values of u. Then, we solve the equation (9) (without adsorption) which represents the second subproblem

$$b(u)\partial_t w + v(u,\nabla u)\nabla w - \operatorname{div}(D(u,\nabla u)\nabla w) = 0. \tag{12}$$

As a result we obtain updated values of w.

In the last step, we get the updated values of w and S from the system

$$b(u)\partial_t w = -\rho\partial_t S, \quad \partial_t S = \kappa(\Psi(w) - S). \tag{13}$$

We consider the boundary conditions

$$u = u^D \quad \text{on } I \times \partial\Omega,$$

$$w = w^D \text{in} \quad I \times \Gamma_1, -D(u,\nabla u)\nabla w \cdot v = 0 \quad \text{on} I \times \Gamma_2,$$

$$\mathbf{J} \cdot v = 0 \quad \text{on} I \times \Gamma_3$$

on the borders of Ω. There is inflow on the boundary Γ_1 and outflow on the boundary Γ_2. The remaining part of the boundary of $\partial\Omega$, Γ_3 is isolated.

The initial conditions are

$$u = u_0, \quad w = w_0, \quad S = S_0 \quad \text{on}\{0\} \times \Omega.$$

In 1D, u^D is represented by $u^{(1)}(t), u^{(2)}(t)$ corresponding to the boundary $x = 0, x = L$ of the sample.

We approximate each subproblem using time and space discretization and we get the following approximation scheme. In the first step, we compute u_i^{k+1} for $i = 1, , n_x - 1$ by using the discrete form of Eq. (11) and known values of u^k (completed with Dirichlet boundary conditions):

$$\begin{aligned} &\frac{b(u_i^{k+1}) - b(u_i^k)}{\tau}\Delta x + F(x_i, u_{i+1}^{k+1}) - F(x_i, u_i^{k+1}) \\ &- \left(A(x_{i+1})D_{i+1}u^{k+1} - A(x_i)D_i u^{k+1}\right) = 0, \\ &u_0^k = u^1(t_k) \quad \text{and} \quad u_{n_x}^k = u^2(t_k), \end{aligned} \tag{14}$$

where $D_{i+1}u^{k+1} = (u_{i+1}^{k+1} - u_i^{k+1})/\Delta x$.

In the second step, we compute the approximation of w, which we denote by $w^{k+1/2}$ at each point x_i, $i = 1, , n_x$. For this purpose we use the discretized Eq. (12) and approximated values w^k, u^{k+1}:

$$
\begin{aligned}
& b(u_i^{k+1})\frac{w_i^{k+1/2}-w_i^k}{\tau}\Delta x+v(u_i^{k+1})(w_i^{k+1/2}-w_{i-1}^{k+1/2}) \\
& -\left(d(u_{i+1/2}^{k+1})D_{i+1}w^{k+1/2}-d(u_{i-1/2}^{k+1})D_i w^{k+1/2}\right)=0, \\
& w_0^k=C_0(t_k) \quad \text{and} \quad w_{n_x+1}^k=w_{n_x}^k,
\end{aligned}
\tag{15}
$$

so that the Neumann boundary condition holds. For simplicity we use the notation $v(u_i^k) := v(u_i^k, \nabla u_i^k)$ and $d(u_i^k) := D(u_i^k, \nabla u_i^k)$.

Finally, in the third step, we use $w^{k+1/2}$ and S^k as the initial condition for $t = t_k$ and compute the values $w^{k+\frac{1}{2}}(t)$, $S^k(t)$ for every point x_i, $i = 1, , n_x$:

$$
\begin{aligned}
& b(u(x_i^{k+1}))\partial_t w_i^{k+\frac{1}{2}}(t)+\rho\partial_t S_i^k(t)=0, \\
& \partial_t S_i^k(t)=\kappa(\Psi(w_i^{k+\frac{1}{2}}(t))-S_i^k(t)).
\end{aligned}
\tag{16}
$$

This last system is a system of ordinary differential equations at each point x_i. Using variation of constants, for S we get

$$
S_i^k(t)=S_i(t_k)e^{-\kappa(t-t_k)}+\kappa\int_{t_k}^{t}\Psi(w_i^{k+\frac{1}{2}}(s))e^{-\kappa(t-s)}ds, \tag{17}
$$

with S_i^k, $w_i^{k+1/2}$ being the initial conditions. For w we get

$$
w_i^{k+1/2}(t)=\frac{\rho}{b_i^{k+1}}\left(S_i^k-S_i^k(t)\right)+w_i^{k+1/2}. \tag{18}
$$

Computing the solution of (18) is very quick and as a result of the third step we set $w_i^{k+1} = w_i^{k+1/2}(t_{k+1})$ and $S_i^{k+1} = S_i^k(t_{k+1})$. Then, we repeat the computations in the next time interval. This scheme describes our numerical approximation of the original problem in the domain $\Omega_T = \Omega \times T$. Our goal is to prove the convergence of such an approximation u^k, w^k and S^k to the variational (weak) solution of the original problem. For simplicity we restrict ourselves to 1D ($\Omega = (0,1)$), but the obtained results and proofs could be extended to the higher dimensional case. For the details see [14].

3 Variational Solution

The convergence results will be obtained in terms of the variational solution in corresponding (standard) functional spaces, see [12]. Let us consider test functions $\Phi(x,t), \zeta(x,t), \vartheta(x,t) \in C^\infty(\Omega \times (0,T))$ with compact support in $\Omega = (0,1)$. We assume $\Phi(0,t) = \Phi(1,t) = 0$, $\Phi(x,T) = 0$; $\zeta(0,t) = 0$, $\zeta(x,T) = 0$ and $\vartheta(x,T) = 0$. Then the variational formulation of the system (8)–(10) (in 1D) is

$$\begin{aligned}&\int_0^T\int_0^1 -b(u)\partial_t\Phi dxdt - \int_0^T\int_0^1 F(u)\partial_x\Phi dxdt\\&+\int_0^T\int_0^1 \mathbf{A}(x)\partial_x u\partial_x\Phi dxdt = \int_0^1 b(u_0)\Phi(x,0)dx, \quad \forall\Phi,\end{aligned} \tag{19}$$

$$\begin{aligned}&\int_0^T\int_0^1 \big(-w\partial_t(b(u)\zeta) + v(u)\partial_x w\zeta + d(u)\partial_x w\partial_x\zeta - \rho S\partial_t\zeta\big)dxdt\\&= \int_0^1 b(u(x,0))w(x,0)\zeta(x,0)dx + \int_0^1 \rho S(x,0)\zeta(x,0)dx, \quad \forall\zeta,\end{aligned} \tag{20}$$

$$\int_0^T\int_0^1 \big(S\partial_t\vartheta + \kappa(\Psi(w)-S)\vartheta\big)dxdt = -\int_0^1 S(x,0)\vartheta(x,0)dx, \quad \forall\vartheta. \tag{21}$$

Here, $u, w, S \in L_2(I, H^1)$ with $\partial_t u, \partial_t S \in L_2(I, L_2)$, where H^1 is the Sobolev space and L_2, L_1, L_∞, $L_2(I, L_2)$, $L_2(I, H^1)$ are standard functional spaces (see [12]).

To prove the convergence of the approximate solutions $\{u^n\}$, $\{w^n\}$, $\{S^n\}$ we follow the ideas from [6, 8]. The convergence is proved under the following assumptions.

4 Assumptions

Let C be a generic positive constant.

(H1) $b(s) \geq \delta_0 > 0$ is Lipschitz continuous, satisfying $0<\delta_0 \leq b'(s) \leq L_b<\infty$ for small δ;

(H2) $F(x, s)$ is continuous and satisfies $|\partial_s F(x, s)| \leq C$ and $F(x, s)$ is decreasing for fixed x;

(H3) $v(v, \xi)$ is a positive continuous function and satisfies $v(v, \xi) \leq C(1 + |v| + |\xi|)$, $D(v, \xi)$ is a continuous, symmetric 3×3 matrix and satisfies

$$||D(v, \xi)|| \leq C(1 + |v| + |\xi|);$$

(H4) $(\mathbf{A}(x)\xi, \xi) \geq C_A|\xi|^2$, $(D(v, \xi)\zeta, \zeta) \geq C_D|\zeta|^2$;

(H5) $|\Psi'(\cdot)| \leq C$, Ψ is monotone;

(H6) initial conditions u_0, w_0, S_0 are smooth and bounded, and also boundary conditions are smooth enough.

5 Convergence

The convergence argument is based on a series of a priori estimates listed in the following Lemmas below (for the detail proofs see [14]). Applying our approximation scheme we obtain also L_2 estimates of derivatives whereas in [8, 13] only L_1 estimates were reached. Consequently, we obtain stronger convergence results and more regular variational solution as in the above cited papers. Using a priori estimates and the Kolmogorov's compactness argument, the convergence in $L_2(I, L_2)$ follows and following the idea in [6], we can prove also the convergence in $L_2(I, H^1)$. Then, following the idea of Majda and Crandal [4] we get the convergence of our approximation to the variational solution defined above. Let's define the vector u^k at time level t_k by

$$u^k = \{u_i^k\}_{i=0}^{n_x}$$

(and similarly also for w and S). To prove the following lemmas 1 and 2 we use the maximum principle in a suitable form.

Lemma 1 The solution u^{k+1} obtained by the operator splitting method is bounded, that is

$$\|u^{k+1}\|_\infty \leq \max(C, |u^1(t)|, |u^2(t)|), \quad t \in \langle 0, T \rangle,$$

where $C \in \mathbb{R}^+$ is a constant independent of time and space discretization and u^1, u^2 are the boundary conditions.

Lemma 2 Let $w^{k+1/2}$ be a solution obtained by the operator splitting method. Then

$$\|w^{k+1/2}\|_\infty \leq \max\{\|w^k\|_\infty, C\}, \quad C = \max_t\{w^D(t)\}, t \in \langle t_k, t_{k+1} \rangle.$$

On the base of the previous facts the following lemma can be proved

Lemma 3 Let w^{k+1} and S^{k+1} be obtained from (16). Then

$$\|w^{k+1}\|_\infty + \|S^{k+1}\|_\infty \leq C,$$

where $C \in \mathbb{R}^+$ is a constant independent of time and space discretization.

A priori estimates for derivatives in terms of L_2 norms are included in the following Lemmas: The a priori estimate

Lemma 4

$$\sum_{k=1}^{n_T} \tau \sum_{i=1}^{n_x} (D_i u^k)^2 \Delta x \leq C$$

holds uniformly for n_x and n_T

The time derivative for $b(u)$ is included in the following estimate:

Lemma 5 The a priori estimate

$$\sum_{k=1}^{r} \tau \sum_{i=1}^{n_x-1} \left(\frac{b_i^k - b_i^{k-1}}{\tau} \right)^2 \Delta x \leq C$$

holds uniformly for n_x and n_T.

Consequently, due to Assumption $(H1)$, we get an a priori estimate for the time derivative of u. An a priori estimate for time derivative of S can be obtained easily from (17), (18) and L_∞ estimates for w, S. The space derivatives for w in L_2 norm are result of the following lemma:

Lemma 6 The a priori estimate

$$\sum_{k=0}^{n_T-1} \tau \sum_{i=1}^{n_x} (D_i w^{k+1/2})^2 \Delta x \leq C$$

holds uniformly for n_x

On the base of this estimate we can prove: The a priori estimate

Lemma 7

$$\sum_{k=1}^{n_T} \tau \sum_{i=1}^{n_x} (D_i w^k)^2 \Delta x \leq C$$

holds uniformly for n_x and n_T

We cannot prove a priori estimates of the time derivative of w in the L_2 norm, but we can guarantee the following weaker estimate (corresponding to $\partial_t^{1/2}$) for the product $b(u)w$. This will be sufficient to obtain compactness of $b(u)w$ and consequently for w, since $b(u) > \delta > 0$. For this purpose we need additional assumption concerning boudness of ∇u. We can guarantee it by a small modification of our model – replacing $D(u, \nabla u)$ by the truncation $D_K(u, \nabla u) := D(u, \nabla u|_K)$, where

$$\nabla u|_K = \begin{cases} \nabla u(x), & |\nabla u(x)| \leq K \\ \operatorname{sgn}(\nabla u(x))K, & \text{otherwise.} \end{cases}$$

The same truncation $v_K(\eta, \xi)$, we shall assume for $v(\eta, \xi)$. We formulate the following Assumption:

(H7) In the place of $D(\eta, \xi), v(\eta, \xi)$ consider $D_K(\eta, \xi), v_K(\eta, \xi)$;

Lemma 8 If, additionally, (H7) is satisfied, then the a priori estimate

$$\sum_{j=1}^{n_T-p} \tau \sum_{i=1}^{n_x} (b_i^{p+j} w_i^{p+j} - b_i^j w_i^j)^2 \Delta x \leq C\tau p,$$

holds, where C is a constant independent on time and space discretization and $p \in \mathbb{N}$.

By means of u_i^k, w_i^k and S_i^k we construct approximate solutions u^ν, w^ν and S^ν ($\nu = (\Delta x, \tau)$) defined on $\Omega \times (0, T)$ by extension (piecewise linear or piecewise constant) from grid points. All modifications of such an extension converge in

$L_2(I, L_2)$ to the same u, w and S due to our a priori estimates and due to the Kolmogorov's compactness argument. Moreover, piecewise linear approximations converge also in the $L_2(I, H^1)$ norm. Due to our strong a priori estimates, $\partial_t u^\nu$ and $\partial_t S^\nu$ converge to $\partial_t u$, and $\partial_t S$ weakly in $L_2(I, L_2)$, respectively.

We have to note that due to the operator splitting method, approximate solutions w^ν and S^ν will be obtained in two steps, each along time interval τ. Moreover, S^ν during the first step was constant (adsorption was not considered). Therefore, we change w^ν and S^ν by compressing the time variable to half and denote the results by $\bar{w}^\nu$, $\bar{S}^\nu$, respectively. This compression in time represents the real approximation of our original problem, since along τ all flow, diffusion with convection and adsorption are realized. The convergence results obtained for w^ν and S^ν hold true also for $\bar{w}^\nu$, $\bar{S}^\nu$ and are not influenced by this compression. Now, following the idea of Majda and Crandal [4], we are able to prove the final convergence result.

Theorem 1 We assume that $(H1) - (H7)$ are satisfied. Then the numerical solution $\bar{u}^\nu, \bar{w}^\nu, \bar{S}^\nu$ obtained by the operator splitting method converges to the variational solution of (19), (20) and (21) (where D, ν are replaced by D_K, ν_K – see (H7)).

We note, that if the flow velocity is bounded (i.e. $\nabla u \in L_\infty$) then in our convergence result we can remove Assumption (H7), i.e., truncation ν_K, D_K.

6 Numerical Experiments

In our numerical experiments we solve a realistic model in 1D, when contaminated water ($w = 1$) from a 3 cm column infiltrates (with gravitational and capillary pressure force) into the 10 cm sample (porous underground). In fact we can expect that along the infiltration a fully saturated zone (in top of the underground) appears with moving interface separating fully saturated zone (governed by steady state Darcy's flow) from the partially saturated zone (governed by the Richard's equation). Moreover, a second moving interface appears between the partially saturated zone and the dry zone. Finally another interface can appear in fully and partially saturated zones, which separates the region where the contaminant is present and where the contaminant is absent. This phenomenon can appear when adsorption is strongly nonlinear. Theoretical results (for "porous media type" equations) prove the existence of these interfaces. On the base of these facts we are facing the difficult task to obtain a precise numerical solution of our complex problem. To demonstrate the practical applicability of the operator splitting method in our problem, we need to compare our results with exact (or very precise) ones obtained without splitting. Last time such a benchmark solution has been obtained in 1D (see [10]) where the mathematical model was reduced to the system of ODEs with exact modeling of evoluting interfaces (also in form of ODEs) and moving grids, based on the movement of interfaces [2, 9]. In our numerical experiments we have used the results concerning the flow (in terms of u, ν, D) from [10] as input data for our contaminant transport with dispersion and adsorption. There the splitting is substantially realized. The flow problem could be

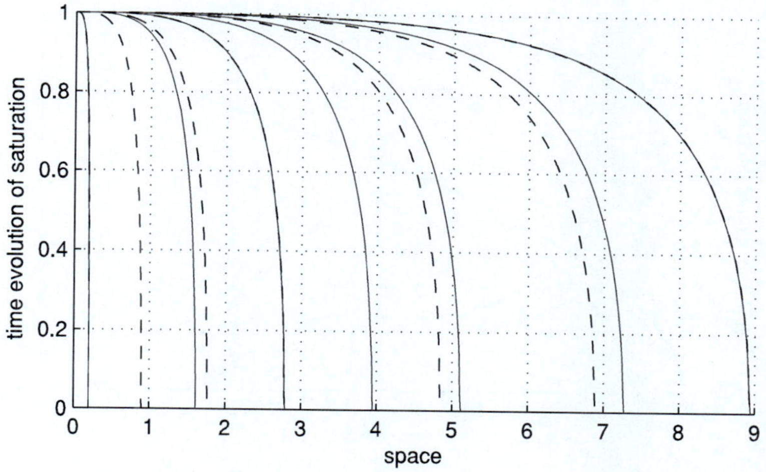

Fig. 4 Saturation in time moments T_2-*dashed* and T_1-*solid*

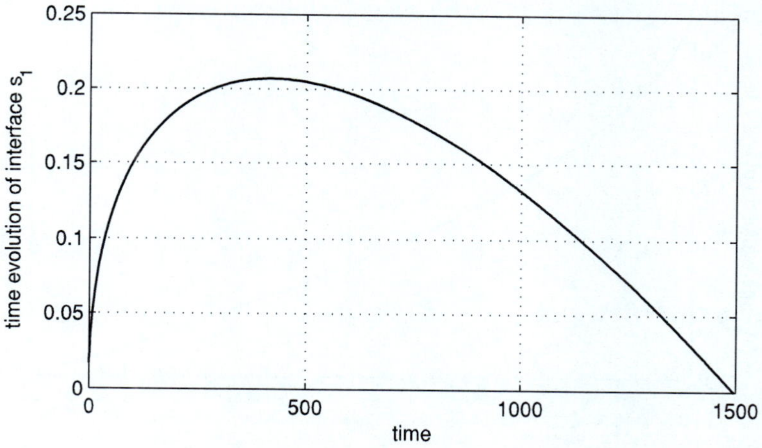

Fig. 5 Time evolution of the interface s_1

solved separately and the splitting (along time $\tau = t_{k+1} - t_k$) is not applied. Along the convection-diffusion and adsorption period we keep u, v, D constant as is assumed in our approximation scheme. Then, we compare our results with the ones in [10] and discuss the magnitude of the error with respect to the length of τ.

We approximate the flow model

$$\partial_t \theta = \operatorname{div}(K(\bar{\theta})\nabla(h+z)), \tag{22}$$

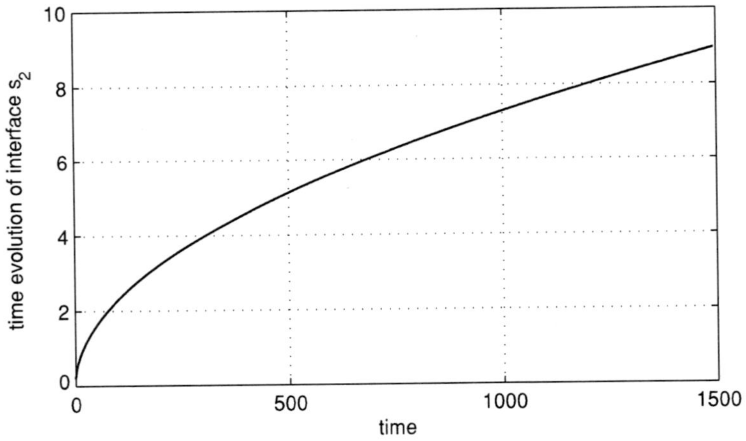

Fig. 6 Time evolution of the interface s_2

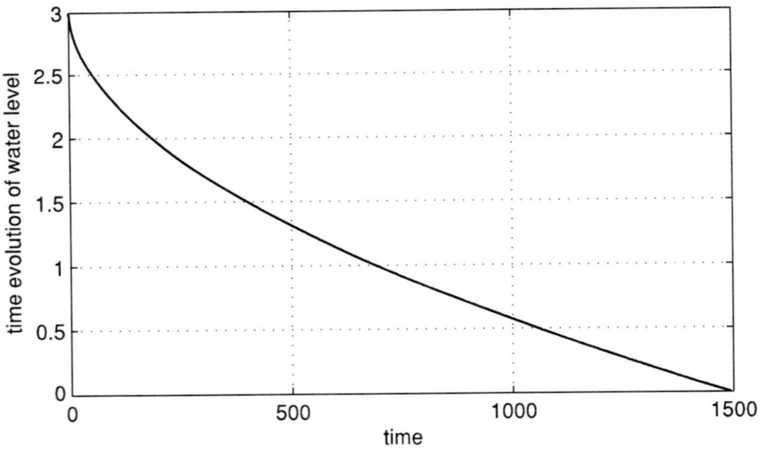

Fig. 7 Time evolution of the water level l

with

$$K(z) = K_s k(z), \quad k(z) = z^{\frac{1}{2}}(1 - (1 - z^{\frac{1}{m}})^m)^2, \; z = \bar{\theta} \tag{23}$$

and soil parameters $K_s = 2.4 \cdot 10^{-4}$, $n = 2.81$, $\alpha = -0.0189$, $\theta_s = 0.38, \theta_r = 0$. Then, the water flux

$$v = -K(\bar{\theta})\nabla(h + z)$$

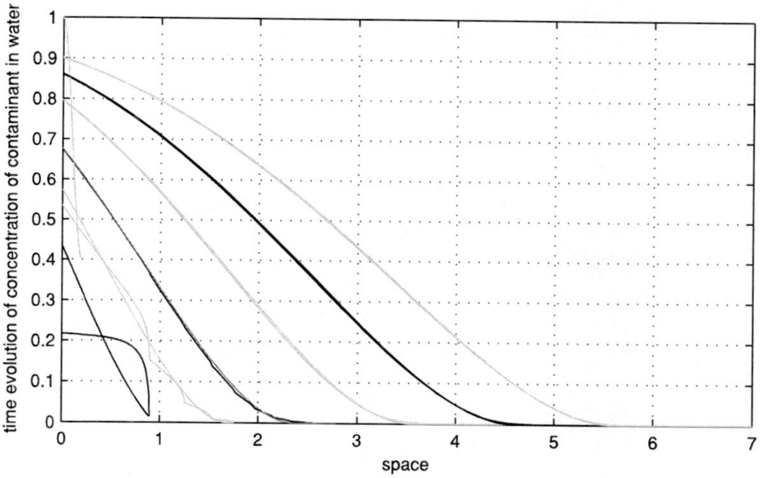

Fig. 8 Time evolution of concentrations w at T_2

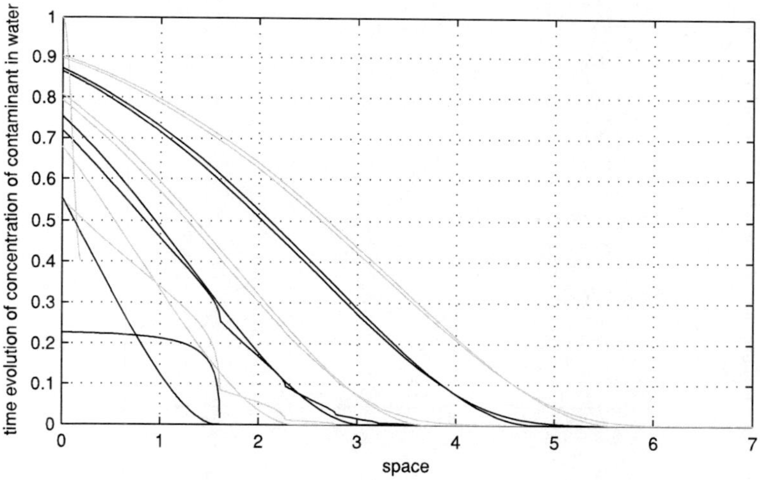

Fig. 9 Time evolution of concentrations w at T_1

is inserted into the contaminant transport with dispersion

$$\partial_t(\theta w) + \operatorname{div}(vw - \theta D \nabla w) = -\rho \partial_t S \tag{24}$$

and also into the adsorption

$$\partial_t S = \kappa(\psi(\theta w) - S), \tag{25}$$

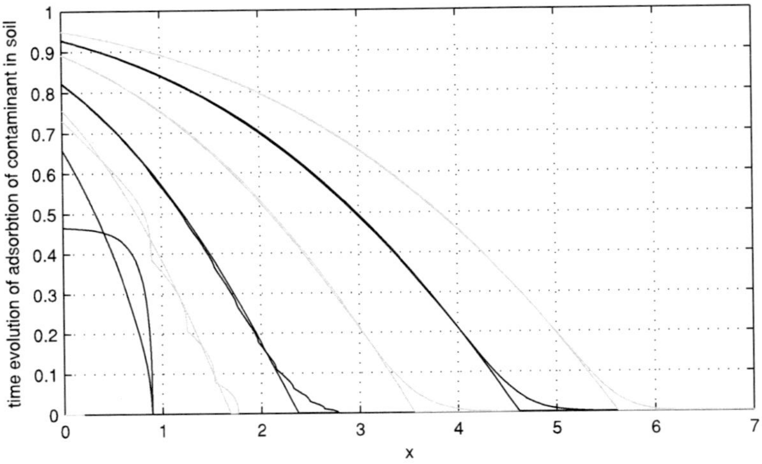

Fig. 10 Time evolution of adsorptions S at T_2

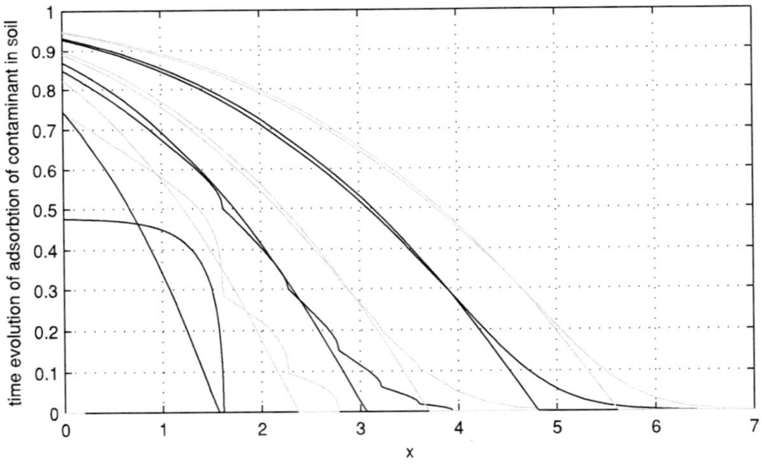

Fig. 11 Time evolution of adsorptions S at T_1

with the adsorption isotherm $\psi(s) = s^{\gamma}, \gamma \in (0, 1]$. Specifically, in our experiments we take $\gamma = 0.75$ and the adsorption kinetic $\kappa = 5$. In the dispersion matrix $\mathbf{D}$ we consider $D_0 = 5 \cdot 10^{-4}$ and $\alpha_L = 1$ since the molecular diffusion is significantly smaller than the dispersion.

The infiltration process will proceed about $1,500$ s. In the following pictures we present the time evolution of the water saturation (θ), contaminant concentration (w), its mass (θw) and adsorbed mass (S) in the solid for both benchmark and approximate solutions at the same time moments. Thus, the comparison is

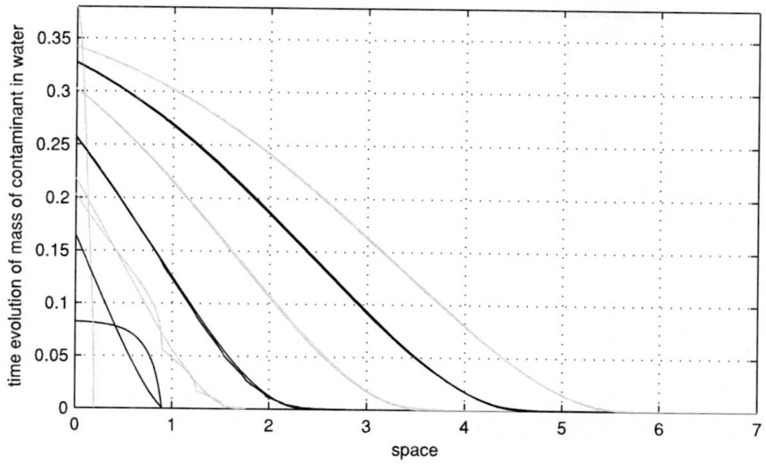

Fig. 12 Mass of contaminant θw at T_2

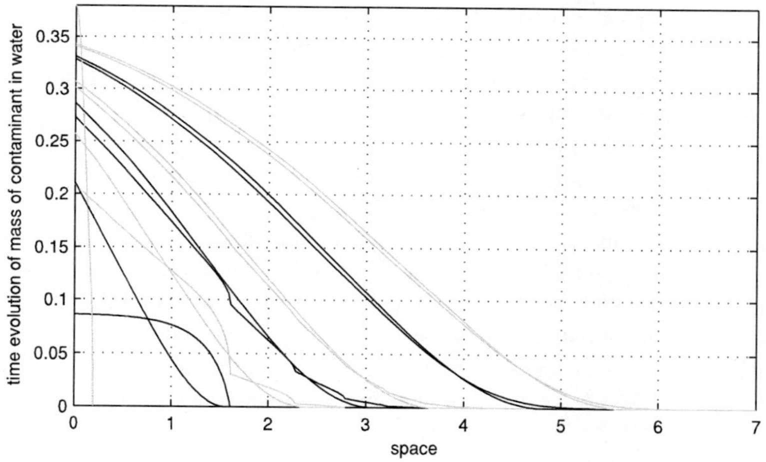

Fig. 13 Mass of contaminant θw at T_1

transparent. We also shortly present the influence of splitting of the time interval on the error by comparing our results with the benchmark solution at two different time splitting intervals $\tau_1 = 50$ and $\tau_2 = 15$. The comparisons will be presented at the time moments $T_1 = 50 \cdot [0, 1, 3, 6, 10, 20, 30]$ for τ_1 and at $T_2 = 15 \cdot [0, 1, 4, 10, 30, 60, 100]$ for τ_2. In our approximation scheme we have used 500 grid points along our sample $x \in (0, 10)$. For the splitting of the interval of the length τ we have used 5 grid points. We have solved the adsorption problem exactly as two ODE equations in each space grid point. To complete the presentation of our numerical results with the "operator splitting" method, we present also the time

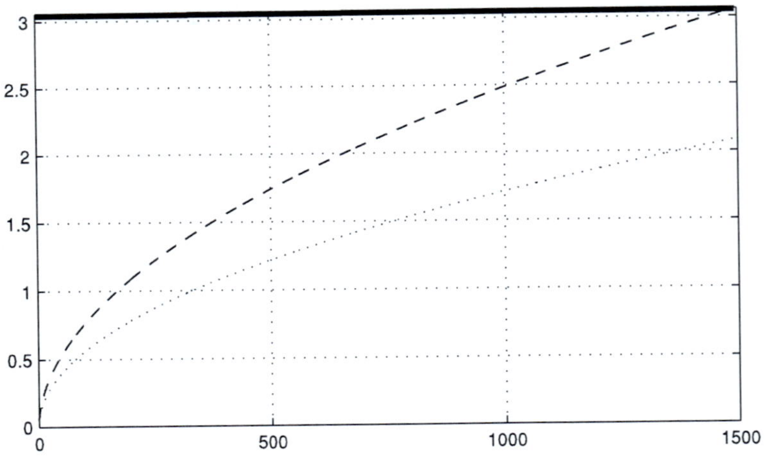

Fig. 14 Time evolution of contaminant: in soil (*dotted*); soil and water (*dashed*); total mass (*solid*)

evolution of flow characteristics from the benchmark solution which we have used in our computations. In Fig. 4 , the water saturation in both time moments T_1 and T_2 is drawn and in Fig. 5–6 we present the time evolution of interfaces s_1, s_2 separating saturated-partially saturated and partially saturates-dry zones, respectively. In Fig. 7 , time evolution of the water level in the column over the sample is plotted. In Fig. Figs. 8 , 9 , contaminant concentrations under the splitting steps τ_2, τ_1 in the corresponding time moments T_2, T_1 are depicted. Similarly, in the following Figs. 10 and 11 , the adsorbed contaminant into the soil are drawn. In Figs. 12 and 13, masses of contaminant dissolved in the infiltrated water per unite volume of the sample (θw) are plot. Finally, in Fig. 14 we present the time evolution of the contaminant mass in the soil (dotted), soil and water (dashed) and total mass (solid) for the benchmark solution. We can see that the mass balance is satisfied up to 1 %. Also the mass balance corresponding to the splitting step $\tau_1 = 50$ is surprisingly very good (less then 3 %). This supports the correctness of our results.

The discrepances of "operator splitting" and benchmark solutions in Figs. 4, 8–13 at the first moments of infiltration are due to the starting process of infiltration. It is because we have started with small nonzero saturation and concentration (at the top of the sample) to avoid the jump effect at the start $t = 0$. In fact we were obligated to take the same initial conditions as in [10] to be able to compare them. The influence of the used initial values at the start were too strong for operator splitting unlike for the method in [10] since ODE solvers there start with significantly smaller time discretization.

Remark. The benchmark solution in [10] is based on the modeling of interfaces s_1, s_2 (in form of ODE) and this is used substantially in construction of moving grid points leading to a very accurate approximation of our problem in terms of

ODEs. Generally, modeling of interfaces in the higher dimensional case is not known and thus the method in [10] can be applied only in 1D.

7 Conclusions

- An efficient numerical method is proposed for infiltration of contaminated surface water into the dry underground.
- The complex model is reduced to the successive solution (along small time intervals) of separate subproblems: unsaturated-saturated flow; convection and dispersion; adsorption.
- Convergence of approximated solutions is discussed and its proof is sketched.
- The approximate solution, based on the proposed "operator splitting" method, is compared to the benchmark solution in 1D.
- The obtained results confirm the applicability of the proposed "operator splitting" method to the solution of the contaminant transport model when polluted surface water infiltrates into the unsaturated-saturated underground in 3D.

Acknowledgments The authors confirm financial support by the Slovak Research and Development Agency under the contracts APVV-0184-10 and APVV-0743-10.

References

1. J. Bear, A. H.-D.Cheng: Modeling Groundwater flow and Contaminant Transport, Springer. ISBN 978-1-4020-6681-8
2. Constales, D., Kacur, J.: Determination of soil parameters via the solution of inverse problems in infiltration. Comput. Geosci. **5**, 25–46 (2004)
3. Constales, D., Kacur, J., Malengier, B.: A precise numerical scheme for contaminant transport in dual-well flow. Water Resour. Res. **39**(10), 1303 (2003)
4. Crandal, M.G., Majda, A.: The method of fractional steps for conservation laws. Numer. Math. **34**, 285–314 (1980)
5. Frolkovic, P., Kacur, J.: Semi-analytical solutions of contaminant transport equation with nonlinear sorption in 1D. Computat. Geosci. **3**((10), 279–290 (2006)
6. Kacur, J., Malengier, B., Van Keer, R.: On the mathematical analysis and numerical approximation of a system of nonlinear parabolic PDEs. J. Anal. Appl. **28**, 305–332 (2009)
7. J. Kacur, B. Malengier, M. Remesikova: Contaminant transport with equilibrium and non-equilibrium adsorption. Comput. Methods Appl. Mech. Eng. **194**, 497-489, 2005
8. Kacur, J., Malengier, B., Remesikova, M.: Convergence of operator splitting method on a bounded domain for a convection-diffusion-reaction system. J. Math. Anal. Appl. **348**(2), 894–914 (2008)
9. J. Kacur, B. Malengier, P. Kison: Numerical modelling of unsaturated-saturated flow under centrifugation with no outflow. arXiv:1001.1070v1[physics.comp-ph], Submitted (2010)
10. J. Kacur, J. Minar: A benchmark solution for infiltration and adsorption of polluted water into the unsaturated-saturated porous media (in preparation)

11. Knabner, P., van Duijn, C.J.: Solute transport in porous media with equilibrium and nonequilibrium multiple-site adsorption, Traveling waves. J. fur die reine Angew. Math. **415**, 1–49 (1995)
12. Kufner, A., John, O., Fucik, S.: Function Spaces. Academia, Prague (1977)
13. M. Remesikova: Numerical Solution of direct and inverse contaminant transport problems with adsorption, PhD Thesis, Faculty of Mathematics, Physics and Informatics, Comenius University (2005)
14. E. Trojakova: Numerical modelling of convection diffusion reaction, PhD Thesis, Faculty of Mathematics, Physics and Informatics, Comenius University (2011). http://hore.dnom.fmph.uniba.sk/ trojakova/PhDThesisTrojakova.pdf
15. Th.van Genuchten, M.: A closed-form equation for predicting the hydraulic conductivity of unsaturated soils. Soil Sci. Soc. Am. J. **44**, 892–898 (1980)